THE MONITOR BOYS

THE MONITOR BOYS

THE CREW OF THE UNION'S FIRST IRONCLAD

John V. Quarstein
Research Assistants Cindy L. Lester & Diana Martin

Published by The History Press
Charleston, SC 29403
www.historypress.net

Copyright © 2011 by John V. Quarstein
All rights reserved

Front cover: View of USS Monitor. *From left to right:* Robinson W. Hands, Louis Stodder, Albert B. Campbell and William Flye. Photograph by J. Gibson, 4 July 1862.
Courtesy of *The Mariners' Museum*.

Back cover: Officers of the USS Monitor. *Standing, left to right:* George Frederickson, Mark Sunstrom, Samuel Dana Greene, L.H. Newman (USS *Galena*) and Isaac Newton. *Seated, second row, left to right:* Louis Stodder, William Keeler, William Flye and Daniel Logue. *Seated, front row, left to right:* Robinson Hands, Albert Campbell and Edwin Gager. Photograph by J. Gibson, 4 July 1862. *Courtesy of The Mariners' Museum.*

First published 2011
Paperback edition 2015

ISBN 978-1-5402-0325-0

Library of Congress Cataloging-in-Publication Data
Quarstein, John V.
Monitor boys : the crew of the Union's first ironclad / John V. Quarstein.
p. cm.
Includes bibliographical references and index.
Hardcover ISBN 978-1-46711-948-1
1. Monitor (Ironclad) 2. Monitor (Ironclad)--Pictorial works. 3. Monitor (Ironclad)--Biography. 4. Sailors--United States--Biography. 5. United States. Navy--Biography. 6. Hampton Roads, Battle of, Va., 1862. 7. United States--History--Civil War, 1861-1865--Naval operations. 8. Seafaring life--United States--History--19th century. I. Title.
E595.M7Q83 2011
973.7'52--dc22
2010052980

Notice: The information in this book is true and complete to the best of our knowledge. It is offered without guarantee on the part of the author or The History Press. The author and The History Press disclaim all liability in connection with the use of this book.

All rights reserved. No part of this book may be reproduced or transmitted in any form whatsoever without prior written permission from the publisher except in the case of brief quotations embodied in critical articles and reviews.

Contents

Foreword by Anna Holloway	7
Preface	9
Acknowledgments	11
1. Technology Is King	15
2. Concept to Construction	23
3. Into Service	33
4. A Day Late	55
5. Showdown in Hampton Roads	71
6. Under a Glass	89
7. The Great Monster *Merrimack* Is No More	111
8. The *Monitor* Was Astonished	123
9. Long, Hot Summer	137
10. Respite and Refit	155
11. Hatteras	167
12. Aftermath	179
Appendix I. USS *Monitor* Chronology	205
Appendix II. USS *Monitor* Statistics and Dimensions	257
Appendix III. USS *Monitor* Casualties	259

Contents

Appendix IV. USS *Monitor* Officers' Assignment Dates	261
Appendix V. USS *Monitor* Officers' and Crew Biographies	263
Appendix VI. USS *Monitor* Purported Crew Members	299
Notes	303
Bibliography	323
Index	333
About the Author	351

Foreword

I'd like to describe a scene for you.

It is early June in 1862. The celebrated ironclad USS *Monitor* is patrolling the James River in Tidewater Virginia on a less than glamorous duty. Entries in the ship's logbook that June are a litany of temperatures that would be familiar to anyone who has ever been near the James River at that time of year.

Acting Lieutenant William Flye wrote on 2 June that "at 1 pm thermometer stood 142° inside the galley, the door being open and the blowers of the engine being in action." The heat continued to be oppressive while the engine driving the blowers became overworked and finally gave up. On 13 June, Fireman George Geer wrote home to his wife Martha, "We took the Impriture [*sic*] of several parts of the ship or rather I did as I have charge of the Thurmomitor [*sic*] and found in my Store Room which is fartherest [*sic*] stern it stood at 110 in the Engine room 127 in the Galley where they Cook and after the fire was out 155 on the Berth Deck where we Sleep 85…so you can see what a hell we have."

Though a thunderstorm later in the week cooled things off a bit, things would get worse. At 1:30 a.m. on 23 June, the men discovered a fire around the stovepipe of the galley. Geer wrote that they were able to extinguish it, but there was enough damage to knock the galley out of commission for a few weeks. He did not feel much like eating, though. He was suffering from jaundice.

Then the photographer showed up.

The photos taken by James Gibson on 9 July 1862, take on new meaning when you know what the men on the *Monitor* had just been through. The

Foreword

battle they had fought recently had not been with Confederate forces, but rather with oppressive heat, swarms of mosquitoes, a broken engine and terrible ennui. Even the imperious genius John Ericsson, who rarely found fault with any of his own creations, voiced his sympathies for the men on board, writing to his friend, Chief Engineer Isaac Newton, that he could not "imagine anything more monotonous and disagreeable than life on board the *Monitor*, at anchor in the James River, during the hot season." The expressions on the men's faces in Gibson's photos, which can be found in Admiral John Worden's personal photo album, bear that out. But you won't find George Geer in any of the photos. He was too sick to come on deck.

We can re-create this scene thanks to the wealth of information on the "cheesebox on a raft" that can be found in The Mariners' Museum Library and Archives and in archives, libraries, historical societies and private collections around the world.

The artifacts recovered from NOAA's *Monitor* National Marine Sanctuary take on new meaning when paired with letters, drawings and photographs. Here in this intersection of archive and artifact, mute metal pieces strive to give voice to the men who used them—and help to tell their stories and paint their pictures about their history, their discovery and recovery.

But they cannot always speak on their own. Historian John Quarstein has taken all of this information and more and brought the self-proclaimed "*Monitor* Boys" to life once again. A master storyteller, John has found their voices and has helped them tell their stories.

So dive into the world of the *Monitor* and meet William Flye, George Geer and the rest of the men who risked everything by going to sea in the celebrated "cheesebox on a raft" and became the hope of a nation wracked by war.

Meet the *Monitor* Boys.

Anna Holloway
Vice-President, Museum Collections & Programs
Curator, USS *Monitor* Center
The Mariners' Museum

Preface

I remember the day the *Monitor*'s turret finally returned to Hampton Roads like yesterday. Positioned atop a barge as she put into shore near The Mariners' Museum Lions Bridge, I declared then that she looked from afar as the *Monitor* would have looked 140 years before. As the turret was moved along a road to her conservation tank, I took a chance and touched her side. I felt an overwhelming connection with this unique warship that I had never felt before. Since that moment, I marveled at all the men who had been associated with the ship and desired to learn all I could about them. It was an urge that brought me to write this book.

When the turret arrived, I was already serving as a consultant on the *Monitor* Center Project developing exhibit names and researching the CSS *Virginia*. When I was younger, I was enthralled by the Battle of the Ironclads and favored the Confederate ironclad. Nevertheless, my work with The Mariners' Museum gave me a greater appreciation for the *Monitor* as I was able to stand inside the turret, touch the Dahlgrens and see unique archaeological discoveries. All of this made me think more about the men who served aboard this novel warship.

Even though many books have been written about the *Monitor*, I realized that no comprehensive study had been accomplished detailing the crew and their lives aboard the *Monitor*. Irwin Berent published his outstanding research for NOAA in 1982. Consequently, I compiled all the available service records and other related documents to produce a biography of each crew member. Simultaneously with this effort, I constructed a detailed chronology of the *Monitor* and her crew's service. All of this information was then organized into appendices.

Preface

When I sat down to write the narrative, I recognized that I had to tell the complete story, however, and, whenever possible, interject what various officers and crew thought about in the life on the *Monitor*.

The end result is *The* Monitor *Boys*. A name the crew gave themselves, it is a reflection of how these hundred-odd men served together through storms, battles, boredom and disaster. Many recognized that they were part of history; moreover, the *Monitor* Boys were agents in the change of naval warfare.

Acknowledgments

The Monitor *Boys* is an amazing story of the officers and crew of the USS *Monitor*. These men deserve the highest compliments, as they served faithfully in an iron ship unlike any vessel ever seen before. Just serving aboard the ironclad took uncommon effort and fortitude: the heat, combat and un-seaworthiness all combined to make each of the *Monitor* Boys an individual who deserves notice.

This volume documenting the *Monitor*'s officers and crew was a six-year effort, and I must first express my appreciation to Anna Holloway, vice-president of Collections and Programs at The Mariners' Museum and the National Oceanic and Atmospheric Administration's *Monitor* National Marine Sanctuary. We conceived the need to utilize the outstanding research work of Irwin Berent as a base upon which I should produce a narrative and genealogical entries presenting a comprehensive overview of the Union ironclad's officers and crew. Berent did an amazing job for NOAA, compiling facts about the crew; however, Anna and I both believed that enough new research had come to light since Berent published his findings in 1982 that the stories of the *Monitor* Boys should be revisited.

In addition to Berent's work on the *Monitor*'s enlisted men, I also must thank and laud former East Carolina University professor Dr. William N. Still for his work documenting the *Monitor*'s officers as well as the Northern industries that made the creation and production of this technological marvel able to change naval warfare in 1862.

This book was truly made possible by my research assistants, Diana Martin and Cindy L. Lester. Diana Martin worked as an intern at the Virginia War

Acknowledgments

Museum while I served as the museum's director. She asked if I had any research projects. Of course I said yes, and she then embarked on a two-year effort collecting all of the raw data from wherever she could discover traces of the *Monitor* Boys. Diana did an outstanding job tracking down all available information and then organizing a file on each crew member. Her work enabled another of my assistants, Cindy Lester, to organize the biographical entries and aid my creation of the appendices from the plethora of records, certificates, newspaper articles and reference books. Cindy also typed the various drafts and formatted the final draft for publishing. She truly supported the production of this book in any way she could.

Numerous others also aided the book's transition into a finished volume. The maps and battle diagrams were produced by Sara Johnston of The Mariners' Museum. The images came from two sources. My son, John Moran Quarstein, has an outstanding collection of ironclad images, and a few of the views are his. The vast majority, however, are from The Mariners' Museum's outstanding collection. Claudia Jew and Megan Evans went out of their way to provide the best photos and prints available for inclusion in *The* Monitor *Boys* book. Several of the crew photographs have never before been published.

I would be remiss not to thank my primary editor, Julie Murphy of Circle C Communications, for her fine work correcting my prose, for ensuring that all of the commas were in the right places and for being a sounding board for my literary concepts. Anna Holloway and J. Michael Moore of Lee Hall Mansion served as my primary readers. Their advice was most helpful. Many others encouraged me to stay on target with this volume, including Ashley Hagen, Alex Bradbury and Corinne D'Amato. They all kept my spirits up and provided various levels of support as I struggled to make all of these stories come together.

Of course, this book would not have been possible without the dedicated service of the *Monitor* Boys themselves. These men served on a warship the likes of which had never been seen before. The ironclad was so novel that many deserted as soon as they saw the ship. Others, however, remained and became determined to test this experiment in warship design. Once aboard, they discovered a half-submerged existence overflowing with inventions. Even though they quickly became the toast of the nation for their service on 9 March 1862, they did not believe that combat was dangerous, as they were so well protected by the *Monitor*'s armor. What did make them heroes is how they survived the intolerable conditions whether at anchor, locked in an iron box during the hot and humid summer on the James River or

Acknowledgments

steaming during a heavy gale at sea in an un-seaworthy vessel. When the *Monitor* Boys looked back at their service, they most vividly remembered the night the *Monitor* went down off Cape Hatteras. They were all heroes that evening to save their ship and survive the rescue effort despite the raging sea. The survivors would never forget that dark storm that consumed their ship and sixteen fellow crew members. They survived "a night of horrors," as George Geer remembered, while others went, as Dr. Grenville Weeks sadly lamented, "gone to a brighter world where storms do not come." Just as Jacob Nicklis "did his duty well until the end," all of the *Monitor* Boys served their nation with a dedication and devotion that must never be forgotten. As Major General John Ellis Wool told William Keeler, "You have made heroes of yourselves."

Chapter 1
Technology Is King

The USS *Monitor*, originally named *Ericsson's Battery*, was not a lone concept that exploded without forethought from the fertile mind of its designer. John Ericsson's vessel was the result of several improvements in naval warfare. The Industrial Revolution introduced new ordnance and motive-power technologies, which caused a major revolution in ship design concepts, construction and composition. This "Age of Inventors" spurred instant changes in the design and manufacturing of ships and armament, which required nineteenth-century navies to continually evolve to utilize, or to counter, these advancements.

The Greek War of Independence (1821–29) witnessed the last battle fought wholly under sail and the first steam-powered warship to serve in battle. Even though the USS *Fulton* was actually the first steam-powered warship, she was primarily a harbor-defense vessel and never witnessed combat. The first purpose-built, steam-powered warship to actually serve in combat was the British-built *Katrina* of the Greek navy. An allied English, French and Russian fleet, fighting for Greek independence, defeated a Turkish squadron on 20 October 1827 at Navarino Bay. Since this conflict introduced steam power to naval warfare, the war truly marked the end of the "Age of Sail." Navies of the world quickly recognized the advantages of steam; however, naval leaders were not totally satisfied with paddle propulsion. Side paddle wheels took up the space that would have otherwise mounted guns. Likewise, both the paddle wheels themselves and their engines mounted on deck were very vulnerable to artillery fire. This circumstance set the stage for the first major nautical design from John Ericsson destined to radically change ship propulsion.

John Ericsson was born in Langbanshyttan, Varmland, Sweden, on 31 July 1803. His father was a mining engineer, and as a youngster, John displayed a genius for mechanics when he designed, and constructed, a miniature sawmill at age ten. He joined the Swedish navy at thirteen and produced technical drawings for the Göta Canal. At age seventeen, Ericsson was commissioned an ensign and served as a topographical engineer. During this time, he developed a heat engine that used the steam from fires as a propellant. He immigrated to England, where his engine eventually proved to be a failure. In 1829, he formed a partnership with John Braithwaite and produced the steam engine called the Novelty. While the Novelty was the fastest engine in the Rainhill Trials competition, it suffered boiler problems and was forced to drop out. Despite this setback, Ericsson continued to produce a series of remarkable innovations, including a steam fire engine, a steam condenser for producing fresh water for a ship's boilers while at sea and a pressure-activated fathometer.

Ericsson became one of several inventors to create a screw propeller in 1839. The screw propeller enabled engine systems to be installed below the waterline, making screw-propelled warship engines virtually shot-proof. Furthermore, tests by the Royal Navy proved that the screw propeller produced more power than the paddle-wheel system. With the financial support of Francis Ogden, the American consul in Liverpool, Ericsson produced a screw-driven steam tugboat, the *Francis B. Ogden*, in 1837. Despite her success (several Royal Navy officers nicknamed the tug the "Flying Devil"), the British Admiralty failed to recognize the innovative power of steam-screw propulsion.

An American naval officer, Robert F. Stockton, had the foresight to recognize the naval application. Stockton, who was from a very well-connected New Jersey family, had dreamed of creating a steam-screw warship for the U.S. Navy. He persuaded John Ericsson to immigrate in 1839 to the United States, where Ericsson received the contract to design and supervise the construction of the USS *Princeton*. The *Princeton* was laid down on 20 October 1842 at the Philadelphia Navy Yard, as a seven-hundred-ton corvette. Not only did Ericsson design the two vibrating-lever engine and the six-bladed screw propeller, but he also created the ship's collapsible funnel, an improved rangefinder and improved recoil systems for the main battery of forty-two-pounder carronades. The *Princeton*, a novel combination of sail and steam power, also featured the most modern advances in artillery: two twelve-inch shell guns.

With the introduction of Admiral Horatio Nelson's tactics of annihilation during the early nineteenth century, equally destructive weaponry became

Technology Is King

essential. Explosive shells and Columbiads had all been introduced to land warfare; however, naval leaders had not recognized their application to ship-to-ship combat. Accordingly, old techniques of warfare ended when Brigadier General Henri-Joseph Paixhans published two books, *Nouvelle Force Maritime et Artillerie* in 1822 and *Experiences Faites sur une Arme Nouvelle* in 1825, in which he advocated a new system of naval gunnery based on standardization of caliber and the use of shell guns. Although even he admitted his concepts were not new, his thoughts did unify a series of ideas that proved to be extremely revolutionary. In 1824, Paixhans tested an eighty-pounder shell gun against an old eighty-gun ship of the line, *Le Pacificateur*, at Brest, France. The battleship was demolished by only sixteen shells. Besides demonstrating the tremendous destructive power of explosive shells, Paixhans argued that modern warships should be steam-powered, iron-plated and armed with like-caliber shell guns.

Shells were far superior to solid shot in terms of naval combat. Whereas solid shot strove to penetrate (and often did not) the wooden sides of warships, shells were designed to explode in a ship's side, tearing an irregular hole that could sink a vessel. Sparks from the explosion could ignite fires on the damaged ship. Furthermore, the resulting wooden splinters and shell fragments had effective anti-personnel properties that could decimate a crew. Initially, the lower velocity required to propel shells against a target meant that shell guns could be lighter. This allowed more powerful guns to be mounted in a ship's battery, thereby increasing the weight of a warship's broadside.

Testing Columbiad at Fortress Monroe, circa 1860. *Courtesy of John Moran Quarstein.*

Other ordnance improvements followed Paixhans' work. Harvard professor David Treadwell introduced cast-iron, smoothbore guns strengthened with wrought-iron bands. The purpose of the banding was to increase projectile weight and velocity. John Ericsson considered all of these factors as he planned the armament for the *Princeton*.

His allegiance to Paixhans' concept was witnessed in the *Princeton*'s main battery composition of twelve 42-pounder carronades. In order to give the ship more effective firepower, he selected a twelve-inch shell gun he had already manufactured, known as "The Oregon." Originally named "The Orator," Ericsson's shell gun was a smoothbore muzzleloader made out of wrought iron and was capable of firing a 225-pound shell using a 50-pound charge. Ericsson had produced it at the Mersey Iron Works in England. The design followed Treadwell's revolutionary concepts, in that it used the "built-up construction" of placing red-hot iron bands around the gun's breech end. This action pre-tensioned the gun and greatly increased the charge the breech could withstand.

The Oregon was shipped to the United States by Ericsson; however, he needed a second pivot gun for the *Princeton*, and Stockton insisted on designing the weapon himself. Relations between Stockton and Ericsson had begun to sour. Stockton had endeavored to force Ericsson off the project and sedulously avoided admitting Ericsson's major role in the ship's design. Therefore, Stockton alone designed and supervised the fabrication of the *Princeton*'s second gun, "The Peacemaker." The gun was fatally flawed, since Stockton did not fully comprehend the design of Ericsson's Oregon. Instead of using Ericsson's hoop construction, he reinforced the breech by simply making it thicker. As a result, this twelve-inch muzzleloader weighed 27,000 pounds. Despite its size, the lack of a reinforced breech made it unable to withstand the pressure of the charge and doomed it to eventually burst.

The *Princeton* made her trial voyage on 12 October 1843. She steamed to New York on 1 January to receive the Oregon and Peacemaker. Stockton failed to advise Ericsson of the correct sailing time, and Ericsson was left in New York when the *Princeton* departed for Washington, D.C. Stockton wished to claim complete credit for the *Princeton*. Even though Robert F. Stockton was an outspoken advocate for the *Princeton*, the design and engineering credit belonged to John Ericsson.

Stockton's warship was an instant success when she arrived in Washington. Several special excursions were made in mid-February; however, the grand event was held on 29 February 1844. It was a trip down the Potomac River with President John Tyler, his cabinet and over two hundred other guests. The champagne flowed, and Peacemaker was fired several times as entertainment. When the huge gun was fired for a final time, it exploded, killing eight attendees, including Secretary of State Abel Upshur, Secretary of the Navy Thomas Gilmer, Chief of the Bureau of Construction, Equipment and Repairs Captain Beverly Kennon and Congressman David Gardiner of New York. The explosion also wounded twenty others, including Robert Stockton himself.

When the smoke cleared and a Court of Inquiry was held, Stockton was somehow exonerated, while the blame entirely fell on Ericsson. Furthermore, Stockton blocked the navy from paying Ericsson for his work on the *Princeton* project. These circumstances resulted in a long-lasting, mutual aura of mistrust between John Ericsson and the U.S. Navy.

The *Princeton* affair turned the U.S. Navy away from the production of large shell guns, as Congress limited funds for ship-design development. Europeans took minimal notice of this and continued to move ahead with startling new concepts. The desire to produce a more accurate and reliable weapon resulted in Major Giovanni Cavalli, of the Sardinian army, introducing the first effective rifled cannon in 1845. An explosive shell could now be hurled at a target with greater velocity and accuracy.

While the Mexican-American War proved the effectiveness of steamers during war for coastal operations, the resupply of an amphibious operation

Awful Explosion of Peace-Maker on Board the US Frigate, Princeton. Currier & Ives, circa 1844. *Courtesy of The Mariners' Museum.*

and for blockade duty, it was the Crimean War in Europe that spurred even greater technological changes. The stunning Russian naval victory at Sinope, Turkey, on 30 November 1853 proved the superiority of the new shell guns when Admiral Pavel Stepanovich Nakhimov's squadron totally destroyed a Turkish fleet. Thereafter, allied navies refused to engage the Russian batteries defending Sevastopol, in the Crimea, fearing the impact of Russian shells on their ships.

A stalemate continued until the French reread Paixhans' books and sought to build floating batteries. Emperor Louis Napoleon called for designs, and John Ericsson responded. Ericsson submitted to the emperor a radical design that was steam powered and completely covered with iron plate. The ironclad's profile was unlike any other submitted. "Ericsson's Impregnable Battery and Revolving Cupola" was a revolutionary design, as the entire living space and engine system were below the waterline for the first time. Only the armored turret was seen above the water. It was an innovative design, but Napoleon apparently did not respond to Ericsson's proposal. Instead, the French developed the *Lave* class of floating batteries.

These vessels featured eighteen sixty-eight-pounder shell guns in a casemate covered by four-inch iron plating. Three of these batteries, *Dévastation*, *Lave* and *Tonnante*, were towed into the Black Sea and were used in the allied assault against the Russian batteries at Kinburn Peninsula in the Ukraine on 17 October 1855. Anchored just eight hundred yards from the forts, the French ironclads were able to withstand four hours of heavy cannonading. The only damage occurred when a Russian shell entered a gun port, killing two French sailors. In turn, the Russian forts were shelled into submission. Kinburn proved the value of armored vessels against fixed fortifications. Both Great Britain and France built oceangoing ironclads in the war's aftermath. The French were first, with the thirty-six-gun *La Gloire*. This screw frigate was covered with four-and-a-half-inch iron plate and could make 13 knots under sail and steam. The British responded with several iron-plated floating batteries and, ultimately, the HMS *Warrior*. The *Warrior* was an all-iron vessel capable of a combined speed of 17.5 knots and mounting forty cannon. The *Warrior* could escape what she could not destroy. By early 1861, the Royal Navy had twelve ironclads under construction.

The U.S. Navy had observers in the Crimean War Theater, and all of the technological and tactical changes were recorded by a team of army officers led by Major Richard Delafield. (Delafield's team included Major Alfred Mordecai and Captain George Brinton McClellan.) This study, commonly called the Delafield Report, recommended modernizing the United States'

Technology Is King

military establishment. The report especially noted that Great Britain and France had used "their greatest exertions to devise the means of destroying sea-coast casemated defenses of their enemy" and advised that the U.S. Navy should construct steam-powered ironclad vessels, armed with the most advanced ordnance, to compete with European navies. Instead, the U.S. Navy focused on the construction of steam-screw wooden warships. While the public questioned the U.S. Navy's reluctance to build ironclads, the desire not to enter into an arms race with Europe was due to several factors. First, U.S. overseas interests did not appear threatened by any European power in the 1850s, and secondly, the U.S. Navy was content to allow the Europeans to complete the costly experiments with ironclads. After all, the U.S. Navy was already heavily invested in the *Stevens Battery*. This revolutionary warship concept was initiated in 1842 but was still incomplete by 1860.

Meanwhile, John Ericsson entered into a business relationship with Cornelius H. Delameter. They became very close friends, despite the strains of business and Ericsson's aloof personality. Together, they produced the *Iron Witch*, the first iron steamboat. Ericsson used Delameter Ironworks

John Ericsson, circa 1862. *Courtesy of The Mariners' Museum.*

for the creation of many of his concepts, as well as to continue to improve his hot-air engine. While Ericsson had already invented this engine system (called "caloric engine") when still living in Sweden, he would eventually be awarded the Rumford Prize of the American Academy of Arts and Science for this design.

Ericsson understood, and participated in, the development of the new technologies that had caused naval warfare to evolve. Screw propellers, engines, cannon production, iron-ship construction and explosive shells were all keen interests of Ericsson. It would take a new conflict in human affairs to enable him to place all of his skills and knowledge into one project that would bring forth an even newer age of naval warfare.

Chapter 2
Concept to Construction

As soon as the guns went silent in Charleston Harbor, President Abraham Lincoln recognized that an important key to the Union victory would be control of the over three-thousand-mile-long Confederate coastline. The commercial link, the "cotton for cannon"[1] trade between Europe and the South, had to be cut. Consequently, the Union declared a blockade of the Southern coast on 15 April 1861 and immediately sought the ships to enforce it. President Lincoln's efforts to strangle the Confederacy via a blockade of Southern ports seemed assured of success. Without a navy to defend its harbors and contest the Federal fleet, the Confederacy appeared doomed to lose its all-important link to European manufactured goods.

When the Confederate government was formed under the leadership of Jefferson Davis, the new nation would be fortunate to secure the services of Stephen Russell Mallory as secretary of the navy. Mallory, from Key West, Florida, had previously served as chairman of the U.S. Senate's Naval Affairs Committee. He immediately realized the Confederacy could never match the North's superior shipbuilding capabilities unless a new "class of vessels hitherto unknown to naval science"[2] was introduced to tip the balance in favor of the South. Mallory knew that iron-cased warships, armed with the most powerful rifled guns, could destroy the North's wooden navy. Since the Federal fleet contained no ironclads, Mallory advocated that "inequality of numbers may be compensated by invulnerability; and thus not only does economy but naval success dictate the wisdom and expediency of fighting iron against wood."[3] Mallory was on target with his concept, but unfortunately, the South did not have the industrial infrastructure to implement his plan.

He believed the Confederacy could eventually acquire an ironclad from Europe, but since this action would take time, the Confederacy needed to act quickly, before the North's blockade could interrupt the flow of goods in and out of Southern ports.

The answer came with Virginia's decision to leave the Union on 17 April 1861. The Commonwealth of Virginia contained two major industrial resources: Tredegar Iron Works in Richmond and Gosport Navy Yard in Portsmouth. Tredegar was the only facility in the South capable of rolling iron plates. Gosport gave the Confederate shipbuilding program an immediate advantage, as the facility was one of the best-equipped yards in America. Even though the Federals burned the yard when they abandoned it on 20 April 1861, they left with such haste that their destructive work was far from complete. Numerous warehouses containing naval supplies survived the blaze. The Federals also abandoned a tremendous array of ordnance, including 1,085 heavy cannon and 250,000 pounds of gunpowder. Many dwellings, as well as the foundry, machine shop and several workshops, remained untouched. More importantly, the retreating Federals failed to destroy the granite dry dock. Overnight, the Confederacy gained the infrastructure to construct the vessels required to challenge the Union blockade. The Richmond press gloated over the abundance of equipment and supplies, stating, "We have material enough to build a navy of iron-plated ships."

All of the wooden warships anchored at Gosport, except for the venerable frigate the USF *United States*, were put to the torch. Of these vessels, the most important was the steam-screw frigate the USS *Merrimack*.[4] The frigate's engines had been condemned, and she was placed in ordinary at Gosport in February 1860, awaiting repairs. Because she had sunk while burning, the *Merrimack*'s hull and engines had not been destroyed. The *Merrimack* was then raised and placed in Gosport's dry dock. The question remained as to what to do with this burned and blackened hulk.

Mallory viewed the *Merrimack* as the best solution to jump-start an ironclad construction program. The Confederate secretary of the navy held a meeting in Richmond on 23 June 1861 to review and plan the *Merrimack*'s conversion. Lieutenant John Mercer Brooke, Naval Constructor John Luke Porter and Chief Engineer William Price Williamson formed a committee to execute the *Merrimack*'s transformation into an ironclad. Porter had actually brought with him a model of an iron-cased, floating harbor-defense battery he had created in 1848, while Brooke provided drawings he had just made at Mallory's request. Both designs featured an inclined casemate, but Brooke's concept

Concept to Construction

submerged the bow and stern of the vessel to enhance speed and buoyancy. Since Mallory unrealistically wanted an oceangoing armored warship, Brooke's design became the plan selected for the *Merrimack*'s reconfiguration.

Work on the *Merrimack*'s revitalization began in June. As Porter supervised the cutting away of the *Merrimack*'s charred timbers, Brooke and Williamson sought to solve the power plant problems. They quickly learned that new engines, provided by Tredegar Iron Works, would take too long to build. Chief Engineer Williamson decided the old, previously condemned engines of the *Merrimack* could be reworked, despite serious corrosion from the salt water of the Elizabeth River. Meanwhile, Porter supervised the removal of all of the upper works and then cut the vessel on a straight line, from bow to stern, at the berth deck level. The main gun deck was laid, and the casemate began to take shape. Porter's plan called for an overall length of 262 feet and a draft of 21 feet.

The casemate was the ironclad's most distinctive feature, extending 28 feet from the bow and reaching aft 172 feet. The fantail continued another 56 feet. The sides were sloped upward at a thirty-six-degree angle to deflect shot. This acute slope allowed only 7 feet of headroom and a beam of 30 feet, which forced the cannon to be staggered along the opposing broadside to accommodate recoil. The roof was grated to provide ventilation to the gun deck. The grating was manufactured of two-inch iron bars, supporting rafters of yellow pine and white oak. Three hatchways were constructed to enable access to the 64-foot-wide deck. At the front of the casemate was a sixty-two-inch-thick iron, conical pilothouse.

The casemate was constructed of four inches of oak laid horizontally, eight inches of yellow pine laid vertically and twelve inches of white pine laid horizontally. It was all bolted together and then eventually sheathed with two- by six-inch iron plate, two inches thick, laid horizontally. A second course of similar iron plate covered the first layer vertically. The deck, designed to be almost awash with the sea, was covered with one inch of iron plate. An additional course of one-inch iron plate extended three feet from the deck to a depth of three feet around the vessel. The joining of the casemate to the hull, where a catastrophic separation of the two sections could occur, was an obvious weak point. Porter had devised a displacement that would submerge the knuckle two feet below the waterline. The casemate eaves were also extended two feet to provide additional protection from shot aimed at the ironclad's hull.

As Tredegar Iron Works rolled the *Merrimack*'s iron plate, John Mercer Brooke worked with the foundry on the production of rifled cannon to arm

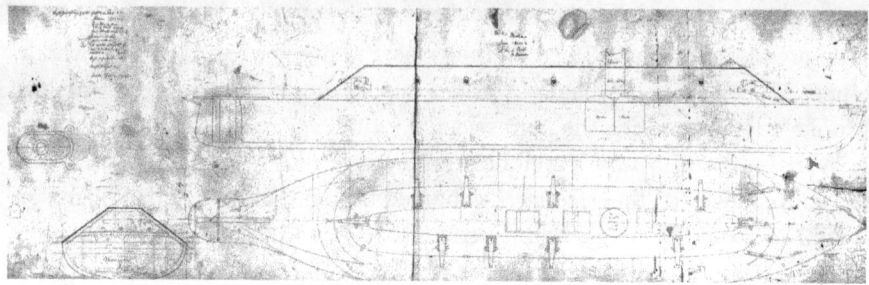

CSS Merrimac. John Luke Porter, 1861. *Courtesy of The Mariners' Museum.*

the ironclad. He developed a brilliant system of converting old smoothbore cannon into rifles by forging bands over their breech to resist the greater pressure of firing rifled projectiles. Brooke invented special explosive shells and, more importantly, an elongated armor-piercing wrought-iron bolt for both the 7-inch and 6.4-inch versions of his rifled cannon. Since Secretary of the Navy Stephen Mallory wanted the *Merrimack* armed with the finest heavy cannon, Tredegar immediately forged ahead with the production of Brooke's rifles and projectiles. In his desire to make the *Merrimack* a deathblow for wooden warships, Brooke decided the ironclad would be armed with a broadside battery of six IX-inch Dahlgren smoothbores and two 6.4-inch rifles. Two of the Dahlgrens were hot shot guns. At each end of the casemate, three gun ports were pierced for the two 7-inch Brooke rifles, which served as pivot guns.

Brooke and Mallory also recognized that a shot-proof, steam-powered, armored ship could revitalize the ancient "ram" as a weapon of modern navies. Ramming, as a decisive offensive tactic, had been virtually abandoned with the rise of large sailing ships mounting artillery, yet steam power made this battlefield technique once again a viable weapon. Mallory was keenly aware that "even without guns the ship would be formidable as a ram" and likened ramming to a "bayonet charge of infantry." Considering the gunpowder shortage in the South, he recognized that Confederate ironclads could, instead, employ the ram as a technological weapon that could punch into the sides of Union wooden vessels. Eventually, the *Merrimack* would be fitted with a 1,500-pound cast-iron, wedge-shaped ram. Unfortunately, this ram was poorly mounted, and one of the flanges holding the ram in place was broken. Nevertheless, the *Merrimack*'s battery and ram combined made the ironclad a weapon that no wooden warship could counter.

Even though the Confederates initiated the *Merrimack*'s conversion in June 1861, it did not proceed as planned. Mallory had initially hoped to unleash his ironclad upon the Union blockade by October 1861. Unfortunately, despite the head start and the vast resources of Gosport, the Confederacy lacked the skilled manpower and infrastructure in order to rapidly complete the ironclad project.

News of the *Merrimack* project reached the North shortly after the conversion was begun. Information leaked, almost daily, to the Federals across Hampton Roads. Escaped slaves, exchanged prisoners of war, Northern sympathizers and other informants kept the Federals informed with up-to-date knowledge about the Confederate ironclad project. The Southern press, which continually boasted about the power of their new ironclad, was another problem.

On 11 August 1861, the *Mobile Register* proclaimed:

> *It would seem that the hull of the* Merrimack *is being converted into an iron-cased battery. If so, she would be a floating fortress that will be able to defeat the whole Navy of the United States and bombard its cities. Her great size, strength, and powerful engines and speed, combined with the invulnerability secured by the iron casting, will make the dispersal or the destruction of the blockading fleet an easy task for her. Her immense tonnage will enable her to carry an armor proof against any projectile, and she could entertain herself by throwing bombs into fortress Monroe, even without risk. We hope soon to hear that she is ready to commence her avenging career on the seas.*[5]

Many Northern naval leaders thought when Gosport was abandoned that the Confederacy could not make use of the demolished yard and scuttled ships. Flag Officer Silas Horton Stringham, commander of the North Atlantic Blockading Squadron in the summer of 1861, advised that the *Merrimack* was "pronounced worthless. Her machinery was all destroyed." Yet stories of the conversion continued to filter northward. The Union secretary of the navy, Gideon Welles, had received copies of his Confederate counterpart, Stephen Mallory's, reports and detailed information about the *Merrimack*'s revitalization. The concept of creating armor-clad vessels, capable of countering any ironclad the South might produce, was not lost upon Gideon Welles. Welles, a former newspaper owner from Hartford, Connecticut, and a Democrat who served in the Navy Department before joining the Republican Party in 1855, recognized that the U.S. Navy was outdated and desperately needed a special shipbuilding program to enforce President Lincoln's blockade of the South.

Secretary of the Navy Gideon Welles, circa 1861. Courtesy of John Moran Quarstein.

Welles had already spent a great deal of effort striving to make the blockade a reality. All vessels that could float, and mount cannon, were required and commissioned into service. While often ridiculed, this "soap box" navy was able to maintain a presence off Southern ports. The need for Union ironclads became apparent as the Confederates worked on the *Merrimack*'s conversion, as well as fortifications defending south-side Hampton Roads. This harbor was the North Atlantic Blockading Squadron's anchorage and was key to the control of the Chesapeake Bay.

On 4 July 1861, Welles asked Congress for an appropriation of $1.5 million to construct three experimental ironclads. He further requested the creation of an Ironclad Board, consisting of three currently serving naval captains, to determine the appropriate design. Congress approved the bill on 3 August, and President Lincoln signed it into law the next day. Flag Officers Joseph Smith and Hiram Paulding and Captain Charles Henry Davis were appointed to the Ironclad Board to select one to three prototype-vessels for construction. Welles solicited bids on 7 August.

By early September, sixteen proposals had been received, of which only two were considered. The first would eventually be known as the USS *New Ironsides*. Merrick and Sons of Philadelphia submitted this proposal. This ironclad was a European-style armored frigate, mounting sixteen guns in broadside batteries. The other selection was a design submitted by businessman Cornelius Bushnell and would be known as the *Galena*.

Bushnell had amassed a fortune in shipping and railroads. He was a prewar acquaintance of Welles and had actively solicited support in Congress for the ironclad appropriation. Consequently, Bushnell had submitted a design

Concept to Construction

by Naval Constructor Samuel Pook, an ironclad displacing 950 tons and mounting six cannons. While the Ironclad Board had approved the design, Davis questioned the ironclad's stability. Bushnell had to provide proof of the *Galena*'s seaworthiness, and knowing little about vessel design, he sought the counsel of one of his partners, Cornelius Delameter, owner of Delameter Ironworks. Delameter advised Bushnell to seek out John Ericsson for advice. During this meeting, Ericsson explained why the *Galena* was seaworthy. He then showed Bushnell his own model of a floating battery design, the one he had offered to France during the Crimean War. Bushnell immediately recognized the brilliance of the concept and offered to promote the idea.

Bushnell went to Gideon Welles's home in Hartford, Connecticut, and shared the design with the secretary of the navy, exclaiming, "The country was safe because I had found a battery which would make us master of the situation as far as the ocean was concerned." Welles agreed, noted that the ironclad was "extraordinary and valuable"[6] and urged him to present the model to the Ironclad Board. Bushnell, who had a keen understanding of politics, secured an appointment with President Lincoln through his connection with Secretary of State William Seward. The president, who was intrigued by the new gadgets of war, saw value in Ericsson's concept and agreed to accompany Bushnell to the Ironclad Board meeting.

On 13 September 1861, Bushnell presented Ericsson's model to the Ironclad Board. The members, despite President Lincoln's presence, were not impressed. Captain Davis simply could not imagine such a warship and told Bushnell to "take the little boat home and worship it as it would not be idolatry, because it is in the image of nothing in the heaven above or on the earth beneath or in the waters under the earth." Once the Ironclad Board members realized that the novel design Bushnell was promoting belonged to John Ericsson, they ridiculed the concept as another "Ericsson's Folly." Yet the president provided a slight reprieve for the plan, commenting, "All I have to say is what the girl said when she stuck her foot into the stocking: 'It strikes me there's something in it.'"[7]

Despite President Lincoln's support, the members of the Ironclad Board remained skeptical, and Bushnell knew the only person who could convince the board members was Ericsson himself. He convinced Ericsson that the board actually liked his concept but the inventor needed to travel to Washington to thoroughly describe the ship's unique features. Ericsson agreed. On 15 September 1861, Ericsson appeared at the Ironclad Board meeting and, after a few questions, realized that his project was only receiving criticism. Ericsson's response to a question about the vessel's

C.S. Bushnell, circa 1860. *Courtesy of The Mariners' Museum.*

stability, however, led one listener to state: "His lengthy and detailed retort thrilled every person present." Ericsson concluded his presentation with the comment, "Gentlemen, after what I have said, I consider it to be your duty to the country to give me an order to build the vessel before I leave the room."[8] A follow-up meeting, with Secretary Welles present, finalized the deal.

Ericsson promptly went to work on transforming the concept into building plans. He had agreed to a one-hundred-day delivery date once the contract was signed, but the inventor realized he needed money to place the ironclad's construction into action in order to meet the obligation. Therefore, a syndicate was formed that included Ericsson, Cornelius Bushnell, Congressman John A. Griswold and John Winslow. These men signed a U.S. Navy contract on 4 October 1861 to build, at the cost of $275,000, an "Iron-Clad Shot-Proof Steam Battery." Because of Ericsson's past troubles with the navy, a refund had to be made if the vessel failed to perform.

Ericsson's genius was evident in virtually every aspect of the project. He selected his partners not only for their wealth and political influence but also for their business connections. If Ericsson was to fulfill the contract's construction timeline, then he needed to subcontract many of the ironclad's

Concept to Construction

components to other firms. John Griswold's firm, Rensselaer Iron Works in Troy, New York, would make the rivets and the bar iron for the pilothouse. The Albany Ironworks, owned by John Winslow, would provide angle iron for the ship's frame, as well as some of the iron plate. Iron plate was also ordered from S. Holdane and Company of New York and H. Abbott & Sons. Even though Abbott was the largest and best-equipped mill in the nation, it could only produce one-inch plate, rather than the four-inch plate specified for the turret. Due to time constraints, Ericsson compromised on eight bands of one-inch plate for the turret.

Several features of Ericsson's original design were discarded or modified due to the rush to complete the vessel to counter the *Merrimack* threat. Ericsson had proposed to arm his ironclad with his own steam gun and proto-torpedoes (also called "hydrostatic javelins"), but these weapons were replaced by a conventional pair of XI-inch Dahlgren smoothbores. He had initially planned a sloping deck; instead, a simpler flat deck was constructed. The turret was originally conceived as a hemispheric turret. It, too, was replaced with a less complex, cylindrical turret. The turret itself was a massive structure, over nine feet high and twenty feet in diameter. The walls were constructed of eight layers of one-inch plate bolted together with overlapping joints. An additional layer was added to the turret's interior to protect the gun crew from flying bolts caused by cannon fire.

The ironclad itself was built at Continental Iron Works, at Greenpoint Ship Yard, in Brooklyn. The New York City firm of Novelty Iron Works, located across the East River from Greenpoint, was contracted to build the ironclad's turret. The Delameter Ironworks, located near Novelty Iron Works, received the critical task of building the vessel's engines. The power plant consisted of two "vibrating-lever" engine designed by Ericsson and two Martin boilers. All of the machinery was located in the hull, below

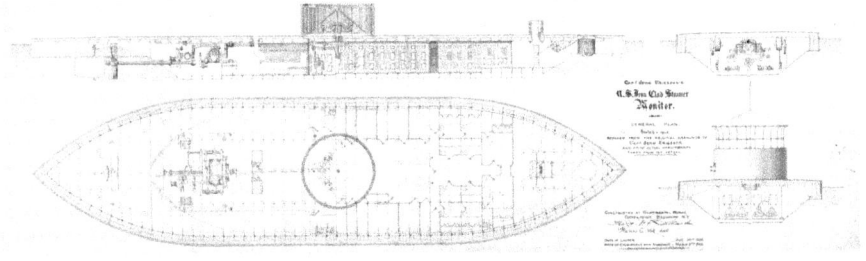

U.S. Ironclad Steamer Monitor. *John Ericsson, 1861. Courtesy of The Mariners' Museum.*

and behind the turret chamber. A watertight bulkhead separated the two sections, providing an airtight seal. The required air supply was brought into the engine compartment via a forced draft system, using two large deck air intakes and blowers. The engine was rated to produce four hundred horsepower under full steam, which enabled the ironclad to make seven knots (she was supposed to have a speed of nine knots). Another firm, Clute Brothers Foundry of Schenectady, New York, was contracted to build the special anchor-hoisting mechanism, engine room grates, gun carriages and the auxiliary steam engines.

The Abbott firm had made special curved one-inch plates that were shipped from Baltimore to New York and then fitted into the turret's frame by Novelty Iron Works. Once the turret was constructed, Novelty's workmen installed the massive port stoppers that had been made at the Niagara Steam Forge in Buffalo, New York. The port shutters were very large, wrought-iron shields, designed to drop down and close the gun ports. Once the shutters were installed, Ericsson discovered that, due to his own error, only one stopper could be raised at a time. Therefore, only one gun could fire at a time, which severely limited the warship's firepower. Ericsson instantly developed a remedying modification for this oversight, but the port stoppers would still prove cumbersome and difficult to operate in service. The problem would not be rectified until 17 April, when Edwin Gager noted in the ironclad's log that the gun crews "succeeded into getting both guns out at once."[9]

The 120-ton turret was far too heavy to transport intact to Greenpoint. Workers at Novelty disassembled it and carefully coded each part. The turret components were then barged across the river to Brooklyn and reassembled on the ironclad's deck at Continental Ironworks.

By early January, the ship was beginning to take its unusual shape. Congressman John Winslow, one of the ship's investors, advised Ericsson on 2 January 1862 that if the ironclad proved successful, he could guarantee twenty more ironclads would be built. As an act of confidence in the project, the U.S. Navy made its fourth payment of $37,500 to the partnership. Yet there was still much to accomplish if this experiment would be finished in time to counter the Confederate ironclad that was reaching completion in Hampton Roads.

Chapter 3
Into Service

Although the one-hundred-day construction timetable for *Ericsson's Battery* had expired on 12 January, the ship had already taken its unusual form and was sure to be completed in a few weeks. Accordingly, Gideon Welles selected Lieutenant John Lorimer Worden as commander of the ironclad. Worden was born in Mount Pleasant Township, Westchester County, New York, on 12 March 1818. He had served in the U.S. Navy for twenty-seven years and had a reputation as a "scientific" officer. Worden had served at sea, first aboard the USS *Erie* and then the USS *Cyane*, before he attended the Philadelphia Naval School. John Worden was named a passed midshipman in 1840 and was detailed to the Pacific Squadron, serving aboard the USS *Relief* and the USS *Dale*. His next assignment would be the start of his "scientific career," as Worden was transferred to the Naval Observatory in Washington, D.C., until the Mexican-American War erupted. He then saw service as the executive officer of the USS *Southampton* along the West Coast. Following other duties with the Pacific Squadron, he returned to the Naval Observatory for a two-year tour. In 1852, Worden cruised aboard the USS *Cumberland* in the Mediterranean, and then the USS *Levant* in the Caribbean, until being detached back to the Naval Observatory in 1855. Worden was transferred to the Brooklyn Navy Yard in 1856 and was named first lieutenant of the USS *Savannah* of the Home Squadron in 1858.

Worden received his first assignment during the Civil War, when he was given the mission to take secret orders to the Pensacola Squadron, telling them to reinforce the Union garrison manning Fort Pickens on Santa Rosa Island, Florida. On his return from this assignment, Worden became the first Union prisoner of war when he was captured on 13 April 1861 by Confederate authorities. Worden was held in a Montgomery, Alabama jail until he was exchanged on 22 November, returned to New York and hospitalized. He was still recovering his health when, on 13 January 1862, the day after the ironclad's one-hundred-day construction deadline expired, a letter arrived from Flag Officer Joseph Smith of the Ironclad Board assigning Worden as commander "of the battery under contract with Captain Ericsson, now nearly ready at New York. I believe you are the right sort of officer to put in command of her."[10]

Although John Worden's family—who called him "Jack"—was concerned that he was still too weak to return to duty, he immediately hurried to Greenpoint to review his new command. The same day, he advised Smith: "After a hasty examination of her," he was "induced to believe that she may prove a success. At all events, I am quite willing to be an agent in testing her capabilities."[11] Worden formally assumed command of the experimental vessel on 16 January.

However, Worden was not the first officer assigned to *Ericsson's Battery*. That distinction belonged to Alban Crocker Stimers, who, though not an official member of the crew, nonetheless was with her from the beginning as an official observer. Stimers was born in Smithfield, New York, on 5 June 1827. At twenty-one years of age, he was appointed a third assistant engineer and assigned to the USS *Water Witch*. He then sailed aboard the all-iron-hulled USS *Michigan*. Stimers was promoted to second assistant engineer on 26 February 1851 and transferred to the USS *Walker*, where he served until assigned to the office of engineer in chief. After service aboard such noted steamers as the USS *San Jacinto* and the USS *Merrimack*, he was given a special assignment in New York. He served briefly on the USS *Arctic* and was promoted to the rank of chief engineer in July 1858. Just before the war, Stimers had served on a board evaluating the completion of *Stevens Battery*. He was detailed as chief engineer of the USS *Roanoke*, and then he was given the task of superintendent of the *Ericsson's Battery* project on 4 October 1861. Stimers was never officially a member of the crew; however, he would remain with the ironclad during her voyage to Hampton Roads, present as an official observer.

While rumors persist that the navy assigned Stimers to the project to keep a "watch" on Ericsson, actually it was Ericsson who requested Stimers to be named as the overseeing agent. Ericsson wanted "an engineer of the highest intelligence,"[12] and he knew Stimers's previous service had well prepared him for this duty. Stimers was a very ambitious man, yet he and Ericsson became project soul mates.

Into Service

Another early appointment to *Ericsson's Battery* was Acting Assistant Paymaster William Frederick Keeler. This early assignment was due to the recently developed role of paymaster, which was equivalent to the rank of lieutenant in the U.S. Navy. The assignment of a paymaster was critical to preparing the ship for service. The position was clerical, and its duties included maintaining the ship's accounts, provisions, supplies and payroll. A paymaster was expected to exhibit strong writing and mathematical skills, all of which Keeler possessed from his business background. Keeler's assignment to *Ericsson's Battery* may have been the result of his wife's father's connection with Cornelius Bushnell. Anna Dutton Keeler's father, Henry Dutton, was head of Yale Law School and would eventually become governor of the state of Connecticut. The new ironclad had already become the "talk of the town," and people were fascinated with this new-styled warship. Keeler wrote in February: "Our vessel has been visited by hundreds of ladies who heartily expressed good wishes we shall carry with us."[13]

Keeler was not the typical sailor. He was born in Utica, New York, on 9 June 1821. He attended school there and eventually moved to Bridgeport, Connecticut. There, he married Anna Elizabeth Dutton in 1846. In 1849, he sailed to California during the Gold Rush aboard the *Anna Reynolds*, cruising around Cape Horn and then on to China and back. Both of his brothers died—James at sea and Edward in California—during this venture. He then returned to La Salle, Illinois, and established a shop on Main

Acting Assistant Paymaster William Keeler, USN, 1862.
Courtesy of The Mariners' Museum.

Street, selling and repairing watches, jewelry and other goods. By 1857, he had established La Salle Iron Works, Founders & Machinists. When the war erupted, Keeler, at age forty, sought service in the U.S. Navy because he considered slavery as a "hideous deformity" and believed the rebellious Southerners to be "traitorous and wretched souls." When he arrived at the Brooklyn Navy Yard on 4 January 1862, the processing officer noticed Keeler was from Illinois, saying, "We don't get a great many sailors from the prairies, out there." Keeler retorted that he had been at sea before but then naïvely asked if he needed a uniform or if he could just wear civilian clothing onboard. He was sternly told, "Get a uniform before you go to sea." Keeler did exactly that.

Meanwhile, the ironclad at Greenpoint was rapidly nearing completion. At the beginning of January, Assistant Secretary of the Navy Gustavus Vasa Fox asked John Ericsson for suggestions for a name for the battery. Ericsson replied, on 20 January:

> *In accordance with your request, I now submit for your approbation a name for the floating battery at Greenpoint. The impregnable and aggressive name of this structure will admonish the leaders of the Southern rebellion that the batteries on the banks of their rivers will no longer present barriers to the entrance of Union forces. The ironclad intruder will thus prove a severe monitor to those leaders. But there are other leaders who will also be startled and admonished by the booming of the guns from the impregnable iron turret. Downing Street will hardly view with indifference this last Yankee notion, this monitor. To the Lords of the Admiralty, the new craft will be a monitor...On these and many similar grounds I propose to name the new battery* Monitor.[14]

Fox could only agree with Ericsson's logic.

Since the *Monitor* was nearing completion, both Keeler and Worden had many tasks to perform. Keeler wrote his wife:

> *I have charge of all the provisions, clothing, stationery, what are called Small Stores, such as Tobacco, soap, candles, thread, buttons, needles, jack knives, & all the thousand and one little things a sailor will stand in need of—besides money...My steward's business is to give out the men's rations daily & render me an a/c* [account]. *Clothing, small stores and everything but the daily rations I issue myself.*[15]

Into Service

Keeler noted that he had nothing to do with arms, ammunition and other aspects of the ship's operation. These duties belonged to Worden or his designee. Keeler was a little taken aback by Worden's appearance when he first met him.

> *Capt. Worden is in the regular service. He is tall, thin, & quite effeminate looking, notwithstanding a long beard hanging down his breast—he is white & delicate probably from long confinement & never was a lady the possessor of a smaller or more delicate hand, but if I am not very much mistaken he will not hesitate to submit our ironsides to as severe a test as the most warlike could desire. He is a perfect gentleman in manner.*[16]

Two of Worden's most important tasks were assembling the officers and crew and securing cannon for the turret. Ericsson's original design, submitted to Emperor Louis Napoleon, featured his "hydrostatic guns." Since time was of the essence, he specified that XII-inch Dahlgrens should be mounted in the turret. Even though Dahlgrens had been produced specifically for the *Monitor*, these shell guns had been used by other vessels. Consequently, Worden could not find any of these large smoothbores available. He was, nevertheless, able to procure two XI-inch Dahlgrens from the USS *Dacotah*. These guns would not be transferred until 30 January.

Worden also began assembling the officers and crew. Lieutenant Samuel Dana Greene was named executive officer on 24 January. Greene, the son of Brigadier General George Sears Greene, was born in Cumberland,

Lieutenant John L. Worden, 1862.
Courtesy of The Mariners' Museum.

Lieutenant Samuel Dana Greene, 1862. *Courtesy of The Mariners' Museum.*

Maryland, in 1840. He graduated seventh in a class of twenty from the U.S. Naval Academy in 1859. He was warranted a midshipman on 9 June 1860. Greene served aboard the USS *Hartford*, on the China Station, and was promoted to lieutenant on 31 August 1861. He was detailed as the ironclad's executive officer on 28 January 1862. Keeler described Greene as "a young man also in the regular service, black hair & eyes that look through a person & will carry out his orders I have no doubt."[17]

The other officers included Louis Napoleon Stodder of Boston, Massachusetts, and John Joshua Nathaniel Webber of Brooklyn, New York, who were the two acting masters. Both men had previous sea service. Keeler remarked that both Stodder and Webber "are volunteers from the merchant service, good steady men."[18] Webber had enjoyed a rather dangerous and daring life at sea before joining the *Monitor*. He was from a seafaring family, his father and uncles having served on privateers during the War of 1812. He went to sea with his father at the age of three, and when he was seventeen, Webber was the third officer of a vessel involved in smuggling opium into China. Eventually, he became first officer of the *Harvest Queen*. The bark was heavily damaged during a strong gale, and twelve passengers were washed overboard. Webber acquired his own schooner just before the war; however, she was rotten and leaked so badly in a storm en route from New York to Florida that he was forced to run her ashore and abandon the vessel.[19]

The ironclad also needed experienced sailors for duty as petty officers. George Frederickson was born in 1834 on the island of Møn, in Denmark, but resided in Philadelphia, Pennsylvania. When Frederickson enlisted on 4 December 1861, in New York, he was immediately assigned as the ironclad's acting master's mate. Frederickson, Keeler believed, was "a good honest Dutchman." Quartermaster Peter Williams was born in Norway in 1831. His prewar home was in California, and Williams had nine years of sea service before enlisting in New York on 27 January 1862. The other quartermaster was Peter Truscott. Truscott's real name was Samuel Lewis. (The use of an alias when enlisting was not an uncommon practice for seamen to follow. It allowed them to desert anonymously if they were dissatisfied with a posting.) Peter Truscott had five years of prior service in the U.S. Navy before enlisting in New York as a seaman on 20 January 1862. He was transferred from the U.S. Receiving Ship (USRS) *North Carolina* on 6 March 1862 as ship's number eleven. (Each crew member was assigned a special ship's number upon being transferred to the ironclad. The number is chronological and reflects the date of muster aboard the warship. Ship numbers one to forty-nine were members of the ship's crew when she eventually left New York.)

John Stocking, the alias of Wells Wentz, was from Binghamton, New York, and enlisted as a boatswain's mate on 25 January 1862. He was transferred from the USS *Sabine* to the *Monitor* by 6 March 1862 as ship's number forty-three. Landsman Francis Banister Butts—known as "Frank"—believed that Stocking was "one of the very best types of an American sailor, and my tutor in seamanship."[20] Joseph Crown of New York City was the gunner's mate. He enlisted on 20 January 1862 and, by 6 March, was transferred from the receiving ship *North Carolina* to the *Monitor* as ship's number ten.

On 27 January 1862, Lieutenant Worden reported to Secretary of the Navy Gideon Welles that he had surveyed the vessel and understood the manpower required to operate the vessel.

> In estimating the number of her crew I allowed 15 men and a quarter gunner for the two guns, 11 men for the powder division, and 1 for the wheel, which I deem ample for the efficient working of her guns in action. That would leave 12 men (including those available in the engineer's department) to supply deficiencies at the guns, caused by sickness or casualties.[21]

The ship would require ten officers: four engineers, one medical officer, one paymaster, two masters, an executive officer and a commander. These individuals would be:

- Lieutenant John L. Worden, commander
- Lieutenant Samuel Dana Greene, executive officer
- Acting Master Louis N. Stodder
- Acting Master J.N. Webber
- First Assistant Engineer Isaac Newton Jr.
- Second Assistant Engineer Albert B. Campbell
- Third Assistant Engineer Robinson W. Hands
- Fourth Assistant Engineer Mark T. Sunstrom
- Acting Assistant Paymaster William F. Keeler
- Acting Assistant Surgeon Daniel C. Logue

Four of the ship's officers were of the line. They were responsible for the handling of the vessel and operating her guns in action. Two of the officers, Keeler and Logue, had other assignments beyond the daily operation of the vessel. The engineering officers were considered a class unto themselves.

Other than Alban C. Stimers, the most experienced engineer aboard was First Assistant Engineer Isaac Newton Jr. Newton held a degree in civil engineering from the University of the City of New York, as well as a State of New York Engineer's Certificate. Isaac Newton had previously served as assistant engineer on the New York to Liverpool packet run. He joined the U.S. Navy on 15 June 1861 as a first assistant engineer aboard the USS *Roanoke*. He eventually was transferred to the *Monitor* on 7 February 1862.

Serving under Newton were three other engineers. Albert Bogart Campbell had joined the U.S. Navy on 26 August 1859 as a third assistant engineer. He was promoted to second assistant engineer in October 1861 and was responsible for actually operating and maintaining the machinery. Robinson Woollen Hands, who was born at sea aboard his father's vessel, was from Baltimore, Maryland, and was a student of mechanical engineering before being named third assistant engineer for the *Monitor*. Robinson's brother, Captain George Washington Hands Jr., was a Confederate infantry officer and considered Robinson a traitor. The junior engineering officer, Mark Trueman Sunstrom, was also from Baltimore. Prior to the war, his occupation was as a bookkeeper.

The ship also required petty officers, and the leading senior enlisted men included:

Into Service

Fourth Assistant Engineer Mark T. Sunstrom, 1862. *Courtesy of The Mariners' Museum.*

- Gunner's Mate Joseph Crown
- Master's Mate George Frederickson
- Master-at-Arms John Rooney
- Boatswain's Mate John Stocking
- Quartermaster Peter Williams
- Quartermaster Peter Truscott
- Captain's Clerk Daniel Toffey

Before Worden could gather these men together and muster a crew, the *Monitor* must be completed.

At 10:00 a.m. on 30 January 1862, the *Monitor* was launched at Greenpoint. There were many skeptics present. One observer told Acting Master Louis Stodder: "You had better take a good look at her now as you won't see her after she strikes water. She's bound to go to the bottom of the East River and stick there, sure."[22] Instead of sinking, the *Monitor* slid down the ways, a defiant Ericsson standing on deck, and floated just as designed. The Union command was elated and ready for the ironclad to see service. G.V. Fox telegraphed Ericsson: "I congratulate you and trust she will be a success. Hurry her for sea, as the *Merrimack* is nearly ready at Norfolk and we wish to send her there."[23] Gideon Welles had hoped to get the *Monitor* to Hampton Roads while the Confederate ironclad was still under construction. Welles believed that the *Monitor* could easily steam up the Elizabeth River past the Confederate batteries and destroy the *Merrimack* as she sat in dry dock.

Worden now needed to redouble his efforts at securing the crew, as the *Monitor* was being readied for commissioning. The fifty-seven men who made up the *Monitor*'s crew were all volunteers. Since the ironclad was an experiment, and different from any other vessel in the U.S. Navy, he did not wish to accept men just arbitrarily assigned from receiving ships at Brooklyn Navy Yard. Instead, Worden went aboard the *North Carolina* and the *Sabine* to ask for volunteers. The response was overwhelming, as more men volunteered than were needed. The chosen enlisted men would be transferred to the *Monitor* from early February through to the day of the *Monitor*'s sailing. Worden needed firemen, coal heavers and ordinary seamen to operate the vessel. Many of the men lacked experience and listed their prewar occupations as farmer, machinist, carpenter, stonecutter, sail maker or none. Nevertheless, several of these volunteers had previous sea service. William Bryan had naval experience before enlisting as a seaman at age thirty-one on 19 July 1861. He had been detailed to the USS *Ohio* but was transferred from the *Sabine* to the *Monitor* on 6 March to serve as yeoman, ship's number thirty-nine. John Ambrose Driscoll, from County Cork, Ireland, had abandoned his wife and children and immigrated to America. Driscoll, who also used the alias "John White," enlisted on 15 February 1862 as a first-class fireman for a 3-year term. By 6 March 1862, he had been transferred to the *Monitor* as ship's number twenty. Driscoll later recalled:

> While we were lying at the navy yard at Brooklyn, NY, prior to starting for Hampton Roads, Va, all manner of uncomplimentary and satirical remarks were made with regards of her fate[.] When she got to sea one declared…the first heavy sea to wash her decks would swamp her for how

could such a mass of iron float if it once got under water[.] *Another old sea dog who had followed the sea all his life remarked that if she got into a fight any ordinary ship would run over her with ease, or if boarded by a strong party they could wedge the turret and work the guns in such cramped quarters*[.] *On one occasion an old seaman said to the writer in a very solemn and prophetic tone that thing you are going in will never stay up long enough to get out of sight of Sandy Hook. You fellows certainly have got a lot of nerve or want to commit suicide one or the other.*[24]

These comments, as well as the very sight of the low-lying iron warship virtually awash with the sea in calm water and the unusual living space below the waterline, prompted several seamen to desert shortly after they arrived on the *Monitor*. Master's Mate George Frederickson noted in the *Monitor*'s log on 4 March, "Norman McPherson and John Atkins deserted taking the ship's cutter and left for parts unknown so ends this day." Coal Heaver Thomas Feeney deserted seven days after he enlisted and the very day he arrived on the *Monitor*. Seaman Francis A. Riddey, also known as Frank Ryeday, was a sailor from Philadelphia. He deserted on 21 February but later reenlisted and served as a gunner's mate on the USS *Princeton*.

Despite these problems, many volunteered for service aboard the ironclad as a sense of duty or as an opportunity to find a place in their new nation. Hans A. Anderson was originally from Gothenberg, Sweden, and had served in the Swedish Merchant Marine. Sometime in the early 1850s, he had immigrated to the United States and served in the American Merchant Marine until enlisting in the U.S. Navy on 29 December 1856. He sailed on vessels like the USS *Falmouth* and the USS *Congress* until discharged with the rank of coxswain. He reenlisted as a seaman and was transferred from the *North Carolina* to the *Monitor*, ship's number nine, as acting quartermaster. Another Scandinavian was Charles, also known as Philip, Peterson, from Norway, who served as quartermaster, ship's number five.

Several other European natives shipped aboard the *Monitor*. Seaman Anton Basting and Carpenter's Mate Derick Bringman were from Germany, while Coal Heaver William (Wilhelm) Durst was a Jew from Austria. Several of the crewmen were born in the British Isles, including Seaman James Fenwick (Scotland); Seaman Daniel Walsh and First-Class Fireman Hugh Fisher (Ireland); and Coal Heaver David Roberts Ellis (Wales).

Many of the men who shipped aboard the *Monitor* were individuals who had joined the service in an effort to help preserve the Union, and little is known of their prior lives. There were two men with the name of Thomas

Carroll among the original volunteers, so they were officially designated as #1 and #2. Thomas Carroll #1, from Boston, had enlisted as a seaman for a 2-year term. He was five feet, six and a half inches tall, with brown hair, blue eyes and a fair complexion. Carroll #1 transferred from the *Sabine* and served as the captain of the hold, ship's number nineteen. The other Thomas Carroll, noted as #2 in the ship's log, was born in Ireland and immigrated to the United States, though for some unknown reason he claimed his birthplace to be Ritchfield Spring, New York, when he enlisted. He stood five feet, six and a half inches tall and was nineteen years old at the time of his enlistment. Carroll #2 was described as having gray eyes, brown hair and a fair complexion. Carroll #2 served as first-class boy, ship's number forty-four, and was noted as being a "strong robust boy." First-Class Firemen John Garety and Patrick Hannan were both from Ireland and both transferred from the *North Carolina* for service on the *Monitor* as ship's numbers thirty-six and twenty-five, respectively. Michael Mooney also hailed from Ireland and was a grocery store clerk in New York City when he enlisted in Company H, Twelfth New York Volunteer Infantry Regiment. Following his five-month enlistment, he returned home only to enlist in the U.S. Navy for a 3-year term as a coal heaver, volunteering to transfer to the *Monitor* as ship's number twenty-seven on 25 February 1862. Two African Americans were among the initial crew. One of them was William H. Nichols, born a freeman in Brooklyn, who enlisted in New York on 13 February 1862 as a landsman for a 3-year term. He was ship's number seventeen, nineteen years old and described as a "dark mulatto."

While crew members like John Rooney had nine years' prior service in the U.S. Navy and assumed important stations (Rooney was master at arms, ship's number fourteen and known as the "funny man" of the berth deck) aboard the *Monitor*, others, such as George Spencer Geer, saw service in the U.S. Navy as an opportunity for advancement. Geer was originally from Troy, New York, where he had worked in his father's stove foundry. He moved to New York City but had lost his job and fallen into debt in order to support his wife and two children. Geer enlisted as a first-class fireman on 15 February 1862 to earn money (this rating earned him thirty dollars per month rather than the eighteen for a coal heaver) and learn a reliable trade. Five days later, he was transferred from the *North Carolina* to the *Monitor* as ship's number twenty-four.

Lieutenant Worden also needed staff to support him and the other officers on the ship. Worden selected his nephew Daniel Toffey of Pawling, New York, as the captain's clerk. Worden's wife was Toffey's aunt, Olivia

Into Service

Toffey Worden. Acting Assistant Paymaster Keeler was overjoyed that Worden had secured a clerk, noting to his wife, "[He] will serve to make my duties still lighter, & more, they give me a servant." Keeler referred to his servant as "contraband,"[25] which meant that Keeler's personal servant was an African American. The word "contraband" was often used to describe runaway slaves, a term based on Major General Benjamin Franklin Butler's 24 May 1861 decision not to return slaves coming into Union lines at Fort Monroe, Virginia, but to view them as contraband—that is, as illegal trade goods. Butler's "Contraband of War" decision ended the Fugitive Slave Act in reference to the Southern states that had seceded and were at war with the Union. Butler reasoned that since slaves were chattel property, and the former states like Virginia were at war with the Union, they could be confiscated by the U.S. Army. Besides his "contraband," Keeler was also detailed a steward, Robert Knox Hubbell of St. Louis, Missouri.

The original wardroom steward was Lawrence Murray. During a dinner served just before the *Monitor* left New York, he "had been testing the brandy & champaine [sic] before putting it on the table," according to William Keeler. The dinner was a failure, he continued, because "the fish was brought in before we had finished the soup & champaine [sic] glasses were furnished us to drink our brandy from & vice versa."[26] The drunk Murray was placed in irons and confined to the anchor well, an ersatz brig, and was released once sober. However, by 4 March, Murray was drunk again. The original captain's steward was David Cuddeback of Port Jervis, New York, who was later promoted to ship's cook.

As the crew was being secured, the *Monitor* was turned over to the navy for testing on 19 February 1862. Overall, the *Monitor* was 173 feet in length, weighed 987 tons and had a beam of 41.5 feet. The ironclad's draft was 10.5 feet with a freeboard of only 1.5 feet. Since the *Monitor* was an experimental vessel and was hastily constructed, numerous problems were discovered. The communication link (a speaking tube) between the pilothouse and turret only worked well when the turret faced toward the pilothouse. This circumstance would certainly compromise communications during battle between the gun crew in the turret and the ship's commander in the pilothouse. Ericsson originally planned to place the pilothouse atop the turret, but the construction timeline prohibited this time-consuming fabrication. Consequently, it was isolated near the ironclad's bow and was at possible risk from an errant broadside. (Ericsson later admitted that this was an "omission.") The pilothouse was accessed through a hatch in the floor and was constructed of iron logs, 9 by 12 inches thick, which were bolted to oak beams below the deck. The ¼-inch gaps offered the only views out in all directions. The pilothouse measured 4 feet by 5 feet and stood 4 feet above the deck. Although there was over 6 feet of vertical space, the horizontal area was limited. Therefore, the pilothouse's interior was very cramped, considering that the quartermaster had to work the wheel as the pilot watched the waters and the ship's commander used it as a command post.

Of even greater concern were problems described during the ironclad's initial trials. Her brief trip on 19 February across the East River from Greenpoint to the Brooklyn Navy Yard, was troublesome. The *Monitor* experienced steering problems, and Chief Engineer Alban Stimers reported that the steering required significant repairs. In addition, the ship could only make three and a half knots instead of the planned nine knots. This problem was corrected with an engine valve adjustment. Still, Ericsson and Stimers could only coax six knots out of the engines.

Nevertheless, the USS *Monitor* was commissioned at the Brooklyn Navy Yard as a third-rate steamer on 25 February 1862. Captain John A.B. Dahlgren called the vessel "a mere speck, like a hat on the surface." Commander David Dixon Porter, who had been assigned the task of assessing the ironclad's combat capabilities, declared the *Monitor* "a perfect success, and capable of defeating anything that then floated."[27] Lieutenant Worden, meanwhile, prepared the ironclad for sea. The crew was mustered aboard, and all were amazed by the vessel. William Keeler commented to his wife, "I shall not attempt a description of her now, but you may rest assured your better half will be in no more danger from rebel compliments than if he was seated with you at home. There isn't danger enough to give us any glory."

Quartermaster Peter Truscott later recounted, "She was a little the strangest craft I had ever seen."[28] The *Monitor* featured over forty of Ericsson's patented inventions and was totally different from anything else afloat. The turret was the dominant feature. It weighed 120 tons, was placed onto a brass ring set into the deck and would turn on high ball bearings, each ten inches in diameter. A small steam engine, connected to a central, vertical drive shaft by four large horizontal gears, placed the turret in motion. This entire system was operated by only one man. The turret, which was the warship's most lasting contribution to naval architecture, was twenty-one feet in diameter and protected by eight layers of one-inch iron plate. It was rounded to help deflect shot. When in proper alignment, the two floor hatches, one next to each gun, allowed access to the deck below. This design feature enabled shot, shell and powder to be passed up to the turret during

Into Service

combat and provided a protected pathway to reach the outside through the turret's roof. Two ladders provided access from the turret's top to the deck and were gathered in whenever the ship was underway. The roof was covered with perforated plate, which permitted smoke and fumes to escape during action and permitted some air and light to enter at other times. The turret was fitted with an awning supported by a wooden center pole and iron stanchions on its edge. When in action or underway, the canopy would be removed and the hatches closed. Then the only access to the outside world was through the top of the turret.

The two XI-inch Dahlgrens from the USS *Dacotah* were fitted into Ericsson's specially designed carriage, with brakes tightened by hand cranks to accommodate recoil. Since the big guns were mounted side by side, the crank housings were placed on the outside of the gun carriages so the crew would not have to stand between the guns to make adjustments. This arrangement seemed to imply that the wheels were mirror images, having to be turned in different directions to tighten the brakes, but both were designed with a left-hand screw and were supposed to be turned in the same direction, counterclockwise. During the 19 February test run, Alban Stimers turned the hand wheel on gun number one as if it were a right-hand screw and then ordered it to be fired. The right-hand turn served to release

Interior view of the turrets of the Monitor *fleet, 1862. Courtesy of The Mariners' Museum.*

the friction gear dissipating the gun's recoil, and the Dahlgren jumped off the carriage, bouncing off the rear wall of the turret. Fortunately, no one was injured. Stimers, believing the second crank wheel should be turned in the opposite direction, again relieved any broking power the gun carriage had and caused the same startling result with gun number two, amazingly again with no injuries.

The Dahlgrens were mounted in the confined space of the turret. Each XI-inch Dahlgren weighed over eight tons, and their massive size left little room in the turret for the gun crews. Solid shot, made by Novelty Iron Works, was stored along the base of the turret beside the guns. The turret was fitted with huge, pendulum port shutters to protect the turret's interior when the guns were being reloaded. In fact, a special loophole was bored into each shutter to allow the cannon to be sponged, wormed and reloaded when the shutter was closed. The guns were retested on 4 March, as Master's Mate George Frederickson reported: "First of firing blank cartridges, second a stand of grape, third with cannister with a full charge of grape."[29] A 15-pound charge of powder would enable the Dahlgrens to hurl a 187-pound solid shot or a 168-pound explosive shell up to 1,700 yards. The *Monitor* was under orders by naval officials not to use the rated 30-pound powder charge when in action, due to their fear of a bursting gun within the turret.

The *Monitor* actually incorporated two separate hulls. The top (called the upper hull by Ericsson) was in the form of a flat iron raft with sharp ends. An armor belt, made of five layers of one-inch graduated iron plate, protected the 5-foot-deep vertical sides. Hence, courses one, two and three ran the full 5-foot course, course four ran for 3 feet and course 5 ran for 2.6 feet. The iron was backed by thirty-inch oak and pine to protect the belt from shot, shell and ramming. The upper hull overhung the lower hull and protected both the propeller and anchor well. The deck itself was covered by two layers of half-inch-thick iron plates laid over the deck beams.

The lower hull was actually a flat-bottomed iron cradle that hung under the raft, below the waterline. The two hulls were joined together by vertical stanchions, brackets and rivets. This seal had to be maintained or else the ship would lose its buoyancy. The lower hull consisted of half-inch-thick iron plates butted together over a wooden frame. Fully loaded, the vessel had a draft of only ten feet, six inches, and its deck reached just eighteen inches above the waterline. The only features seen above the deck were the turret amidships, with its rounded shape to help deflect shot, and the pilothouse, near the bow. All other features of the ship—including bollards, ventilator boxes and smoke boxes—could be removed before going into combat.

Into Service

The *Monitor* was virtually a submarine. All crew spaces and the ship's operations were located below the waterline. There, the crew lived and worked, often unaware of the world above. Everything mechanical was protected from both the sea and enemy shot, and this made for a unique experience.

The vessel's interior was divided into two halves by an iron bulkhead that supported the turret. The rear section contained the ship's machinery, the galley and the heads, or toilets. Two large iron hatchways controlled access to the ironclad's interior. The crew hatch was located abaft the midship bulkhead. The forward hatch was primarily for use by the officers. When closed, these hatches established an airtight seal, required to maintain draft to the furnaces firing the boilers. Air was forced in by belt-driven blowers. To guard against fire, the floor for the engine and galley areas was covered with diamond-patterned cast-iron plates. The galley itself was located at the front of the rear section. The large cast-iron stove backed onto the boilers. The galley area also held the heads, "water closets" designed by Ericsson. The *Monitor* was the first warship with below-the-waterline flushing toilets and required the operation of several valves to remove the waste. Acting Assistant Surgeon Daniel Logue suffered the indignity of being blown off his seat by a jet of water when he operated the valves in the wrong order. Two water closets for enlisted men were on the starboard side, and one for officers was on the port side.

The two Martin boilers were situated behind the galley stove. They were fourteen by nine feet and provided all of the steam necessary to power the ship's machinery. Ericsson's four-bladed propeller was powered by a "vibrating-side-lever" engine. Ericsson had designed the engine to operate in a confined space. The thirty-ton, four-hundred-horsepower engine, with pistons three feet in diameter, stood on a raised platform, which enabled the engineers to have access to all of the engine's moving parts. At the rear of each engine was the valve chest used to operate the machinery. Along each side of the engine area were the bunkers, designed to hold an eight-day supply of anthracite coal. The pumps and other auxiliary machinery were located to the rear. The *Monitor* had three pump systems. The Worthington Steam Pump, the Adams Centrifuge Pump and a bilge injection pump were installed to meet any emergency the ship might encounter at sea. The steam system also heated the vessel. The remaining space was consumed by steering mechanisms connected to the rudder.

Forward of the main bulkhead could be found the berth deck, officers' quarters, magazine and shell room. The open area beneath the turret was the enlisted berth. It was crowded and dark and used oil lamps for lighting. The walls were painted white to reflect light. George Geer believed that his new ship was "so much more comfortable than the old *North Carolina*,"[30] while John Driscoll recalled that he and his fellow shipmates "were made all manner of fun by those on bard [*sic*] for going to sea in a tank as most of them term it."[31]

The powder magazine was located on the starboard side of the berth deck. It measured nine feet square and was lined with lead to prevent any sparks. Copper tanks were provided for the storage of loose powder, cartridges and explosive shells. A light room was attached to this room, from which an oil lamp illuminated the powder room through a thick glass porthole. In case of fire below deck, the magazine could be flooded by a seacock. The iron structure had a foyer-style entrance with a heavy iron door; during action, the entranceway was covered by a thick, canvas curtain. On the port side was the iron-walled shell room, which held projectiles, fuses and cannon primers. The solid shot were stored in the middle of the berth deck, ready to be hoisted up into the turret via the main hatch.

Forward of the berth deck, and separated by a wooden bulkhead, were the officers' quarters and the wardroom, where the officers socialized and ate their meals. Ericsson had personally underwritten the cost for all of the officers' furnishings, including an oak table, hardwood chairs, lanterns and shelves. The rooms were finished with canvas floor coverings, oriental rugs, goat-skin mats, lace and damask. Blowers at the stern of the vessel ventilated the berth deck, officers' quarters and wardroom with fresh air. Eight cabins for the officers lined the wardroom. These spaces were furnished with black walnut berth, drawers and closets. Lighting for each living area was by oil lamp, candles and skylight. The skylights were six inches in diameter and set into the deck with thick glass. Iron plates covered these lights when the ship was made ready for combat. The daylight was often filtered by seawater that washed across the deck. William Keeler noted, "I have seen no room as handsomely fitted up as ours. The only objection is that they are too dark." He later wrote to his wife, "Whenever I write, day or night in my state room, I have to use a candle. My little deck light lets in light enough for all purposes except reading & writing." Nonetheless, he called his room his "little snuggery."[32]

The captain's cabin and stateroom were of similar size—sixteen feet square—and were located across a narrow passageway from each other. Both were equally well furnished, and the cabin had its own flushing head. The stateroom was for the captain to entertain important guests and to conduct ship's business. The passageway between the captain's spaces led to the chain locker at the ship's bow. The anchor system was sited just forward of the ladder leading up to the pilothouse. The anchor could be raised or lowered without exposing crew members to the outside world.

The *Monitor* indeed featured an undersea world. Paymaster William Keeler accounted: "Nothing would strike a stranger with more surprise after walking our cheerless, wave-washed iron deck than to go below and see our bright, well-lighted, cozy wardroom with the officers grouped around the table reading, writing, or talking." He added: "The dash of the waves as they roll over our heads is the only audible sound that reaches us from the outer world. One would hardly suppose from the quiet stillness that pervades our submarine abode that a gale was raging around us...Our life I assure you is getting monotonous enough."[33]

When author Nathaniel Hawthorne visited the *Monitor* in March, he was surprised by the new world beneath the sea: "It was like finding a palace, with all its conveniences, under the sea...[the crew] hermetically seal themselves and go below; and until they seem fit to reappear, there is no power to man in hereby they can be brought to light. A storm of cannon-shot damages them no more than a can of peas."[34]

As the *Monitor* was being readied for sea, William Keeler took great pride in escorting dignitaries, curiosity seekers and the media on tours of the new ironclad. Ericsson's ironclad epitomized the strengths of American industry, and this fired an interest among the public. Keeler was proud of his new uniform and noted to his wife that it brought great attention to him:

> *Our vessel has been visited by hundreds of ladies who heartedly expressed good wishes we shall carry with us. The duty devolved upon me to show most of these around the vessel. You can imagine your polished & accomplished husband <u>shining</u> in this new sphere—I believe I got along well enough. I rubbed up my antiquated & somewhat indistinct ideas of etiquette & bright buttons & shoulder straps made up any deficiency.*
>
> *I find that they (buttons &c) are a sure passport to the notice of the weaker sex & I rather enjoy the idea of handling those around who, if I were dressed in other clothes, would scarcely notice me.*[35]

Keeler gave tours of the vessel to his father-in-law, dignitaries like the former consul to Lisbon and the editor of the *Scientific American*.

Once the *Monitor* was commissioned, the crew was mustered onboard and everything was being made ready for the voyage south. The "men were at work at every part of the boat where it was possible to work," Keeler wrote his wife. He also noted that the entire crew was "getting impatient &

want to get alongside the *Merrimac.*"[36] News about the Confederate ironclad, under construction at Gosport Navy Yard in Portsmouth, Virginia, ranged from false propaganda published in the *Norfolk Day Book* indicating that the *Merrimack* was a complete failure and "she will make an invaluable floating battery for the protection of Norfolk, better good for something than nothing,"[37] to reports from Union commanders in Hampton Roads that the ironclad would soon strike at the Federal fleet. Gideon Welles wanted to get the *Monitor* immediately underway and gave orders to Worden to "proceed with the USS *Monitor*, under your command, to Hampton Roads, Virginia."[38]

However, the *Monitor* was still experiencing technical difficulties. On 27 February, the *Monitor* went on a trial run during a storm. The ironclad steamed into the East River, but the steering became extremely erratic. "We ran first to the New York side then to the Brooklyn and so back and forth across the river, first to one side then to the other, like a drunken man on a sidewalk,"[39] Keeler recalled. Eventually the *Monitor* crashed into a dock and had to be ignominiously towed back to the Brooklyn Navy Yard. Stimers, in his report, stated: "The man at the wheel had not sufficient command over the rudder to enable him to steer the vessel."[40]

Flag Officer Hiram Paulding, commandant of the Brooklyn Navy Yard, suggested that the *Monitor* be placed back into dry dock to correct the steering problem with a new rudder. John Ericsson flatly refused, declaring, "Put on a new rudder. The *Monitor* is mine and I say it shall not be done.

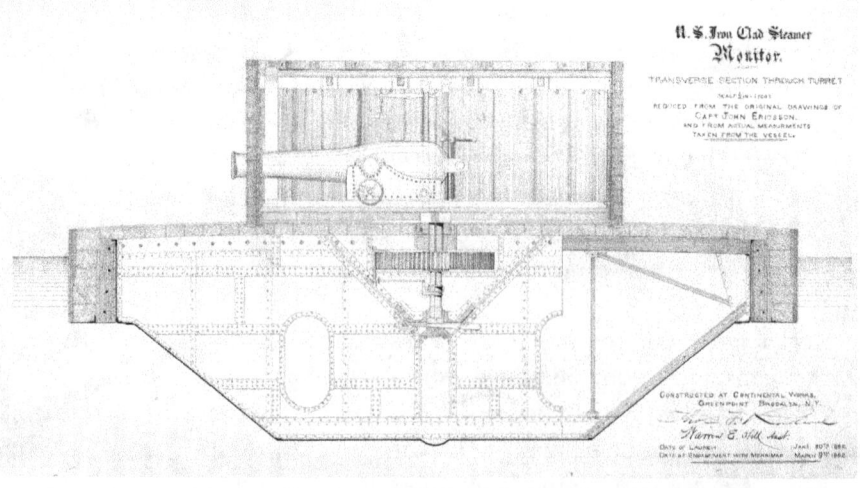

U.S. Ironclad Steamer Monitor. John Ericsson, circa 1861. *Courtesy of The Mariners' Museum.*

Into Service

They would waste a month in doing it. I will make her steer easily in three days."[41] The rudder was seriously overbalanced, but Ericsson and Stimers corrected the problem by installing a series of pulleys between the tiller and the steering wheel drum. The result enabled the *Monitor* to turn in three times her length of 173 feet within five minutes. Worden was still worried about how the *Monitor* would handle in bad weather and insisted upon trials off Sandy Hook under the review of a board including Flag Officer Francis H. Gregory, Chief Engineer Benjamin F. Garvin and Naval Constructor Edward Hartt. The *Monitor* passed this inspection on 4 March 1862 and was deemed ready for sea. Some were still unsure. Seaman David R. Ellis made perhaps the most telling remark about the ironclad as it readied to leave New York, commenting, "She had not been pronounced seaworthy, and no one could safely judge of her fighting qualities."[42]

Nevertheless, Flag Officer Hiram Paulding ordered Worden, "When the weather permits, you will proceed with the *Monitor* under your command to Hampton Roads and on your arrival report to the senior officer there… wishing you a safe passage."[43] A powerful coastal storm moved up the coast the very day Worden received his orders, and the ironclad's commander decided to delay his departure. Worden recognized, despite Ericsson's claims to the contrary, that the *Monitor* was not really designed to be an oceangoing vessel, and he felt it best to wait for good weather.

Chapter 4
A Day Late

On the morning of 6 March 1862, Worden ordered the *Monitor* to cast off from the quay at the Brooklyn Navy Yard. Crew member John Driscoll remembered:

> *It was customary at that time when a craft was going on the blockade for the crew of the receiving ship* North Carolina *to man the netting and give the departing vessel three cheers, as she passed down the Bay, also for the ferry boats and other steam craft to toot there* [sic] *whistles as a parting salute. No*[t] *so when the* Monitor *started out. As we passed the* North Carolina *not a head was seen above the rail not a whistle sounded to cheer us as we went out. Those we passed seemed to think it would be better to have played the funeral dirge than to have given us the customary cheer.*[44]

Since the *Monitor*'s engines did not make the specified nine knots, and due to questions regarding the ironclad's seaworthiness, it was deemed best to tow her southward. The gunboats were assigned to the flotilla, not for protection (each ship mounted only four thirty-two-pounders and one twenty-pounder rifle), but rather they were intended to serve as rescue vessels. Once into the bay, Worden ordered the four-hundred-foot towline secured to the steam tug *Seth Low*. Then she joined her escorts, two steam-screw gunboats, the USS *Currituck* and the *Sachem*, and by 4:00 p.m., the ships had passed Sandy Hook and turned south toward the Chesapeake capes.

During the first evening at sea, the officers congratulated themselves on the good weather. Stimers noted in a report to Ericsson: "I never saw a vessel more buoyant or less shocked…There has not been sufficient movement to disturb a wine glass setting on the table." William Keeler remembered standing atop the turret with "the water smooth & everything seems favorable…Not a sea has yet passed over our deck, it is as dry as when we left port."[45]

When Keeler awoke the next morning, he "found much more motion to the vessel & could see the green water through my deck light." He noted that several of the crew, particularly Captain Worden and Dr. Logue, were suffering from extreme seasickness and had been moved atop the turret. A gale had worked its way up the coast, and as Alban Stimers noted, "The sea commenced to wash right across the deck."[46] Even though Ericsson had designed the heavy turret to form a watertight seal fitting into the brass ring on the deck, Brooklyn Navy Yard officials had insisted on jacking the turret on wedges and caulking it with oakum, loose fiber used to fill the seams of a boat to make it watertight. This modification failed to stop water entering the ironclad; in fact, the water gushed through the turret like a waterfall. William Keeler commented that this flood of water "drowned out the Sailors whose hammocks were hung on the berth deck immediately below. The water was coming down this morning from under the tower & from the hatches & deck lights & various other openings making it wet & very disagreeable below."[47]

Executive Officer Samuel Dana Greene wrote that the angry, ten-foot waves "would strike the pilot-house and go over the turret in beautiful curves, and it came through the narrow eye-holes in the pilot-house with such force as to knock the helmsman completely around from the wheel."[48] Paymaster Keeler noted "the top of every sea that breaks against our side rolls unobstructed over the deck dashing and foaming at a terrible rate."[49]

Ericsson had originally designed the *Monitor* to have no stacks; however, he had fitted the ventilation ducts and smokestack with temporary six-foot funnels to protect against high seas. Nevertheless, water soon began to come down the blower pipes, leaking onto the leather belts operating the blowers circulating air throughout the vessel. The soaked belts soon failed, and the blowers stopped. Without this necessary airflow, the coal fires were extinguished, and fumes quickly overcame the engineers and firemen. They were saved from suffocation only by being taken to the fresh air at the top of the turret by a rescue team led by Lieutenant Greene. Second Assistant Engineer Albert Campbell remembered the scene:

> *The next afternoon we broke both our blower belts which spoiled the draft of our fires and drove all the gas into the engine room. This of course was rather inconvenient, for carbonic acid gas and hydrogen is not calculated to*

support animal life. I found myself getting weak and lost all consciousness and did not know any more until I found myself on top of the turret with a couple of engineers lying alongside of me, looking more dead than anything else.[50]

The engineers opened the engine room doors, flooding the entire ship with gas. Keeler went below to investigate and met one of the engineers "pale, black, wet, and staggering along gasping for breath." Chief Engineer Stimers ordered everyone out of the engine room in an effort to fix the problem himself, until he found he "was also getting very much confused in head and very weak in the knees."[51] He was forced to leave.

John Driscoll recalled the incident:

I was on duty in the fire room at that time[,] the fire and engine room being both in one[,] at eight bells. 4 P.M. I was relieved by the next watch[.] At that time there seemed to be no danger of a serious nature threating [sic] [.] There is on board ship, what is known as dog watches[.] The first runs from six to eight P.M. Coming off duty at four I was scheduled to go on again at eight[.] Realising [sic] my need of sleep I retired to a loft under the turret where the hammocks were stowed and then I went to sleep and that sleep saved me from the terrible fate that befell all the other 18 engineers and firemen who were suffocated with gas which came from the furnaces[.]

...I had scarcley [sic] gotten asleep when the belt on the port side blower flew off. Engineer R.W. Hands, who was on duty at that time[,] with the assistance of the crew shortened and laced the belt. By this time the fan box was full of water[.] Consequentley [sic] on every attempt to start the engine the belts would fly off[.] While attempting to get the port blower started[,] the star-board [sic] belt...blew off and since all draft was cut off the gas soon filled the engine room, suffocating all who was [sic] in there at that time. The other firemen on the berth deck smelling the gas, rushed in a body to the engine room...and dragged out those who were overcome[.] The last man to be carried out was Chief Engineer Newton.

The ladder leading to the turret was very close to where I was asleep, and when Newton...was being carried up the noise awakened me. So I rushed like the others to the engine room[.] The only means of reaching the engine room was by a narrow passage leading from the berth deck and passing between the boilers and coal bunkers. As the pressure of gas was so strong I was forced to retreat but by tyeing [sic] a silk handkerchief over my mouth and nose and keeping so close to the floor as possible, I succeeded

> *in reaching the engine room and it was thick with gas*[.] *Like the others I tried to start one of the blowers but the belt flew off…Rushing into the store room I procured a hammer and chisel and knocked a hole through the sheet box*[.] *While I was working the water from the blower was rushing over me but it helped to expell* [sic] *the gas from about me*[.] *When the water was all out there was nothing to prevent the fan from starting*[.] *5 minutes had not elapsed since the time I had entered the engine room untill* [sic] *I got the blower started. With one blower started the gas was gradually expelled*[.] *At this time the crew were all up into the turret while the engineer and fireman were being revived by the use of brandy administered by doctor Logue on top of the turret* [sic][.] *I had scarcley* [sic] *gotten one blower started when two seman* [sic] *came into the engine room*[.] *They had wet cloths over their mouth*[.] *They informed me that they had been sent by Captain Worden to find me and bring be* [sic] *up on the turret supposing that like the others I had been overcome by the gas and was overlooked… When I got on top of the turritt* [sic]*…it was then for the first time I thought that the predictions of some of the croakers was about to come to pass*[.] *I informed Capt. Worden of what I had done and the condittion* [sic] *of things in the engine room and requested that I get some help from among the seamen and then I received a glass of brandy which relieved me of my troubles a great deal.*[52]

Despite Driscoll's efforts, the *Monitor*'s mechanical systems were still failing, and the ironclad was in serious trouble.

Without the ability to properly repair the blower belts, the boiler fires went out. Consequently, the engines ceased, stopping the steam-powered, large-capacity Worthington and Adams pumps. Greene later wrote: "The water continued to pour down the smoke-stacks and blower-pipes in such quantities that there was imminent danger that the ship would flounder."[53] Since Worden was seasick, Greene set about saving the ship. The *Monitor* was rapidly filling with water. Hand pumps were tried, but they were not powerful enough to pull the water out through the hatch of the turret. Greene organized a bucket line to bail out the water. While this did little to combat the rising water, it did steady the crew. William Keeler wrote about the foundering ship, "Things for a time looked pretty blue, as though we might have to give up the ship."[54] The *Monitor*'s flag was flying upside down, indicating distress, but only when Greene hailed and ordered the *Seth Low* to tow the *Monitor* toward calmer waters near Fenwick Island, Delaware, was the ironclad saved.

A Day Late

Seth Low. James Bard, circa 1860. *Courtesy of The Mariners' Museum.*

By morning, the storm had abated, and the engineers were organized to go back into the engine room for short periods of time in order to complete specific repairs before the fumes overcame them. Eventually, the engine room was vented and the boilers restarted. This enabled the *Monitor* to continue her voyage down the coast.

The evening of 7 March was clear and beautiful. Lieutenant Greene laid down for a much-needed nap about midnight. As the ironclad passed Chincoteague Island, Virginia, he was startled awake by "the most infernal noise I ever heard in my life." Greene knew the ironclad was again in serious trouble, as he later recounted:

> *We were just passing a shoal and the sea suddenly became rough and right ahead. It came up with tremendous force through our anchor well and forced the air through our hawse-pipe where the chain comes, and then the water would rush through in a perfect stream, clear to our berth deck, over the wardroom table. The noise resembled the death groans of twenty men, and was the most dismal, awful sound I have ever heard.*[55]

The hawsepipe had not been plugged before the *Monitor* left the Brooklyn Navy Yard. The storm forced the water through the pipe where the anchor chain passed. Greene and Worden went forward and were able to stop the flow of water.

The sea's fury increased and, once again, came into the vessel via the blower stacks. Greene remembered: "We began to think that the *Monitor* would never see daylight again."[56] While the engineers and firemen struggled to keep the boilers operating, the wheel ropes jumped off the steering wheel and the ship began to sheer, stressing its towline with the *Seth Low*. The *Monitor* began to face broadside to the sea and rolled erratically. Fortunately, the crew of the *Seth Low* noticed the *Monitor*'s distress and towed her near the shore. When the storm lessened, the tiller ropes were repaired, and the vessel was able to steam on toward the Chesapeake capes.

The *Monitor* had survived two close encounters with the angry sea, and by late afternoon on 8 March, Cape Charles, Virginia, was sighted. Once she entered the Chesapeake Bay, many crewmen thought they heard firing in the distance. As the *Monitor* slowly made her way toward Hampton Roads, Keeler wrote his wife:

> *Clouds of smoke could be seen hanging over it in the direction of the Fortress [Fort Monroe], & as we approached still nearer little black spots could be occasionally seen suddenly springing into the air remaining stationary for a moment or two then gradually expanding into a large white cloud—these were shells & tended to increase the excitement. As darkness increased, the flashes of guns lit up the distant horizon & bursting shells flashed in the air...As we reached the harbor the firing slackened & only an occasional gun lit up the darkness—vessels were leaving like a covey of frightened quails & their lights danced over the water in all directions.*[57]

When the *Monitor* took on a pilot, they learned the Confederate ironclad had already wreaked havoc upon the Federal fleet. The *Monitor* had, indeed, arrived one day too late.

Just two days before, a stalemate had existed in Hampton Roads. Both navies were rushing their ironclads into service to attain naval superiority, and the Hampton Roads harbor was the key to success. Formed by the confluence of the James, Nansemond and Elizabeth Rivers, Hampton Roads is the largest natural harbor in the world and just eighteen miles from the Virginia capes. Control of this harbor was considered critical. The Confederates needed to defend the James River approach to Richmond, as well as the Norfolk-Portsmouth industrial center. The Federals, in turn, had already proven the harbor's immense value for launching expeditions against the southern coast. More importantly, in March 1862, Major General George Brinton McClellan had presented an ambitious plan to "take Fort Monroe as a base, and operate with complete security…up the Peninsula"[58] to capture the Confederate capital at Richmond. The campaign's success hinged upon the U.S. Navy's control of the harbor and its ability to support the Army of the Potomac's move up the James River. President Lincoln was concerned about the campaign's feasibility, due to the threat of the *Merrimack*, but was assured by Assistant Navy Secretary Gustavus Vasa Fox: "You need not give yourself any trouble about that vessel."[59]

Flag Officer Louis Malesherbes Goldsborough had replaced Flag Officer Silas Horton Stringham as commander of the North Atlantic Blockading Squadron in September 1861. Goldsborough was born in Washington, D.C., and entered the U.S. Navy as a midshipman at the tender age of seven. A veteran of the Seminole and Mexican-American Wars and a former superintendent of the U.S. Naval Academy in Annapolis, Goldsborough was a huge (reports indicate he weighed well over three hundred pounds) and intimidating man with a powerful temper. One naval officer noted that Goldsborough possessed "manners so rough, so that he would almost frighten a subordinate out of his wits." Since assuming command of the squadron, Goldsborough continuously bombarded Gideon Welles with reports about the Confederate ironclad. On 17 October 1861, the flag officer wrote Welles, noting: "I have received further minute reliable information with regard to the preparation of the *Merrimack* for an attack on Newport News and these roads, and I am quite satisfied that unless her stability is compromised by her heavy top works of wood and iron and her weight of batteries, she will in all probability prove to be exceedingly formidable."[60]

To counter any possible threat from the Confederate ironclad, Goldsborough turned the Hampton Roads blockading force into a strong complement of wooden warships. At the fleet's heart were the two sister steam-screw wooden frigates, the forty-seven-gun USS *Minnesota* and the forty-two-gun USS *Roanoke* of the *Merrimack* class. The station also included three sailing wooden warships: the fifty-gun USS *Congress*, the forty-four-gun USS *St. Lawrence* and the twenty-four-gun sloop of war USS *Cumberland*. The squadron was supported by the steamer *Cambridge*, the U.S. Storeship *Brandywine*, three coal ships, a hospital ship, five tugboats (including the one-gun USS *Zouave* and one-gun USS *Dragon*), a side-wheel steamer and a sailing bark. Goldsborough, before leaving for North Carolina with Brigadier General Ambrose Everett Burnside's expedition to capture Roanoke Island, had planned to confront the Confederate warship when it entered Hampton Roads by surrounding the ironclad in crossfire. "Nothing, I think," wrote Goldsborough, "but very close work can be of service in accomplishing the destruction of the *Merrimack* and even of that a great deal may be necessary."[61]

CSS Virginia *in dry dock*, circa 1862. *Courtesy of The Mariners' Museum.*

There were concerns that the Union ships, without the support of an ironclad of their own, might be unable to confront the Confederate ironclad. Lieutenant Joseph B. Smith of the *Congress* noted that his frigate "had been a model in her day." Nevertheless, he worried that since all of his cannon were older smoothbores, "we should only be a good target for them, as none of our guns could send a shot to them."[62] Obviously, Welles knew that only the *Monitor* could stop the *Merrimack*, and his constant inquiries about when the Union ironclad would be ready epitomized the sense of urgency felt by the Union command. Captain Gershom Jacques Henry Van Brunt, commander of the USS *Minnesota*, noted: "The *Merrimack* is still invisible to us, but report says she is ready to come out. I sincerely wish she would; I am quite tired of hearing of her." Van Brunt added: "The sooner she gives us the opportunity to test her strength, the better."[63]

The Confederate ironclad project was plagued with delays. The *Merrimack*'s conversion had been a true challenge to the Confederate industrial and transportation infrastructure. She was launched, commissioned and christened as the CSS *Virginia* on 17 February 1862. Flag Officer Franklin Buchanan, as commander of the James River defenses, was placed in command of the ironclad a week later. Buchanan, a hero of the Mexican-

A Day Late

Captain Franklin Buchanan, circa 1855. *Courtesy of The Mariners' Museum.*

American War and former superintendent of the U.S. Naval Academy at Annapolis, advised Confederate Secretary of the Navy Stephen Russell Mallory on 4 March that the Confederate ironclad was ready for service. Mallory then ordered Buchanan: "The Virginia is a novelty in naval construction, is untried and her powers as a ram are regarded as formidable, and it is hoped that you may be able to test them. Like a bayonet charge of infantry, this mode of attack, while the most distinctive, will command itself to you in the present scarcity of ammunition."[64]

Mallory also suggested that if the ironclad could "pass Old Point [meaning Fort Monroe] and make a dashing cruise on the Potomac as far as Washington, its effect upon the public mind would be important to our cause." Such a bold move could surely bring victory at a time when the Confederacy was reeling from defeats in Tennessee and along the North Carolina sounds. Mallory was convinced "that the opportunity and the means for striking a blow for our Navy are now for the first time presented. I congratulate you upon it, and know that your judgment and gallantry will meet all just expectations." He concluded his letter by stating: "Action, prompt and successful action—now would be of serious importance to our cause."[65]

Mallory's orders for prompt action had not been lost on Franklin Buchanan. He planned to surprise the Federal fleet with an early morning attack on 7 March, but a severe gale—the same one that almost sank the *Monitor*—delayed the foray. The Confederate ironclad required calm waters to safely operate. Even though the ironclad was still considered "by no means ready for service,"[66] Buchanan was determined to take his ship into battle the next day.

On 8 March 1862, the weather cleared, and Buchanan prepared his ship for action. The casemate was smeared with a thick coating of "ship's grease." As the executive officer, Lieutenant Catesby ap Roger Jones, noted, it would "increase the tendency of the projectiles to glance." At 11:00 a.m., Buchanan ordered his crew to cast off from the Gosport Navy Yard quay. Workmen dashed off the ship without completing many minor details, and the ironclad began her trip down the Elizabeth River, joined by the wooden gunboats the CSS *Beaufort* and the *Raleigh*.

The banks of the Elizabeth River thronged with thousands of cheering citizens. Surgeon Dinwiddie Phillips commented: "Most of them, perhaps, attracted by our novel appearance, and desirous of witnessing our movements through the water. Few, if any, entertained an exalted idea of our efficiency, and many predicted a total failure."[67] Midshipman Hardin Beverly Littlepage remembered one man shouting from the shore, "Go on with you, old metallic coffin! She would never amount to anything else!"[68] The *Norfolk Day Book* reflected the overwhelming hope that the Confederacy would secure a great victory with its untried weapon:

> *It was a gallant sight to see the ironclad leviathan gliding noiselessly through the water, flying the red pennon of her commander at the forestaff and the gay Confederate ensign aft. Not the least impressive thought which she suggested was that her gallant crew, under a commander and officers worthy to direct their destiny and defend the flag she bore, went thus boldly with smiles and huzzas to solve a new problem in maritime warfare—to make the trial trip of the* Virginia *the trial of battle.*[69]

The Confederate ironclad may have appeared fearsome to many onlookers, yet there were still problems onboard the ship. The broadside port shutters had not been installed, making the ship vulnerable to enemy shot coming in through these gun ports. The warship was slow, making only

A Day Late

CSS Virginia *passing Craney Island batteries en route to attack the Union fleet.* J.O. Davidson, circa 1880. *Courtesy of John Moran Quarstein.*

five knots, and ran so close to the bottom of the Elizabeth River (she had a twenty-two-foot draft) that a towline from the CSS *Beaufort* was needed to help the huge ironclad around a bend in the river. Lieutenant John Taylor Wood commented later, "She steered so badly that, with her great length, it took thirty to forty minutes to turn…She was as unmanageable as a waterlogged vessel."[70] Major William Norris, of the CS Signal Corps, believed that the *Virginia* "was in every respect ill-proportioned and top heavy; and what with her immense length and wretched engines, she was a little more manageable than a timber raft."[71]

As the *Virginia* passed Craney Island, Flag Officer Franklin Buchanan informed the crew, "Sailors, in a few minutes you will have the long-awaited opportunity to show your devotion to your country and our cause…The Confederacy expects every man to do his duty. Beat to quarters." Buchanan then pointed to the Union fleet in Hampton Roads and exclaimed, "Those ships must be taken…so to your guns!" Midshipman Littlepage recalled that Buchanan "also told us that the Confederates had complained that they were not taken near enough to the enemy and assured us that there should be no complaint this time, for he intended to head directly for the *Congress*."[72]

The ironclad dropped its towline from the *Beaufort* at 1:30 p.m. and entered Hampton Roads at high tide. The crew could see the entire Federal fleet, arrayed in a line that stretched from Newport News Point to Fort Monroe on Old Point Comfort. Two French warships, the *Gassendi* and *Catinet*, under the overall command of the Marquis de Montaignac, had come to Hampton

Roads, awaiting the events that were soon to unfold. The Confederates appeared to have achieved a tactical surprise, as drying clothing hung from the rigging of the Union ships. "Nothing indicated," John Taylor Wood remembered, "that we were expected."[73]

The Federals quickly noticed, according to Acting Master Henry Reaney of the tug *Zouave*, "what to all appearances looked like the roof of a very big barn belching forth smoke as from a chimney on fire."[74]

"Suddenly, huge volumes of smoke began to pour from the funnels of the frigates *Minnesota* and *Roanoke* at Old Point," recalled Chief Engineer Ashton Ramsay of the *Virginia*. "They had seen us and were getting up steam. Bright colored signal flags were run up and down the masts of all of the ships of the Federal fleet." Ramsay continued, "The *Congress* shook out her topsails, down came the clothes-lines on the *Cumberland* and boats were lowered and dropped astern."[75] Brigadier General Joseph King Fenno Mansfield, commander of Camp Butler at Newport News Point, telegraphed Major General John Ellis Wool at Fort Monroe, "The *Merrimack* is at hand."[76]

The Confederate ironclad steamed on toward the *Cumberland*; however, as she passed the *Congress*, the two ships traded salvoes. Shot from the *Congress* bounced off the *Virginia* like "pebble stones." In turn, the Confederate warship unleashed her broadside of four guns against the *Congress*. Hot shot and shell ignited two fires on the hapless frigate. The *Congress* appeared critically damaged.

The *Virginia* did not stop to complete the destruction of the *Congress* but continued on a course against the *Cumberland* "like some devilish and superhuman monster, or the horrid creature of a nightmare." The *Cumberland* kept up her fire against the oncoming ironclad, but her shot "struck and glanced off, having no more effect than peas from a popgun."[77] The ironclad then rammed the sloop of war in her starboard quarter. The *Cumberland* was mortally wounded, the ramming only made worse by a simultaneous shot from the *Virginia*'s bow rifle, which killed ten men. The Union warship immediately began to sink, with the *Virginia*'s ram trapped within the *Cumberland*'s hull. As the weight of the *Cumberland* rested upon the ram, the Confederate ironclad began to settle. The poorly mounted ram broke off, freeing the *Virginia*. Ashton Ramsay noted: "Like a wasp, we could sting but once, leaving the sting in the wound."[78]

The ironclad floated fifty yards away from the *Cumberland*, as the two ships continued to bombard each other. Acting Master's Mate Charles O'Neil remembered the "shot and shell from the *Merrimack* crashed through the wooden sides of the *Cumberland* as if they were made of paper, carrying

A Day Late

Virginia *attacking the* Cumberland. Alexander Stuart, circa 1890. *Courtesy of John Moran Quarstein.*

huge splinters with them and dealing death and destruction."[79] The *Cumberland* turned into a slaughterhouse, and the sloop's death toll was 121. The Confederate warship was also wounded from her encounter with the *Cumberland*. The three broadsides from the sloop damaged the ironclad's outer works. The *Cumberland*'s gunners aimed at the *Virginia*'s gun ports, hoping to send a shot inside the casemate. Lieutenant Catesby Jones reported, "Our aft 9-inch gun was loaded and ready for firing when its muzzle was struck by a shell, which broke it off and fired the gun. Another gun also had its muzzle shot off; it was broken so short that at each subsequent discharge its port was set on fire."[80]

The *Cumberland* was sinking, and her commander, Lieutenant George Upham Morris, ordered his men to abandon ship. Morris then commanded the remaining gun crews to "give them a broadside, boys, as she goes."[81] The *Cumberland* sank in 55 feet of water. "She went down bravely, with her colors flying,"[82] Catesby Jones remembered.

The *Virginia*, due to her deep draft and poor steering, was forced to steam up the James River to turn around. While this maneuver was executed, Lieutenant Wood struck the *Congress* with several shells from the seven-inch Brooke rifle. The ironclad also destroyed two Union transports and captured one other anchored along a wharf. She then steamed within two hundred yards of the stranded *Congress*. The frigate's stern was quickly demolished,

Gun deck of the Cumberland. *J.O. Davidson, circa 1880. Courtesy of John Moran Quarstein.*

and the main deck was "reeking with slaughter."[83] The *Zouave* was struck by several shells, and fled the scene. The acting commander of the *Congress*—Lieutenant Joseph B. Smith, the son of Flag Officer Joseph Smith of the Ironclad Board and commandant of the Brooklyn Navy Yard—was killed by a shell fragment. The ship's command evolved onto the shoulders of Lieutenant Austin Pendergrast. The *Congress* then hauled down her flag.

As the Confederates attempted to accept the frigate's surrender and remove the wounded (Franklin Buchanan's brother, Paymaster Thomas McKean Buchanan, was aboard the *Congress*), musket fire from Camp Butler disrupted the operation. Buchanan, standing atop the ironclad and enraged by the Union actions under a flag of truce, shot at the Federal soldiers on the shore with a musket. He was then shot himself and severely wounded in the thigh. Buchanan was carried below and ordered his executive officer, Catesby Jones, to "plug hot shot into her and don't leave her until she is afire."[84] Jones assumed command. The *Virginia* soon left the frigate burning "stem to stern" and reentered Hampton Roads.

Three Union frigates, the *Minnesota*, the *Roanoke* and the *St. Lawrence*, had run aground attempting to support the *Congress* and the *Cumberland*. Jones wanted to destroy the grounded frigates and shelled them from afar.

A Day Late

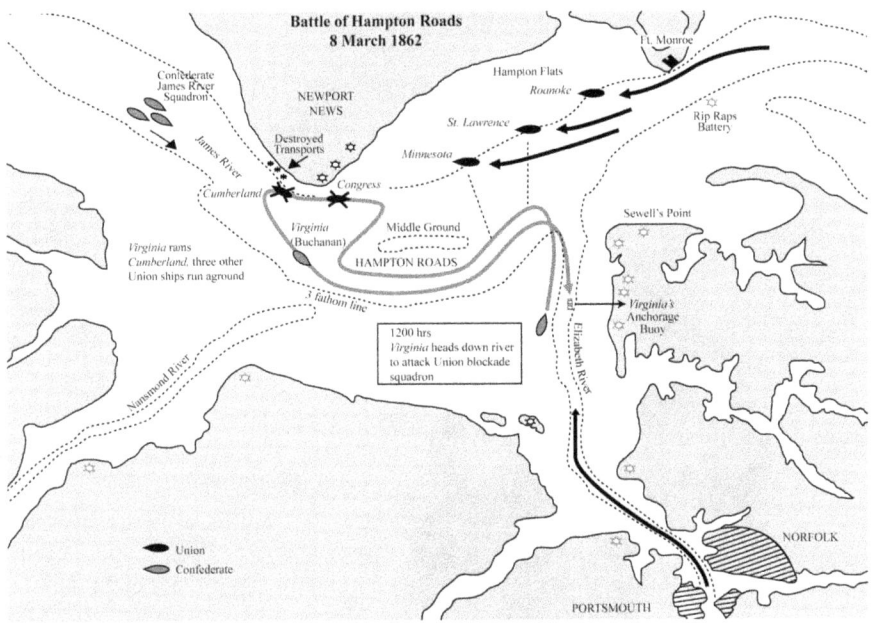

Battle of Hampton Roads—8 March 1862. Illustration by Sara Johnston.

The *St. Lawrence* had a single shell lodged in her main mast, but it failed to explode; meanwhile, over a dozen shells struck the *Minnesota*. Darkness and the receding tide compelled Jones to finally steam his ship to her mooring at Sewell's Point. As the *Congress*'s spars and ropes "glittered against the dark sky in dazzling lines of fire,"[85] Jones vowed to destroy the Federal fleet on the morrow.

The Union command was stunned by the events of 8 March 1862. General Wool reported the destruction of the *Congress* and the *Cumberland* to Washington, D.C., via telegraph. Wool's chief of staff, Colonel LeGrand B. Cannon, noted that the land-based armament at Fort Monroe was "as useless as musket-balls against the ironclad."[86] Major General George McClellan telegraphed, "The performances of the *Merrimac* places a new aspect upon everything, and may very probably change my whole plan of campaign, just on the eve of execution."[87] President Abraham Lincoln viewed the defeat as the greatest Union calamity since Bull Run. Secretary of War Edwin Stanton became "almost frantic," according to Gideon Welles's observations, stating: "The *Merrimac*…would destroy every vessel in the service, could lay every city on the coast under contribution, could take Fortress Monroe; McClellan's mistaken purpose to advance must be abandoned." Stanton

Sinking of the Congress.
J.O. Davidson, circa 1880.
*Courtesy of John Moran
Quarstein.*

feared, as Welles noted, that the Confederate ironclad would soon "come up the Potomac and disperse Congress, destroy the Capitol and public buildings or she might go to New York and Boston and destroy those cities."[88]

Gideon Welles endeavored to allay everyone's fears. He advised that the *Merrimack* could not go everywhere at once. Furthermore, he noted that the ironclad could not attack Washington because the *Merrimack*'s draft was far too great to reach up the Potomac. Welles reminded all that the Union's newest ironclad, the *Monitor*, was then en route to the Chesapeake Bay. When Stanton asked about the Union ironclad and her armament, and when Welles replied, Stanton's "mingled look of incredulity and contempt cannot be described; and the tone of his voice, as he asked if my reliance was on that craft with her two guns, is equally indescribable." Despite Stanton's fears, the *Monitor* appeared to be the only hope to save the Union.

Chapter 5
Showdown in Hampton Roads

The *Monitor* finally entered Hampton Roads at about 9:00 p.m. on 8 March 1862. The crew was shocked by the destruction and chaos left behind by the Confederate ironclad. "Our hearts were so very full," Executive Officer Samuel Dana Greene wrote, "and we vowed vengeance on the *Merrimac*."[89] Commander Worden, as ordered, anchored his ironclad near the *Roanoke* and reported to Captain John Marston, acting commander of the Union naval forces in Hampton Roads. Worden learned at that time of his new orders that had failed to reach him before the *Monitor* left New York: Gideon Welles wanted the ironclad to proceed to Washington, D.C. Marston rescinded the orders from the secretary of the navy. He realized that the best way to stop any Confederate ironclad assault against Washington was to defend the wooden frigates in Hampton Roads. Marston commanded Worden to station the *Monitor* near the grounded *Minnesota* and to protect that frigate from the *Merrimack*. Worden reflected upon the scene and wrote his wife, "The *Merrimac* has caused sad work amongst our vessels. She can't hurt us."[90]

Before Worden could do anything, he needed to secure a pilot to guide his ironclad through the treacherous shoals of Hampton Roads. Considering all the damage wrought by the Confederate ironclad that day, few qualified pilots were willing to volunteer to go aboard the strange-looking *Monitor* and fight the seemingly unstoppable *Merrimack* the next day. After a two-hour search, Acting Master Samuel Howard of the bark USS *Amanda* agreed to pilot the *Monitor*.

As the *Monitor* steamed over to the *Minnesota* and anchored next to the stranded frigate, the death throes of the *Congress* provided a spectacular conclusion to the events of 8 March. The *Congress* had been burning since late in the afternoon, and the frigate's conflagration sent an eerie glow across Hampton Roads throughout the evening. Onlookers commented about how the red-tongued flames danced along the shrouds, masts and stays. When the *Congress* finally exploded after midnight, Samuel Dana Greene commented, "Certainly a grander sight was never seen, but it went right to the marrow of our bones."[91] William Keeler recalled:

> *The night we arrived I was on deck and witnessed the explosion of the burning* Congress, *a scene of the most terrible magnificence. She was wrapped in one sheet of flame, when suddenly a volcano seemed to open instantaneously, almost beneath our feet and a vast column of flame and fire shot forth till it seemed to pierce the skies. Pieces of burning timbers, exploding shells, huge fragments of the wreck, grenades and rockets filled the air and fell sparkling and hissing in all directions. It did not flash up and vanish in an instant, but seemed to remain for a moment or two, an immense column of fire, one end on the earth the other in the heavens. It soon vanished and a dense thick cloud of smoke hid everything from view. We were about two miles from the wreck and the dull heavy explosion seemed almost to lift us out of the water.*[92]

It was an impressive, yet somber, conclusion to the day.

Meanwhile, Worden and Greene went aboard the *Minnesota* to confer with her commander, Captain Gershom Jacques Henry Van Brunt. Van Brunt expected to free his ship at the next high tide, at 2:00 a.m., and required no assistance. The veteran officer had little faith in the *Monitor*'s ability to stop the Confederate ironclad and indicated that if all else failed, he would destroy his ship rather than allow it to be captured. Keeler concluded that the men of the *Minnesota* thought the "idea of assistance or protection being offered to the huge thing by the little pigmy at her side seemed absolutely ridiculous."[93] One of the *Minnesota*'s crew members noted how "insignificant she looked, she was but a speck on the dark blue sky at night, almost a laughable object by day." Others were jubilant that the *Monitor* had arrived. The steam frigate's chief engineer, Thomas Rae, remembered that when he shouted down to the engine room, "The *Monitor* is alongside," the crew "gave a cheer that might have been heard in Richmond."[94] As Worden left the *Minnesota*, he advised Van Brunt, "I will stand by you to the last if I can

Showdown in Hampton Roads

USS Monitor *and USS* Minnesota. J.O. Davidson, circa 1880. *Courtesy of John Moran Quarstein.*

help you."[95] Worden knew that his ironclad was the only thing that could save the *Minnesota* from the *Merrimack*.

There was little opportunity for the *Monitor's* crew to rest during the early morning of 9 March. At 2:00 a.m., Worden mustered his men to their stations, in case the *Minnesota* was able to float free with the high tide, and the *Monitor* needed to move out of the way of the frigate. Nothing worked to free the *Minnesota* from the shoal, however, and she seemed destined to await her fate.

On board the *Monitor*, preparations were underway for the expected engagement with the *Merrimack*. Patrick Hannan, William Durst and John Driscoll were on deck before dawn "to screw the iron plates over the deck lights in the deck and to take down blower pipes and smoke stacks."[96] About 6:00 a.m., it was noted that the Confederate squadron near Sewell's Point was getting their steam up. At 7:00 a.m., Worden ordered the men to breakfast. He then, according to John Driscoll, gave the men a brief speech, reminding them:

> *We had all volunentered* [sic] *to go with him*[,] *that now haveing* [sic] *seen what the* **Merrimac** *had done and from all appearances was now capable of doing and that the fate of the* **Cumberland** *may soon be ours*[,] *that if any one regretted the step he had taken he would put him on board the* **Roanoke**[.] *He was answered by every man jumping to his feet and giving three cheers.*[97]

Driscoll also added details on that special breakfast, their first real food for nearly two days: "We had canned roast beef[,] hard tack and coffee."[98]

Many of the men aboard the *Monitor* had also not slept in forty-eight hours. Their nerves must have been on edge, as they had almost sunk twice and were now witness to all of the havoc wrought by the Confederate ironclad upon the Federal fleet in Hampton Roads. They knew that their ironclad was an experiment, and they had been subject to jeers by other Union sailors about their little ship's abilities. Nevertheless, a different feeling was felt by the crew as they awaited the *Merrimack*, as William Keeler recounted:

> *Everyone was at his post, fixed like a statue, the most profound silence reigned—if there had been a coward heart there its throb would have been audible, so intense was the stillness.*
>
> *I experienced a peculiar sensation, I do not think it was fear, but it was different from anything I ever knew before. We were enclosed in what we supposed to be an impenetrable armour* [sic]*—we knew that a powerful foe was about to meet us—ours was an untried experiment & our enemy's first fire might make it a coffin for all.*[99]

Meanwhile, the officers and crew of the Confederate ironclad had spent a more jubilant evening. The day before, they had achieved a tremendous victory, but the *Virginia* had also suffered significant damage during her engagement with the *Cumberland*. Catesby Jones, the ironclad's executive officer, who had assumed command of the *Virginia* following Franklin Buchanan's wounding, inspected the ship. Jones discovered a small leak in the bow but did not notice that the ram was missing because of the darkness. He thought the ram was merely twisted from its collision with the *Cumberland*. So, despite the missing anchors, boats, flagstaffs and railings, Jones believed that the *Virginia* was ready, once again, to venture out against the Federal fleet of wooden ships. The Confederate ironclad had stood the test of combat rather well on 8 March. "Our loss in killed and wounded was twenty-one," wrote John Taylor Wood. "The armor was hardly damaged, though at one time our ship was the focus on which were directed at least one hundred heavy guns, afloat and ashore."[100] When Surgeon Dinwiddie Phillips returned from delivering Flag Officer Buchanan to the hospital in Portsmouth, he "found all her stanchions, iron railings, boat davits, and light work of every description swept away, her smokestack cut to pieces, two guns without muzzles, and ninety-eight indentations on her plating, showing where heavy shot had struck, but glanced off without doing any injury."

Showdown in Hampton Roads

Phillips later added that the smokestack was so riddled that it "would have permitted a flock of crows to fly through without inconvenience."[101]

When daylight came on Sunday, 9 March 1862, Catesby Jones ordered his men to breakfast. "We began the day with two jiggers of whiskey," William Cline commented, "and a hearty breakfast."[102] They got underway at 6:00 a.m., accompanied by the CSS *Patrick Henry*, the *Jamestown* and the *Teaser*. Due to heavy fog, the small squadron delayed entering Hampton Roads until nearly 7:20 a.m. The day turned "as bright and beautiful as the day preceding it," remembered *Virginia* crew member Richard Curtis. "The broad waters of Hampton Roads were as smooth as glass, not a ripple on its surface, an ideal day to go to church, but alas it was soon to be broken by the roar of cannon and angry men seeking each other's lives."[103]

Lieutenant Jones saw that the *Minnesota* was still stranded on the shoal as the Confederate ironclad closed within range. At 8:00 a.m., Lieutenant Hunter Davidson's forward Brooke rifle sent the first shot of the day, at a range of one thousand yards, through the frigate's rigging. Another shot quickly followed, "exploding on the inside of the ship causing considerable destruction, setting the ship on fire." The ironclad's crew expected to make short work of the *Minnesota*. Ashton Ramsay, chief engineer of the *Virginia*, recounted, "We approached her slowly, feeling our way cautiously along the edge of the channel, when suddenly, to our astonishment, a black object that looked like the historic description, 'a barrel-head afloat with a cheese box on top of it' moved slowly out from under the *Minnesota* and boldly confronted us."[104]

The Confederates were thoroughly amazed by the sight of the Union ironclad. "We could see nothing but the resemblance of a large cheese box," Lieutenant John Randolph Eggleston remembered, and thought it was "the strangest looking craft we had ever seen before."[105] Catesby Jones, who had followed the construction of the Federal ironclad in the Northern newspapers, calmly noted to his officers there that "was an iron battery near" the *Minnesota*. Lieutenant Hunter Davidson, commander of the forward seven-inch Brooke rifle, thought at first that "the *Minnesota*'s crew are leaving on a raft." But he soon realized, "By George, it is the *Ericsson Battery*, look out for hot work."[106] When the Union ironclad moved to intercept the *Virginia*'s course toward the *Minnesota*, Lieutenant John Taylor Wood, commander of the stern seven-inch Brooke gun, reflected, "She could not possibly have made her appearance at a more inopportune time."[107]

Worden, Surgeon Daniel Logue and William Keeler were watching the approach of the *Merrimack* when "a shell howled over our heads and crashed into the side of the *Minnesota*." Keeler added: "Captain Worden…came up

& said more sternly than I ever heard him speak before, 'Gentlemen, that is the *Merrimac*, you had better go below.'" Keeler said that he "did not wait for a second invitation but ascended the tower & down the hatchway, Capt. W. following. The iron hatch was closed over the opening & all access to us cut off. As we passed down through the turret the gunners were lifting a 175 lb. shot into the mouth of one of immense guns. 'Send them that with our compliments, my lads,' says Capt. W."[108] Lieutenant Worden then steered his ship toward the *Merrimack* in an effort to engage the Confederate ironclad as far away from the *Minnesota* as possible.

Almost everyone onboard the *Monitor* had a duty to perform, and all were at their stations as she closed for battle. Executive Officer Samuel Dana Greene recalled:

> *Worden took his station in the pilot-house, and by his side were Howard, the pilot, and Peter Williams, quartermaster, who steered the vessel throughout the engagement. My place was in the turret, to work and fight the guns; with me were Stodder and Stimers and sixteen brawny men, eight to each gun. John Stocking, boatswain's mate, and Thomas Lochrane* [sic— Loughran], *seaman, were gun-captains, Newton and his assistants were in the engine and fire rooms, to manipulate the boilers and engines... Webber had charge of the powder division on the berth deck.*[109]

At first, the *Merrimack* did not appear to take notice of the *Monitor*. Van Brunt was amazed at how the *Monitor*, "much to my astonishment, laid herself right alongside the *Merrimac*, and the contrast was that of a pygmy to a giant."[110]

Worden then stopped his engines and gave the order to Greene to commence firing. The first shot missed, according to E.V. White of the *Virginia*:

> *We didn't have long to wait before she fired. Her first shot fell a little short and sent up a geyser of water that fell on our top and rolled off. We then fired our forward rifle and scored a direct hit on her turret, but with no apparent effect. Her next shot was better and caught us amidships with a resounding wham, but while the old boat shuddered, there seemed to be no appreciable damage.*[111]

When the first shots struck the *Monitor*, the turret rang like a bell; however, the men knew that their ship was impregnable. As Greene remembered, "A

Showdown in Hampton Roads

Monitor & Merrimac *in Action*, from John Worden's photograph album, sketch by "Walker," circa 1862. *Courtesy of The Mariners' Museum.*

look of confidence passed over the men's faces, and we believed the *Merrimac* would not repeat the work she had accomplished the day before."[112]

During the first two hours of the engagement, it became apparent that neither ship's gunnery could do damage to the other vessel. The fire control system on board the *Monitor* was not effective. Samuel Dana Greene later reported:

> *My only view of the world outside of the tower was over the muzzles of the guns, which cleared the ports only by a few inches. When the guns were run in, the port-holes were covered by heavy iron pendulums, pierced with small holes to allow the iron rammer and sponge handles to protrude while they were in use. To hoist those pendulums required the entire gun's crew and vastly increased the work inside the turret. The effect upon one shut up in a revolving drum is perplexing, and it is not a simple matter to keep the bearings. White marks had been placed upon the stationary deck immediately below the turret to indicate the direction of the starboard and port sides, and the bow and port sides, and bow and stern; but these marks*

> *were obliterated early in the action. I would continually ask the captain, "How does the* Merrimac *bear?" He replied, "On the starboard-beam," or "On the port-quarter," as the case might be. Then the difficulty was to determine the direction of the starboard-beam, or port-quarter, or any other bearing. It finally resulted, that when the gun was ready for firing, the turret would be started in its revolving journey in search of the target, and when found it was taken "on the fly," because the turret could not be accurately controlled.*[113]

Greene decided he could not fire his Dahlgrens directly toward the bow, in fear of hitting the pilothouse, nor toward the stern, in case of gunpowder going into the blowers.

Communication between the turret and pilothouse was another serious problem. The very design of the *Monitor* detached the ship's commander and pilot from the gun crew. A speaking tube was installed to provide a constant link between the critical combat functions of command and fire control. Ericsson later admitted that the pilothouse should have been placed atop the turret; however, the *Monitor*'s rapid construction timeline caused it to be placed near the bow. Ericsson had thought that the speaking tube would prove adequate for the exchange of commands. Unfortunately, the tube malfunctioned early in the battle, and it became necessary to use crew members to transfer orders as to what, where and when to fire. Greene, who commanded the gun crews in the turret, later recalled:

> *The drawbacks to the position of the pilothouse were soon realized...Keller* [the paymaster] *and Toffey* [Worden's nephew and clerk] *passed the captain's orders and messages to me, and my inquiries and answers to him, the speaking tube from the pilothouse to the turret having been broken early in the action. They performed their work with zeal and alacrity, but, both being landsmen, our technical communication sometimes miscarried.*[114]

Keeler, who had no official duties during combat, was perfect for this task, and he would later receive a commendation from Gideon Welles for "carrying the orders between the pilothouse and turret, bringing them always in a cool distinct manner that added greatly to the complete understanding between these two positions." Keeler felt that he was hidden in the fog of battle while exchanging information between the two critical command posts, as he later wrote:

Showdown in Hampton Roads

> *With exception of those in the pilothouse and one or two in the turret, no one of us could see her* [the *Virginia*]. *The suspense was awful as we waited in the dim light expecting every moment to hear the crash of the enemy shot…Mr. Greene says "Paymaster, ask the captain if I shall fire." The reply was "Tell Mr. Greene not to fire till I give the word."… Below we had no idea of the position of our unseen antagonist, her mode of attack, or her distance from us, except what was made known through the orders of the captain. "Tell Mr. Greene that I am going to bring him on our starboard beam close alongside."*[115]

Despite all of Ericsson's mechanical genius, the design and machinery failed at a critical time, which lessened the *Monitor*'s combat abilities. According to Greene: "The situation was novel: a vessel of war was engaged in desperate combat with a powerful foe; the captain, commanding and guiding, was enclosed in one place, and the executive officer, working and fighting the guns, was shut up in another, and communication between them was difficult and uncertain."[116]

In addition to the communication and vision problems, the XI-inch Dahlgrens lacked the punch to pierce the *Virginia*'s armor. These Dahlgrens could fire a 187-pound solid shot; however, even though an XI-inch Dahlgren was tested to fire using 30 pounds of powder, the *Monitor* was ordered to use only 15 pounds, due to the fear that a gun could burst in the turret. The *Monitor* could fire a shot every six to eight minutes. Loading was accomplished with the guns turned away from the enemy, and then the turret would revolve toward the target. Because of limited vision and the inability to effectively stop the turret, of the forty-one shots fired by the *Monitor* during the engagement, only twenty struck the Confederate ironclad. William Norris noted that "any three [shots] of them properly aimed would have sunk us, and yet the nearest shot to the waterline was over four feet."[117]

"Not a single shot struck us at the waterline, where the ship was utterly unprotected," John Taylor Wood commented, "and where one would have been fatal. Or had the fire been concentrated on any one spot, the shield would have been pierced; or had larger charges been used, the result would have been the same. Most of her shot struck us obliquely," Wood added, "breaking the iron of both courses, but not injuring the wood backing. When struck at right angles, the backing would be broken, but not penetrated."[118]

Regardless of the communication and fire-control problems, Worden knew that he had to keep his ship in the fight. Even though it appeared that the *Virginia* could not damage the *Monitor* with her guns, she was still a deadly

Monitor—Virginia. Alexander Charles Stewart, circa 1890. *Courtesy of The Mariners' Museum.*

threat against the *Minnesota*. The *Monitor*'s speed, quickness, maneuverability and light draft gave her the ability to block the Confederate ironclad from reaching her wooden prey.

The engagement was fought in concentric circles, as each ironclad endeavored to discover the other's weakness. Onboard the *Monitor*, even without seeing their enemy, the crew was well aware of the desperate combat going on around them. William Keeler noted:

> *Our two heavy pieces were worked as rapidly as possible, every shot telling—the intervals being filled by the howling of shells around & over us, which was now incessant.*
>
> *The men at the guns had stripped themselves to their waists & were covered with powder & smoke, the perspiration falling from them like rain…*
>
> *The sounds of the conflict at this time were terrible. The rapid firing of our own guns amid the clouds of smoke, the howling of the* Minnesota's *shells, which was firing whole broadsides at a time just over our heads (two of her shot struck us), mingled with the crash of solid shot against our sides & the bursting of shells all around us. Two men had been sent down from the turret, who were knocked senseless by balls striking the outside of the turret while they happened to be in contact with inside.*[119]

Showdown in Hampton Roads

Chief Engineer Alban Stimers, whose duty it was to operate the turret during the battle, concurred with Keeler about the force of the *Virginia*'s shot and shell, as he wrote his father, "The crash against the turret was tremendous when their heavy shot struck it and if a man happened to lean against the inside of where a shot struck it knocked him down and stunned him for a couple of hours."[120] Stimers was knocked to the floor when a shot hit the turret as he was resting his hand against it. While Stimers recovered quickly, Louis Stodder and Peter Truscott were both knocked unconscious when a shell hit as they were leaning against the wall of the turret, and they were sent to the surgeon for treatment. Stodder recovered within one hour, but Truscott did not return to duty until several hours later.

After almost two hours of combat, Worden ordered his ship to break off action and steam onto a shoal, away from the Confederate ironclad. The supply of shot in the *Monitor*'s turret was exhausted and had to be resupplied from storage bins on the berth deck. To accomplish this, the turret had to

Acting Master Louis Napoleon Stodder, circa 1862. *Courtesy of The Mariners' Museum.*

be stationary so that the 187-pound solid shot could be hoisted up through the two scuttle holes, one in the deck and the other in the floor of the turret. It was an extremely difficult, and time-consuming, task. Worden ordered Keeler to provide the powder division with a ration of spirits. One crewman, Moses M. Stearns, received a strain that caused "a hernia of the left side," and he was excused from duty. George Geer was then relieved of his "duty in the fire room and went to my station hoisting up shot and shell to the tower guns."[121]

Worden took this time to venture out on deck in order to inspect the damage. While the *Virginia*'s shot had indented the turret's iron plates, none was cracked, and no other injury to the ironclad could be found.

Meanwhile, Catesby Jones headed his warship toward the *Minnesota*. The Confederate ironclad, leaking at the bow due to the loss of her ram from the day before, now ran aground and was unable to defend herself. She was in serious danger. The *Monitor* approached and fired several shots at almost point-blank range. The *Virginia*'s surgeon, Dinwiddie Phillips, later remembered that the *Monitor* took "a position very close to us, and where none of our guns could be brought to bear upon her, she directed a succession of shots at the same section of our vessel, and some of them striking close together, started the timbers and drove them perceptibly in, but not enough to do any serious damage."[122] Ashton Ramsay recognized the serious situation the Confederate ironclad was in, as he later noted that the Union ironclad

> began to sound every chink in our armor—every one but that which was actually vulnerable, had she known it. The coal consumption of the two days' fight had lightened our prow until our unprotected submerged deck was almost awash. The armor on our sides below the waterline had been extended but about three feet, owning to our hasty departure before the work was finished. Lightened as we were, these extended portions rendered us no longer an ironclad, and the Monitor *might have pierced us between wind and water had she depressed her gun.*[123]

Somehow, the *Virginia*'s unreliable engines were able to drag the heavy ironclad off the shoal.

Catesby Jones had become frustrated by his ship's ineffectual fire against the Union ironclad. The Confederates had the wrong ammunition, as they were prepared to fight only against the Federal wooden warships. Jones tried to concentrate on the *Minnesota*; however, the quick and nimble *Monitor* continued

Showdown in Hampton Roads

Lieutenant Catesby ap Roger Jones, circa 1880. *Courtesy of The Mariners' Museum.*

to block her nemesis's approach. The Confederates tried to fire at the *Monitor*'s gun ports, yet the turret turned too quickly. Finally, Jones decided to ram the *Monitor*, unaware that he had lost the ram the day before in the *Cumberland*. As the unwieldy ironclad took the half an hour to get into ramming position, Jones noticed, on the gun deck, that Lieutenant John Randolph Eggleston was not firing his gun at the *Monitor*. "Why are you not firing, Mr. Eggleston?" Catesby Jones asked. Eggleston replied, "Why, our powder is very precious, and after two hours' incessant firing I find that I can do her about as much damage by snapping my thumb at her every two minutes and a half." Jones retorted, "Never mind, we are getting ready to ram her."[124]

The *Virginia* began her half-mile run at the *Monitor*. Worden recognized the Confederate ironclad's motive. He told Keeler to tell Greene to "give them both guns." As Keeler raced through the ship with this command, he remembered thinking, "This was the critical moment, one that I feared from the beginning of the fight—if she could so easily pierce the heavy oak beams of the *Cumberland*, she surely could go through the ½-inch iron plates of our hull."[125]

Just as the Confederate ironclad was about to strike the *Monitor*, Jones reversed the engines, which lessened the impact. Also, as Worden saw the *Virginia* about to strike his vessel, he ordered Quartermaster Peter Williams

Monitor—Merrimack. J.O. Davidson, circa 1880. *Courtesy of John Moran Quarstein.*

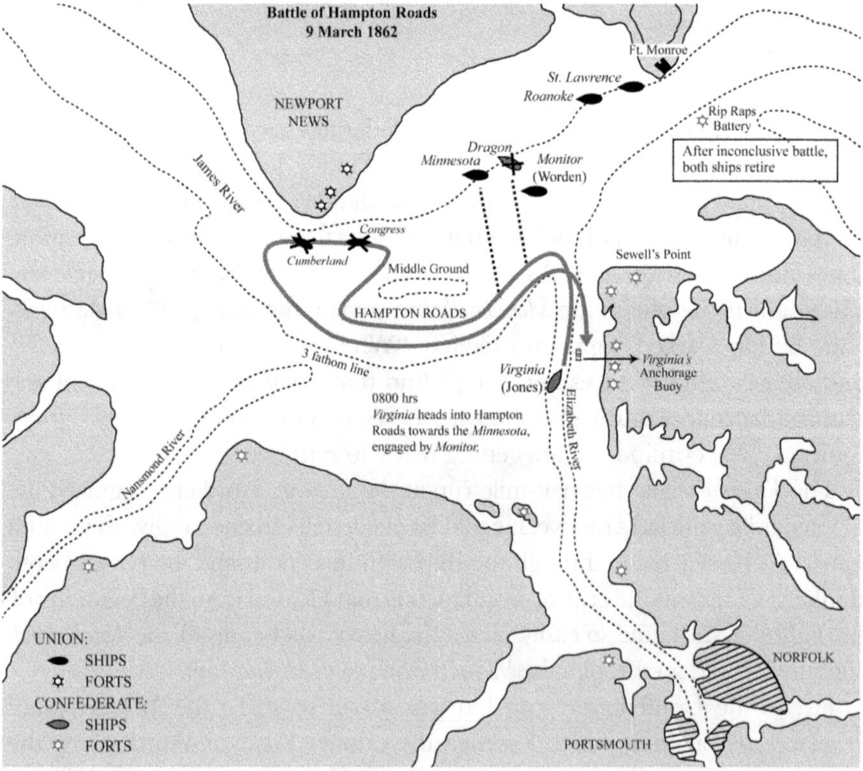

Battle of Hampton Roads—9 March 1862. Illustration by Sara Johnston.

to veer the nimble ironclad off to starboard. These actions caused the *Virginia* to hit the *Monitor* with only a glancing blow. The impact, according to Keeler, was "a heavy jar nearly throwing us from our feet."[126] John Driscoll noted that the "blow was a glancing one yet with such force as [to] jar the chimney off every lamp below and to slightley [*sic*] start one of the deck plates."[127] The only evidence of the ramming, other than the additional damage to the *Virginia*'s bow, were several wooden splinters on a bolt head on the *Monitor*'s deck and a small indentation in her iron plating.

As the *Virginia* passed by, Greene fired his Dahlgrens at the huge ironclad. One shot struck the *Virginia*'s casemate, just above the stern pivot gun port, forcing the shield in two to three inches. "All the crews of the after guns were knocked over by the concussion, and bled from the nose or ears. Another shot at the same place," John Taylor Wood noted, "would have penetrated."[128] Greene knew his shots had hit home. "Had the guns been loaded with 30 pounds of powder, which was the charge subsequently used with similar guns," the executive officer later noted, "it is probable that this shot would have penetrated her armor; but the charge being limited to 15 pounds, in accordance with peremptory orders to that effect from the Navy Department, the shot rebounded without doing any more damage than possibly to start some of the beams of her armor-backing."[129]

As the *Monitor* lay alongside the *Virginia*, Catesby Jones considered the feasibility of boarding the *Monitor*. Jones knew that the *Monitor* did not have any small arms aboard. A group of volunteers, organized by John Taylor Wood, was ready to leap onto the *Monitor*'s deck, cover the pilothouse with a coat to blind the ship and toss specially prepared grenades into the turret and down the funnels. Worden anticipated this move by the Confederates and ordered Greene to double-shot the Dahlgrens with cannister. The *Monitor* quickly slipped by the *Virginia*, and the Confederates were unable to launch their bold, but desperate, attack.

The *Monitor*'s evasive action during the *Virginia*'s ramming attack enabled Jones to, once again, maneuver toward the *Minnesota*. As the Confederate ironclad approached, the stranded frigate sent a broadside from her entire port battery, which, according to her commander, Van Brunt, "would have blown out of the water any timber-built ship in the world."[130] The shot had no impact on the *Virginia*. Even though the pilots would not let Jones move his ironclad any closer than a mile from the *Minnesota*, nonetheless, several shots were sent against the stranded frigate, starting a fire on the *Minnesota*. One shell struck the *Dragon*. The *Dragon*'s boiler burst, and the tug, which had been alongside the *Minnesota* to tow the frigate to safety, sunk.

Battle Between the Monitor *and* Merrimac, *circa 1862. Courtesy of John Moran Quarstein.*

Worden was, once again, able to steer his ironclad between the Confederate ship and the Union frigate. He now decided to ram the *Merrimack*, seeking to disable the larger ironclad's vulnerable fantail. William Norris later noted that this was the Confederate ironclad's Achilles's heel: "Our rudder and propeller were wholly unprotected, and a slight blow from her stern would have disabled both and ended the fight."[131] The *Monitor*'s captain recognized that an effective strike could disable the propeller or rudder, thereby leaving the Confederate ship adrift and subject to capture. The *Monitor* rushed toward the *Virginia*'s fantail but missed her target at the very last moment because of a malfunctioning steering system.

As the Union ship passed the stern of the *Virginia*, Lieutenant John Taylor Wood fired his seven-inch Brooke rifle at the *Monitor*'s pilothouse. Wood's shell blew off one of the wrought-iron bars that formed the pilothouse just as Worden was peering out of an observation slit. The explosion created "a flash of light and a cloud of smoke," which blinded Worden. Worden fell back from the damaged slit and exclaimed, "My eyes, I am blind!"[132] Despite his blindness, Worden could sense the bright light and cool air now coming into the pilothouse and believed that the command center was destroyed. The *Monitor*'s commander, with an amazing presence of mind, ordered the helmsman, Peter Williams, to turn the ironclad to starboard, away from the *Virginia*, and the ship veered onto a shoal.

Showdown in Hampton Roads

Worden's Blinding, circa 1862. *Courtesy of John Moran Quarstein.*

John Worden was helped out of the pilothouse by Keeler and the pilot, Samuel Howard, as Daniel Toffey was sent to the turret to get Lieutenant Greene. Greene placed Alban Stimers in charge of the turret and rushed forward to the pilothouse. When he arrived, he found Worden, holding onto the ladder below the pilothouse, looking "a ghastly sight with his eyes closed and the blood apparently rushing from every pore in the upper part of his face."[133] Greene, along with Dr. Logue, William Keeler and several crew members, including William Durst, helped Worden to his cabin. Worden was laid on a sofa in his cabin as Logue immediately began removing iron particles from the ironclad captain's blackened eyes. Worden told Greene that he was seriously wounded and directed the young lieutenant to take care of the ironclad. Samuel Dana Greene asked Worden what he should do. Worden replied, "Gentlemen, I leave it with you, do what you think best. I cannot see, but do not mind me. Save the *Minnesota* if you can."[134]

As Samuel Dana Greene moved quickly back to the pilothouse, the tactical situation in Hampton Roads was changing rapidly. When the *Monitor* veered away from the *Virginia*, the *Minnesota*'s captain immediately noticed that there was something seriously wrong with the *Monitor*. Van Brunt feared the worse, believing that the "*Monitor* had given up the fight and run into shoal water," and began making preparations to scuttle his ship. Catesby Jones saw that this was another opportunity to attack the *Minnesota*. Unfortunately, the pilots warned him that the tide was falling fast and the huge ironclad should move back into the Elizabeth River. Jones reviewed, with his officers, the circumstances of the *Monitor*'s apparent retreat and the inability of their ship, due to her deep draft, of getting any closer to the *Minnesota*. He summarized the situation by stating, "This ship is leaking from the loss of her prow; the men are exhausted by being so long at their guns…I propose to return to Norfolk for repairs." A majority of the officers agreed with Jones's assessment. Jones, noting that "had there been any sign of the *Monitor*'s willingness to renew the contest, we would have to fight her,"[135] turned the Confederate ironclad away from the *Minnesota* and steamed toward Sewell's Point.

By this time, Greene had returned to the pilothouse. When he realized that the damage was minimal, Greene ordered the Union ironclad back into action. It had taken nearly half an hour for Greene to assume command, and the lull had prompted the Confederate ironclad to break off action. Greene mistook the *Virginia*'s course toward Sewell's Point as a sign of defeat and proclaimed, "We had evidently finished the *Merrimac*."[136] When the crew learned about the *Virginia*'s withdrawal, the entire crew cheered vigorously. A few more shots were fired by the *Monitor* at her antagonist. Several men were so excited by the turn of events that, according to William Keeler, "Our iron hatches were slid back & we sprang out on deck which was strewn with fragments of the fight. Our foe gave us a shell as a parting fire which shrieked just over our heads & exploded about 100 feet beyond us."[137] The first battle between ironclad ships of war was over, and "Ericsson's Folly" had proven her worth.

Chapter 6
Under a Glass

Once Samuel Dana Greene realized the battle was over, he anchored his ironclad next to the *Minnesota*. Several crew members wished that the *Monitor* would have continued the fight. William Durst, Patrick Hannan and Hans Anderson all reflected, in postwar comments, that the crew urged Lieutenant Greene to continue the fight. Anderson later wrote the "*Merrimac* was in sinking condition and was ready to hoist the white flag but found it unnecessary since the *Monitor* had already withdrawn. It was a shame that Captain Worden was wounded, since otherwise we would have sunk the *Merrimac* or perhaps been sunk ourselves."[138] All this was probably going through Greene's mind when he made the decision to act on the defensive and positioned his ironclad near the stranded frigate. After all, the pilothouse was badly damaged and probably could not withstand being struck by another shell. Furthermore, Worden was seriously injured and needed more advanced medical attention. While this decision haunted him for the rest of his life, Greene believed it was the prudent action to take, as he later explained:

> *We had definite orders to act on the defensive, and protect the* Minnesota… *Therefore, after the* Merrimac *retreated, we went to the* Minnesota *and remained by her side until she was afloat. General Wool and Secretary Fox both have complimented me very highly for acting as I did and said it was the strict military plan to follow. This is the reason we did not sink the* Merrimac *and every one on her says we acted exactly right.*[139]

As soon as the battle was over, Alban Stimers immediately telegraphed John Ericsson with the message: "You have saved this place to the nation by furnishing us with the means to whip an ironclad frigate that was, until our arrival, having all her way with our most powerful vessels."[140] The realization that the *Monitor* had saved the day was not lost on any Unionist in Hampton Roads. Numerous small boats flocked to the *Monitor* as soon as she anchored to congratulate the crew on their splendid victory.

Of utmost importance was the *Monitor*'s commander, John Worden. Since Dr. Logue was not an eye specialist, there was only so much that he could do for the blinded Worden. A small tug came alongside, and Worden was brought up from his cabin. Crew members cheered him wildly as he left the *Monitor*. Worden was taken to Fort Monroe. There he was placed in the charge of a friend, Lieutenant Henry A. Wise, who escorted Worden aboard a steamer to Washington, D.C., for treatment.

Meanwhile, the *Monitor*'s crew was ready for a meal after so long under the strain of battle. Keeler noted: "Our stewards went immediately to work & at our usual dinner hour the meal was on the table, much to the astonishment of visitors who came expecting to see a list of killed & wounded & a disabled vessel, instead of which was a merry party around a table enjoying some good beef steak, green peas, &c."[141]

Assistant Secretary of the Navy Gustavus Vasa Fox, who observed the entire engagement from the *Minnesota*, went onboard the Union ironclad and told the *Monitor*'s officers, "Well, gentlemen, you don't look as though you just went through one of the greatest naval conflicts on record." "No, sir," Greene replied, "we haven't done much fighting, merely drilling the men at the guns a little."[142] Greene's boldness was merely an expression of all of the nervous energy felt by the men of the *Monitor*. After all, these iron men had not rested in over two days. In that time, they had twice survived sinking conditions and had fought a fierce engagement with a determined foe. Despite the euphoric feelings of victory, the men had been pushed to the limits of human endurance. George Geer noted in a letter to his wife after the battle that he felt "well with the exception of want of sleep"[143] and told her little about the engagement. The exhausted Greene wrote home later that evening, "I had been up so long, and been under such a state of excitement, that my nervous system was completely run down…My nerves and muscles twitched as though electric shocks were continually passing through them…I laid down and tried to sleep—I might as well tried to fly."[144]

Neither ship had been seriously damaged during the four-hour battle. The *Monitor*'s shot struck the *Merrimack* only twenty-one times out of forty-one fired. Hans Anderson, assigned to Gun Number 1, later contended that the odd amount of shot fired was because he and one of the other Scandinavians onboard, Charles Peterson (the two Swedes onboard were Anderson and Charles Sylvester; Peterson was from Norway), agreed to double-shot their gun. Anderson believed that this shot made "holes in the *Merrimack* and almost delivered her 'the COUP-DE-GRACE.'"[145] Throughout the battle, gun crew members often thought that various shots had inflicted significant damage to the *Virginia*; however, since they could not see their antagonist, they never realized that their Dahlgrens had minimal effect on the Confederate ironclad. Nor did they inflict any casualties among the *Virginia*'s crew. Most balls just bounced off the *Virginia*'s sloped sides and tumbled into the sea. A few iron plates were cracked, and the wood backing to the shield was damaged where two solid shot struck simultaneously when the Confederate ironclad attempted to ram the *Monitor*. The *Monitor*'s guns had limited impact on the *Virginia* because of the navy's directive to only use fifteen-pound powder charges. Had the *Monitor* used a standard powder charge of thirty pounds, or perhaps a double charge, as tests would later prove the Dahlgrens could handle without bursting, the *Virginia* would have suffered significant damage. Furthermore, visual and communication problems made fire control difficult and ineffective.

The Confederate ironclad's shots struck the *Monitor* twenty-two times, hitting the "pilothouse twice, turret 9 times, side armor 8 times, deck 3 times." Confederate E.V. White, who served on the *Virginia*'s gun deck manning the speaking tube and engine room bell, recollected that "balls from the *Merrimack*, especially those fired almost muzzle to muzzle, produced some results. Three cylindro-conical balls fired from rifled guns made an indentation nearly four inches deep in the armor plating. Two of them made an equally deep indentation on the inside of the turret…The other shots which reached the *Monitor*, and were otherwise for the most part round, did not appear to me to have produced a very great effect, those especially which struck the edge of the deck, lifting and tearing it, causing iron plating to give way and breaking three of them. The others only produced insignificant effects."[146] The Confederate ironclad simply had the wrong ammunition with which to significantly damage the *Monitor*. Although John Mercer Brooke had already created armor-piercing shot known as "Brooke bolts," these projectiles were not available on 9 March. Nevertheless, five crew members were injured and subsequently treated by Surgeon Daniel Logue: Moses Sterns (hernia), Alban Stimers (slight concussion), Louis Stodder (concussion), Peter Truscott (severe concussion) and John Worden (facial burns, eye injury, concussion). The *Monitor* had proven herself to be impregnable.

Over twenty thousand soldiers, sailors and civilians had witnessed the battle, and everyone recognized the engagement as a "day of stirring events."[147] "Our ship is crowded with officers of all grades both army and navy. They are wild with joy and say if any of us came to the Fort we can all have all we want free, as we have saved 100s of lives and millions of property to the Government."[148] While the *Monitor* Boys were all proclaimed heroes for fighting the Confederate ironclad to a standstill, William Keeler thought otherwise, as he wrote his wife: "I think that we get more credit for the mere fight than we deserve, any one could fight behind an impenetrable armor—many have fought as well behind wooden walls or behind none at all. The credit, if any is due, is in the daring to undertake the trip & go into the fight, in an untried experiment & in our unprepared condition."[149] It was almost a miracle that the *Monitor* had even made it to Hampton Roads, and a tremendous tribute to the crew and officers for saving the Federal fleet from further destruction.

"The next morning at 8 o'clock we got underway and stood through the fleet," Samuel Dana Greene later wrote about the *Monitor*'s victory procession in Hampton Roads. "Cheer after cheer went up from the frigates and small craft for the glorious little *Monitor* and happy, indeed, we did all feel. I was the captain of the vessel that had saved Newport News, Hampton Roads, Fortress Monroe (as General Wool himself said) and perhaps your Northern ports."[150] Greene, despite his heroism on the trip from New York and his leadership in the turret, was considered far too inexperienced for command. Assistant Secretary Fox detailed two new officers to help Greene manage the *Monitor*.

Acting Lieutenant William P. Flye reported to the *Monitor* at 10:00 p.m. on the evening of the ninth to serve as executive officer to Greene. Flye was a seasoned naval officer, having first served as a professor of mathematics at the U.S. Naval Academy. After service aboard the USS *Jamestown* and at the Naval Observatory from 1848 to 1854, he had resigned from the U.S. Navy. He rejoined the navy as an acting lieutenant on 6 December 1861. He commanded the USS *R.B. Forbes* until that vessel was wrecked and burned on 25 February 1862. He was reassigned to the USS *Roanoke* and was aboard that frigate during the Battle of Hampton Roads.

Acting Master Edwin Velie Gager was detailed to the *Monitor* the next day, on 10 March. Gager was a prewar sailor who enlisted in Brooklyn, New York, in April 1861. He was assigned to the USS *Monticello* in Hampton Roads and was master of the *Monticello* during the war's first amphibious operation against Hatteras Inlet, North Carolina. Gager was able to bring

Under a Glass

Acting Master Edwin Velie Gager, circa 1862. *Courtesy of The Mariners' Museum.*

his ship close in to the shore, despite the dangerous shoals, to rescue a stranded Indiana regiment.

These new assignments reflected upon the Union's recognition of the *Monitor*'s importance to ensuring Federal control of Hampton Roads. Even though the North was euphoric about the *Monitor*'s victory on 9 March, the Confederates also perceived the battle as their victory. The battle was actually a drawn one. The *Monitor* had indeed won a tactical victory, as the Union ironclad had stopped the *Virginia* from destroying the *Minnesota*, relieving Federal fears of a broken blockade and an attack on Northern cities. However, since the *Monitor* had not captured, severely damaged or sunk the *Virginia*, the undefeated Confederate ironclad still effectively controlled Hampton Roads. The *Virginia* blocked the entrance to the James River, thereby defending the water approach to Norfolk and Richmond. The mere existence of the *Virginia* would have a powerful influence on Major General

George McClellan's strategic initiative to capture the Confederate capital in Richmond by way of the Virginia Peninsula. Brigadier General John Gross Barnard, chief engineer of the Army of the Potomac, lamented that the "*Merrimack*...proved so disastrous to our subsequent operations."[151]

The Confederate strategic naval victory effectively closed the James River and most of Hampton Roads to the Federals just when General McClellan decided to march toward Richmond by way of Fort Monroe and the Virginia Peninsula. It was a brilliantly conceived concept. The Union general's original plan entailed the use of both the James and York Rivers. Gunboats could guard his flanks, while steamers could transport supplies as McClellan's troops marched up the Peninsula. The entire campaign was now in jeopardy, all because of the Confederate ironclad. McClellan wired Assistant Secretary G.V. Fox at Fort Monroe: "Can I rely on the *Monitor* to keep the *Merrimack* in check, so that I can make Fort Monroe a base of operations?" Fox replied, "The *Monitor* is more than a match for the *Merrimack*, but she might be disabled in the next encounter...The *Monitor* may, and I think will, destroy the *Merrimack* in the next fight; but this is hope, not certainty."[152] McClellan held a council of war with his corps commanders, and it was agreed to proceed with the campaign using only the York River. The U.S. Navy was expected to neutralize the *Virginia* and help destroy the Confederate batteries on the York River.

Major General John Ellis Wool, commander of the Union Department of Virginia at Fort Monroe, concurred with Fox's opinion. He had already received orders, on 9 March, from McClellan to hold Fort Monroe "at all hazard"[153] against the Confederate ironclad. Wool visited the *Monitor* on 10 March and advised the ironclad's officers that the *Virginia* could have destroyed all the naval forces and shipping in the harbor, as well as, according to William Keeler, "shelled out the Rip Raps (Fort Wool), & Fortress Monroe itself would have been at their mercy...The[y] would have attacked Gen. [Joseph K.F.] Mansfield's army at Newport News in front while [General John Bankhead] Magruder took them in the rear. This & still more extensive plans of operations had been laid by them when our appearance blocked the game."[154] The Confederate ironclad's existence immediately prompted Wool to block the *Virginia*'s access into the Chesapeake Bay by mounting a fifteen-inch Rodman Gun, nicknamed the "Lincoln Gun," next to the twelve-inch rifled "Union Gun" on the beach near the Old Point Comfort Lighthouse. The Union general was sure that the *Monitor*, supported by these two powerful guns, would effectively close the mouth of Hampton Roads to any sortie by the *Virginia*.

Under a Glass

Regardless of Wool's preparations, the Confederate ironclad's role as a "fleet in being" continued to influence the U.S. Navy's ability to support McClellan's campaign against Richmond. McClellan later wrote:

> *The James River was declared by the naval authorities closed to the operations of their vessels by the combined influence of the enemy's batteries on its banks and the Confederate steamers* Merrimac, Yorktown, Jamestown *and* Teaser. *Flag Officer Goldsborough…regarded it (and no doubt justly) as his highest and most imperative duty to watch and neutralize the* Merrimac, *and as he designed using his most powerful vessels in a contest with her, he did not feel able to attack the water batteries at Yorktown and Gloucester. All this was contrary to what had been previously stated to me and materially affected my plans. At no time during the operations against Yorktown was the Navy prepared to lend us any material assistance in its reduction until after our land batteries partially silenced the works.*[155]

McClellan's grand plan to quickly capture Richmond was altered and delayed, despite the *Monitor*'s presence. Welles advised McClellan that "the Confederate ship was an ugly customer." Even though the secretary of the navy believed the *Monitor* to be the superior vessel, Welles thought the Union ironclad "might easily be put out of action in her next engagement and that it was unwise to place too great a dependence on her."[156] Consequently, Flag Officer Goldsborough feared that if the *Monitor* were damaged in another battle with the *Virginia*, it would enable the Confederate ironclad to then strike and destroy the Federal fleet. Until the other Union ironclads under construction arrived in Hampton Roads, Goldsborough was content to remain on the defensive.

The *Monitor* was recognized as the only real defense against the Confederate ironclad. William Keeler noted: "We shall remain here as guardians of Fortress Monroe, and the small amount of shipping which will remain in harbor have been ordered off in apprehension of the reappearance of the *Merrimac*. We shall remain here to meet her. We are very willing and anxious for another interview."[157] The *Monitor*'s crew might have been ready to reengage the *Virginia*, but the Federal command was not. Gideon Welles sent a telegram on 10 March stating, "It is directed by the President that the *Monitor* be not too much exposed, and that in no event shall any attempt be made to proceed with her unattended to Norfolk."[158]

The *Monitor*'s importance prompted a change in command. Greene was considered too young and inexperienced to permanently replace the wounded Worden. Lieutenant Thomas Oliver Selfridge Jr. replaced Greene on the late afternoon of 10 March. Son of a career naval officer who later became an admiral, Selfridge was born at Charlestown Navy Yard in Boston on 6 February 1836, entered the U.S. Naval Academy at Annapolis in 1851 and graduated at the head of his class in 1853, the first officer to receive a diploma under the permanent Naval Academy system. He was detailed as midshipman to the USS *Independence* and was assigned to Coastal Survey tasks. In 1856, he was warranted a passed midshipman and detailed to the USS *Nautilus*. On 22 January 1858, Selfridge was promoted to master and transferred to the USS *Vincennes*, off the coast of Africa, to suppress the slave trade. After an extended sick leave, Selfridge was ordered to the USS *Cumberland* and was chosen to lead a detail to destroy equipment and facilities of the Gosport Navy Yard when the yard was abandoned on the evening of 20 April 1861. He was then placed in command of the gunboat the USS *Yankee* and, while on reconnaissance duty, engaged with the Confederate batteries manned by the Richmond Howitzers on 7 May 1861. On 8 May, he was transferred back to the *Cumberland* and commanded the forward gun battery during that sloop's unfortunate engagement with the *Virginia* on 8

Chief Engineer Alban C. Stimers, circa 1862. *Courtesy of The Mariners' Museum.*

Under a Glass

March 1862. He survived the *Cumberland*'s sinking, saving himself by jumping from a gun port and swimming to a rescue launch. He then witnessed the battle between the *Monitor* and the *Virginia* on 9 March from the shore of Hampton Roads.

Selfridge would only command the *Monitor* for two days, but during his brief command, he ordered repairs to the pilothouse, considered the *Monitor*'s greatest weakness. Alban Stimers and Isaac Newton reconfigured the pilothouse sides from perpendicular to a slope of thirty degrees to deflect shot. The shell of solid oak was covered with three inches of wrought iron, laid in three layers, to further deflect shot.

As the pilothouse repairs were initiated, the crew joyfully learned from Mrs. Worden that their former captain, according to William Keeler, "was very weak & nervous, but the doctor says he will recover his sight though he may be confined for some time." Worden was considered a true hero, and on 10 March, President Abraham Lincoln, escorted by Gideon Welles and Lieutenant Henry Wise, visited the injured Worden. Wise had already given the president an eyewitness account of the battle, but Lincoln insisted on seeing Worden. On entering Worden's recovery room, Wise said, "Jack, here's the President, who has come to see you." Worden, lying in bed with his head swathed in bandages, reached for Lincoln's hand and said, "You do me great honor, Mr. President, and I am only sorry that I can't see you." Lincoln respectfully replied, "You have done me more honor, sir, than I can ever do to you."[159] Lincoln sat on the edge of the bed and asked Worden to tell him the story of the battle.

Worden's tale did not differ from the others, except for his conclusion. The *Monitor*'s former commander acknowledged the shot-proof qualities of the ironclad yet explained that the ship's greatest weakness was that she could be captured by the enemy boarding her. While the crew was safe inside the ironclad, boarders could block the turret, pour water down the funnels to shut down the engines and disable the crew by tossing hand grenades into the turret's gun ports. The Confederates had actually considered such an action during the battle. Worden advocated that the *Monitor* would be safe as long as she had sea room in which to maneuver and the crew had small arms available. The *Monitor* was quickly issued weapons, and orders not to expose the ironclad were reiterated.

Meanwhile, the *Monitor* was cheered everywhere she steamed in Hampton Roads. William Keeler remembered, a few days after the fight, steaming close by Camp Butler on Newport News Point:

The whole army came out to see us, thousands & thousands lined the shore, covered with the vessels at the docks & filled the rigging. Their cheers resembled one continuous roar. Each regiment had its band, the nearest of which as we approached struck up "See the Conquering Hero Comes," and then the "Star Spangled Banner," & so it passed from one band to another as we slowly steamed along in front of them. All our national airs were given, when a lull in the tumultuous cheering would allow them to be heard.

It was laughable to hear the great variety of names applied to us by the soldiers, for we passed so near we could readily converse. Says one, "You're the boys," another "Bully for you," "No sand bag Batteries there," "You're our Saviours [sic]," "Ironsides & iron hearts," "No back down to you," "You're trumps everyone," & etc, etc, etc.[160]

The *Monitor* became a major attraction, and everyone wished to visit the vessel that had saved the Union. Keeler, because of his limited duties, became

Crew of the USS *Monitor*. Photograph by J. Gibson, 4 July 1862. *Courtesy of The Mariners' Museum.*

the primary tour guide. He enjoyed escorting people through the ironclad. He noted all of the distinguished visitors who had come to see the *Monitor*: "Gen. Mansfield & staff & by any quantity of foreign nobles, counts &c who are serving in our army. Among them quite a number of Swedes, who being countrymen of Capt. Ericsson's naturally feel a pride in his invention & its triumphant success." Keeler appears to have truly enjoyed being the "official tour guide," as well as having all the laudatory comments. He advised his wife about a tour that he had presented to a Mrs. Wall:

> *The request that I take care of her for a time—as she was young, handsome, & intelligent of course I couldn't refuse. I asked her if she had been to the turret to see the guns. "Oh, yes," she said. "I kissed them too. I feel as if I could kiss the deck we stand on," &, continued, one of her female friends who was standing near, "I would like to kiss all who were on board during the fight if I thought they would let me."*[161]

Keeler and other crew members noted that while their warship had become an iconic, industrial marvel, they were treated as heroes of the highest order. Albert Campbell was overjoyed to discover that "people ashore could not say enough in our praise and would not take our money for anything." Keeler was overwhelmed by the feeling within the Union positions along Hampton Roads as "the *Monitor* is on everyone's tongue & the expressions of gratitude & joy embarrassed us they are so numerous." While the paymaster was securing fresh provisions for the crew, a sloop owner said, "Tell your ward room officers to come and see me whenever they come ashore. I have first rate quarters & they are always welcome without expense. The safety of all I have is due to them."[162]

Keeler would also give mementoes—shell fragments from the *Virginia*— found on the *Monitor*'s deck to many of her distinguished guests. He collected them right after the battle concluded. In fact, as he was picking up pieces of shell, the last shell from the *Virginia* exploded overhead. "One of the men who had been working the guns, touching his hat, said very coolly, 'Paymaster, there's some more pieces.' I confess I looked rather anxiously to see if any more were coming." Keeler sent fragments to his family and also gave the end of an exploded shell to General Wool's chief of staff, Colonel LeGrand Cannon, to send to President Lincoln "with the respects of the officers of the *Monitor*."[163]

The fragments were holy relics of the industrial age, a time when men became secondary to the machine. Many saw the battle as the dawn of a new era of naval weapons, and the *Monitor* was its great symbol of the victory of technology. This ironclad duel was considered, at least by one Union soldier, as "one of the greatest naval engagements that has ever ocurred [*sic*] since the Beginning of the world."[164] The battle was seen as a "revolution in naval warfare, and henceforth iron will be king of the seas."[165] The battle received international attention. The *London Times* reported: "Whereas we had available for immediate purposes one hundred and forty-nine first-class warships, we now have two, these two being the *Warrior* and her sister *Ironside*." Great Britain immediately committed itself to building a fleet of ironclad vessels, as Lord Clarence Paget, secretary of the admiralty, noted in Parliament: "No more wooden ships would be built."[166] The showdown in Hampton Roads between these two experimental ships with inexperienced crews proved once and for all the power of iron over wood.

Even though many would continue to call the *Monitor* "the strangest looking craft," she was the only protection the Union had against the *Virginia*. Since General McClellan had decided to move forward with his Peninsula Campaign using Fort Monroe as a base, a more experienced technical officer was required to properly manage the Union's technological wonder. Consequently, a short, somewhat rotund officer, Lieutenant William Nicholson Jeffers III, boarded the *Monitor* on 12 March and informed Selfridge that he was the *Monitor*'s new commander.

William Jeffers was born on 16 October 1824 in Swedesboro, New Jersey. Jeffers entered the navy as an acting midshipman on 25 September 1840, and by 1841, he was detailed to the U.S. Frigate (USF) *United States*, the flagship of the Pacific Squadron, commanded by Commodore Thomas ap Catesby Jones. During his service under Jones, Jeffers participated in the seizure of Monterey, California. He also sailed to Hawaii. While in Honolulu, Herman Melville, who later wrote *Moby Dick* as well as an article about the *Monitor*, enlisted as an ordinary seaman aboard the *United States*. In 1845, Jeffers was sent to the newly established Naval Academy at Annapolis, where, in 1846, he graduated fourth in a class of forty-seven as a specialist in ordnance and gunnery. While still a midshipman, Jeffers published his first book, *The Armament of Our Ships of War*, in 1846.

When the Mexican-American War erupted, Jeffers was detailed to the light-draft steamer the USS *Vixen*. He participated in several notable engagements while aboard the *Vixen*, including the capture of Tabasco, Tuspan and Laguna de Terminos, as well as the siege and capture of Vera Cruz and the castle of San Juan d'Ulloa.

At the war's conclusion, Jeffers was sent to Annapolis as an instructor of mathematics. While on this assignment, Jeffers wrote his second and third

Under a Glass

books, *Nautical Routine and Stowage, with Short Rules in Navigation* and *A Concise Treaty on the Theory and Practice of Naval Gunnery*. On 17 September 1850, he married the daughter of a U.S. Army surgeon, Lucie LeGrand Smith. Between 1852 and 1854, he worked on the Coast Survey until he took a leave of absence to survey Honduras for an "Inter-Oceanic Railway" under the employ of E.G. Squires. In Squires's subsequent book, *Notes on Central America...and the Proposed Inter-Oceanic Railway*, published in 1855, Jeffers was recognized as a member of the expedition "with acknowledged scientific and practical ability." He would return to that region in 1857 to conduct a second survey, but the railway was never constructed.

William Jeffers returned to the service in 1854 and was assigned to the Brazil Squadron as acting master of the USS *Allegheny* and then the USS *Germantown*. In January 1855, he was promoted to lieutenant and named the executive officer of the *Water Witch*. This small, side-wheel steamer was commanded by Lieutenant Thomas Jefferson Page and became engaged with Paraguarían forces when surveying the La Plata and Parana Rivers in South America. Page commanded the expedition to chart these waters; however, he fell into a disagreement with Paraguay's president, Carlos Antonia Lopez. President Lopez decreed that no "foreign vessels of war" should enter the rivers of his nation. Lieutenant Page ordered Jeffers to do so anyway, which resulted in an exchange of fire between the *Water Witch*,

Lieutenant William Jeffers. Photograph by J. Gibson, 4 July 1862. *Courtesy of The Mariners' Museum.*

armed with three small brass howitzers, and the Paraguayan Fort Itapiru. The *Water Witch* was damaged, and the event "seriously damaged diplomatic relations between the two countries." While both Page and Jeffers were chided for their involvement in the affair, neither officer was reprimanded by the Navy Department for his action. Despite his problems in Paraguay, Jeffers would be commended for rescuing the Spanish warship *Cartagenera* from being wrecked. In 1858, Congress passed a resolution allowing Jeffers to accept a gold commendation sword from the Queen of Spain for his actions in saving the ship.

Jeffers's interest in gunnery brought him an assignment on the sloop of war the USS *Plymouth* in 1858. The *Plymouth*, with Lieutenant Commander John Dahlgren as commander and Lieutenant Catesby ap Roger Jones as executive officer, was an experimental ordnance ship. Jeffers became one of Dahlgren's protégés and was considered an ordnance expert after he concluded this assignment.

In August 1859, he was detailed to the steam sloop the USS *Brooklyn*, commanded by Captain David Glasgow Farragut. Unfortunately, Jeffers became embroiled in a bitter dispute with Farragut due to an incident when Jeffers failed to acknowledge, or salute, Farragut. No formal charges were brought after Jeffers apologized to the captain.

The *Brooklyn* was then sent to Panama to land an expedition to explore a route across the Isthmus of Chiriqui. Jeffers served as hydrographer for this survey. By the fall of 1860, Jeffers was discovered to be ill with rheumatism, and he was on sick leave when the Civil War erupted.

Jeffers immediately applied for active service and was appointed ordnance officer of the Gosport Navy Yard. The yard, however, fell into the hands of the Confederacy before Jeffers could reach his new post. After a brief stint as commander of the iron-hulled steamer the USS *Philadelphia*, Jeffers was assigned to the steam-screw frigate the USS *Roanoke* at Hampton Roads.

In November 1861, Jeffers was detailed as commander of the four-gun, side-wheel steamer the USS *Underwriter*. After assuming command of that vessel, Jeffers was quickly ordered to take the *Underwriter*, with two towboats hauling stone-filled hulks, to sink the hulks in Ocracoke Inlet, thereby blocking that entrance to the North Carolina sounds. The *Underwriter* was then assigned to Flag Officer Louis M. Goldsborough's task force to capture Roanoke Island, South Carolina. The *Underwriter* was then attached to commander S.C. Rowan's task force to close the canals and capture Elizabeth City, North Carolina. The Confederate squadron, commanded by Flag Officer William F. Lynch, was destroyed, and the entrance to the Dismal Swamp Canal was blocked. Jeffers then took the *Underwriter*, along with a small force, and blocked the entrance to the Chesapeake & Albemarle Canal. Rowan praised Jeffers for his "unusual zeal and intelligence" when closing the canal.

When Goldsborough learned about Worden's wounding, he instructed Jeffers to take command of the *Monitor*. Goldsborough wanted an experienced officer—Jeffers was thirty-eight years old, with over twenty years of sea service—as well as a commander with intelligence and knowledge in gunnery. While Assistant Secretary of the Navy G.V. Fox had appointed Selfridge to the *Monitor*'s command, it was actually Goldsborough's prerogative as squadron commander to name Worden's replacement; therefore, his choice prevailed.

Jeffers met his officers that evening after dinner in the wardroom. They all appeared rather impressed. Samuel Dana Greene wrote his parents: "Mr. Jeffers is everything desirable, talented, educated, and energetic and experienced in battle."[167] Keeler also seemed pleased as he wrote his wife: "Lieut. Jefers [sic] has been in most of the fights along the coast and it is very interesting as we sit at the table to hear him give his experience in the different fights—some of them he sets out in a very amusing light."[168]

Jeffers would quickly lose any esteem held by the officers and crew for him. Keeler would later write:

> *Things don't go as smoothly and pleasantly on board as when we had Capt. Worden. Our new Capt. is a rigid disciplinarian, quick imperious temper and domineering disposition…so far I have got along smoothly enough with Capt. Jeffers, but I am expecting every day that I may forget to touch my hat, or give him the deck in passing or grieviously* [sic] *offend him in some little point of etiquitte* [sic], *when I shall get a blast.*[169]

George Geer called Jeffers "a dam'd old hog,"[170] and the other crew members actually wrote Worden on 24 April, addressing "Our Dear and Honored Captain." The letter continued: "These few lines is [sic] from your own crew of the *Monitor* with there [sic] kindest Love to you[,] Hoping to God that they will soon have the pleasure of Welcoming you back to us again Soon…[S]ince you left we have had no pleasure on Board of the *Monitor*." This passionate yet rough-hewn request was signed:

> *We remain until Death your Affectionate Crew*
> *The* Monitor *Boys*[171]

It appears that Jeffers lacked, according to Keeler, "that noble kindness of heart and quiet unassuming manner to both officers and men which endeared Captain Worden to all on board."[172]

The Confederate ironclad also needed a new commander to replace the wounded Franklin Buchanan. Lieutenant Catesby Jones remained the ironclad's executive officer, and Flag Officer Josiah Tattnall was named commander of all Confederate naval forces in Virginia's waters. Tattnall had joined the U.S. Navy in 1812 and fought his first battle on Craney Island in June 1813. He served with distinction in the Mexican-American War, as well as in China as commander of the East India Squadron. Josiah Tattnall was well known as the "beau-ideal of a naval officer." Many believed that Tattnall possessed all the traits that are found in heroic characters. Tattnall resigned his U.S. Navy commission when Georgia left the Union and served as commander of the Savannah Squadron until assigned to command in Hampton Roads. This directive from Confederate Secretary of the Navy S.R. Mallory was to make the Confederate ironclad

Flag Officer Josiah Tattnall, CSN, circa 1863. Courtesy of John Moran Quarstein.

"as destructive and formidable to the enemy as possible." Mallory added: "Do not hesitate or wait for orders, but strike when, how and where your judgment may dictate."[173]

Since the Confederate ironclad was actually incomplete when she sailed forth into Hampton Roads on 8 March, and was damaged during the two-day engagement, she was immediately placed into dry dock for repairs and improvements. John Mercer Brooke developed new armor-piercing bolts, designed to penetrate the *Monitor*'s armor. He also prepared a new ram that was twelve feet long and featured a steel point designed to strike below the *Monitor*'s armor belt into her one-inch iron hull. Other repairs and improvements included gun port shutters, the replacement of several damaged iron plates and additional armor below the eaves of her casemate to protect the ironclad's knuckle. Time limited this work because, as her former commander Franklin Buchanan concluded, the Confederate ironclad was "the most important protection to the safety of Norfolk."[174]

By late March, most of Major General George McClellan's army had arrived on the Peninsula, and the Union general was preparing to begin his march toward Richmond. The men of the *Monitor* were amazed by all of the soldiers and equipment assembling at Fort Monroe, Camp Hamilton and Camp Butler, just as those arrivals marveled at the *Monitor*. William Keeler commented about the massive influx of troops and equipment: "Steamers have been arriving with troops through the day. How many have been landed here within the last few days I do not know, but a Col. of one of the Penn. regiments stationed at Newport News told me that over one hundred thousand men have passed there within the last few days."[175]

Visitors, both civilian and soldiers, continued to see the *Monitor* and to cheer the "brave, gallant & victorious band of *Monitors*." One lady told Keeler, "You cannot think how pleased I am to stand on the deck of this ship that I have heard so much about & have felt so proud of. You have saved us all."[176] General McClellan even visited the ironclad, but without the pomp and circumstance one might have expected. When the general arrived in Hampton Roads to visit the *Monitor* the next day, all was made ready for his visit. The expected time for the "hero's arrival" had passed when, according to George Geer, "an old row Boat came along side and an officer came on Board and introduced himself as Genl. McClelan [sic]."[177] Everyone was surprised, but McClellan needed to see the warship that had made his campaign feasible and that would continue to protect his supply line.

Besides all of the men and cannon arriving, some new technology also appeared in early April. McClellan had brought with his army the noted aeronaut Thaddeus S.C. Lowe's two gas balloons, *Intrepid* and *Constitution*, aboard the balloon barge USS *G.W.P. Custis*. The balloon was immediately used for reconnaissance in Hampton Roads, and on 30 March, William Keeler noted to his wife, "Just before dark tonight I noticed a huge balloon rearing its head."[178] In addition to Lowe's balloons, another ironclad arrived in Hampton Roads, the U.S. Revenue Marine Ship (USRMS) *Naugatuck*, also known as *Stevens Battery*. The *Stevens Battery* was the U.S. Navy's first ironclad project designed by Robert L. Stevens of Hoboken, New Jersey. The vessel was laid down in 1844; however, its design was modified in 1854. The ironclad was never completed due to lack of funds and an outmoded design. R.L. Stevens's son John offered to complete the ship and donate it to the U.S. Navy. His offer was declined. Eventually, the vessel was accepted by the U.S. Revenue Service and then loaned to the U.S. Navy for service during part of 1862. The *Naugatuck* was an experimental warship. The iron-hulled ship would have her forward and aft compartments flooded to partially submerge the hull. This ironclad's primary armament was a one-hundred-pounder Parrott rifle.

On 4 April 1862, McClellan's army began its movement toward Richmond, only to be stopped the next day by Confederate General John Bankhead Magruder's Warwick River defensive line, anchored on Yorktown and Gloucester Point. Army of the Potomac Chief Engineer John G. Barnard called the Confederate position "certainly one of the most extensive known to modern times,"[179] while Army of the Potomac, IV Corps, commander Brigadier General Erasmus Darwin Keyes noted, "No part of this line as discovered can be taken without an enormous waste of life."[180] McClellan, his path to Richmond blocked, hesitated, and this delay set the stage for a carefully organized ruse. Magruder began shuttling his soldiers up and down his fortifications to create an illusion of many troops arriving and moving into positions of great strength. Magruder earned the title of "Master of Ruses and Strategy" for his make-believe show of strength. "It was a wonderful thing," recorded Confederate diarist Mary Chesnut, "how he played his ten thousand before McClellan like fireflies and utterly deluded him."[181] Since the U.S. Navy refused to support any action against the Confederate water batteries because of the existence of the *Virginia*, McClellan felt that he had no other option than to besiege Magruder's defenses.

Many of the *Monitor*'s crew, such as George Geer, commented that "the Booming of Cannons could be heard all day in the direction of Yorktown and was renewed this morning."[182] The *Monitor* and *Naugatuck* were stationed in the channel between Fort Monroe and Fort Wool. These ironclads and forts were the primary defense for McClellan's transport in the York River.

Under a Glass

Meanwhile, the *Monitor*'s crew had become impatient waiting for the *Virginia* to reappear. Rumors persisted about the Confederate ironclad's return to Hampton Roads. George Geer wrote his wife, "We had a Contraband on board yesterday that ran away from Norfolk…He says he saw the *Merrimack* on Monday and that she was in dry dock repairing…I don't think we will get another chance at her unless we go up there." Keeler noted in a letter to his wife, "We still vainly watch the mouth of Elizabeth River, no *Merrimac* makes her appearance—nor in my opinion will she."[183]

The *Virginia* left dry dock on 4 April 1862, the very day McClellan began his march up the Peninsula. Bad weather and mechanical problems delayed any foray, despite the Confederate high command's desire for their ironclad to strike at McClellan's transports. Such an attack could not be made, as long as the *Monitor* served as a counter-threat to the Confederates' control of Norfolk.

On 11 April, at 6:00 a.m., Flag Officer Josiah Tattnall's squadron moved down the Elizabeth River to Sewell's Point. Tattnall was determined to take on the *Monitor*, as he declared to his crew, "I will take her! I will take her even if hell's on the other side of her."[184] The Confederate ironclad entered Hampton Roads at 7:10 a.m., and the Federal transports scattered to the protection of Fort Monroe "like a flock of wild fowl in the act of flight." The *Monitor* and *Naugatuck* stayed in the channel between Fort Monroe and Fort Wool, on the Rip Raps. The Union ironclad had strict orders not to engage the *Virginia* unless the Confederate ironclad moved out of Hampton Roads into the open waters of the Chesapeake Bay. Tattnall refused to take his ironclad out of Hampton Roads, and the *Monitor* would not accept the *Virginia*'s challenge. "Each party steamed back and forth before their respective friends until dinner time," wrote William Keeler, "each waiting for the other to knock the chip off his shoulder." Keeler summarized the standoff between the two ironclads as: "She had no desire to come under fire of the Fortress and all the gunboats, to say nothing of the rams, while engaged with us, neither did the *Monitor* with her two guns desire to trust herself to the tender mercies of the gunboats and Craney Island and Sewall's [sic] point batteries while trying the iron hull of the monster. I had a fine view of her at a distance of about a mile through a good glass and I tell you she is a formidable-looking thing. I had but little idea of her size and apparent strength until now."[185]

"Had the *Merrimac* attacked the *Monitor* where she was and still is stationed by me, I would instantly have been down before the former with all my force," Goldsborough commented about the *Virginia*'s excursion. "The

The Second Trip of the Merrimack. Theodore R. Davis, 3 May 1862. *Courtesy of John Moran Quarstein.*

salvation of McClellan's army, among other things," he continued, "greatly depends upon my holding the *Merrimac* steadily and securely in check and not allowing her to get past Fort Monroe and so before Yorktown. My game therefore is to remain firmly on the defense unless I can fight on my own terms."[186] Goldsborough's terms, as interpreted by William Keeler, were "to get the *Merrimac* in deep water where the larger steamers fitted up as rams can have a chance at her."[187] Tattnall understood the Union plan to "get me in close conflict with *Monitor*…to seize the opportunity to ram into me the *Vanderbilt* and other vessels."[188] The USS *Vanderbilt*, donated by railroad and shipping magnate Cornelius Vanderbilt, and three other fast steamers had been modified into rams. The *Vanderbilt* had a speed of fourteen knots, and her bow was reinforced with steel specifically to run down the *Virginia*.

    Meanwhile, the Confederates had concocted their own plan to destroy or capture the *Monitor*. Information gleaned from an issue of *Scientific American*, which contained a detailed report on the Union ironclad, indicated that the *Monitor* could be boarded and captured by disabling the crew. This scenario was what Worden had warned President Lincoln and Secretary Welles about as the *Monitor*'s greatest weakness. "We had four of our small gunboats," wrote Midshipman R.C. Foute of the *Virginia*, "ready to take the party, some

Under a Glass

of each division in each vessel. One division was provided with grappling irons and lines, another with wedges and mallets, another with tarpaulins, and fourth with chloroform, hand grenades, etc. The idea was for all four vessels to pounce upon the *Monitor* at one time, wedge the turret, deluge the turret by breaking bottles of chloroform on the turret top, cover the pilothouse with a tarpaulin and wait for the crew to surrender." Foute added that the "plan was very simple, and seemingly entirely practical, provided we were not blown out of the water before it could be executed."[189]

The Confederate ironclad steamed around in Hampton Roads from 9:00 a.m. to 4:00 p.m., hoping that the *Monitor* would dare attack. While the *Virginia* held the attention of the entire Federal fleet, the CSS *Jamestown*, commanded by Lieutenant Joseph N. Barry, captured two brigs, the USS *Marcus* and the *Salbea*, and an Accomac schooner, the USS *Catherine A. Dix*, off Newport News Point and towed them to Norfolk. The *Virginia*, flying the captured transports' flags upside down under her own colors as an act of disdain, fired several shells at the *Naugatuck* and returned to Gosport Navy Yard. Tattnall was praised for his prudent yet gallant actions on 11 April. The Northern press, however, lambasted the Union response to the *Virginia*. A correspondent of the *New York Herald* wrote, "The public are justly indignant at the conduct of our navy in Hampton Roads."[190] Many crew members were disturbed by the lack of action. Geer wrote his wife:

> *We have been very much provoked for the last two days by the* Merrimack. *She comes down in the Roads in plain sight but under cover of Sewel's* [sic] *Point Battery and lays there as much as to dare us to attack her, but our own orders are such that we cannot go after her, but must lay here like an old coward and look at her. One of her consorts, the* Jamestown, *ran over by Hampton and cut out three schooners right under our nose when a fiew* [sic] *shot from us would have sunk her, but our orders are so strict we cannot make a move unless the Comodore* [sic] *signalizes* [sic] *us to. If ever a man was cursed I think that Comodore* [sic] *Golesborrow* [sic] *has been by our crew for the last two days.*[191]

William Keeler concluded this sense of frustration over the order not to engage the *Virginia* in Hampton Roads with "I believe the Department is going to build a big glass case to put us in for fear of harm coming to us."[192]

Chapter 7
The Great Monster *Merrimack* Is No More

The stalemate in Hampton Roads was destined to end in May. The officers and crews of both ironclads were ready for action; however, other events would soon change the tactical situation and open the way for the Federal fleet to ascend the James River.

The last weeks of April had witnessed the siege of the Yorktown-Warwick Line increase in its intensity. Confederate General John Bankhead Magruder's bluff that stopped the Federal advance on 5 April prompted General George McClellan to besiege the Confederate positions. McClellan had his men construct massive fortifications for his heavy artillery. The Union general intended to use weapons, such as one-hundred-pounder Parrott rifles and thirteen-inch seacoast mortars, to pound the Confederates into submission and open the land route to Richmond. The siege was the talk of the *Monitor*'s crew. George Geer wrote his wife, "Every day we can hear heavy firing from the direction of Yorktown. I suppose when the fight commences in good earnest we will hear it very plain."[193] When General William Buell Franklin visited the *Monitor*, William Keeler asked him about the siege operations and then wrote his wife: "There is no grumbling here at McC's want of energy or slowness, everyone is perfectly satisfied thus far & thinks he has done all that could be expected. We have heard occasional heavy firing in that direction all day."[194]

As Goldsborough seemed content to await the arrival of a third ironclad, the USS *Galena*, the Federal timidity amazed the Confederates. Lieutenant John Taylor Wood wrote how "frightened they must be, with all of their forts and 3 or 400 vessels in their Navy to be afraid of our vessel."[195] General

Robert Edward Lee, meanwhile, continued to press for the *Virginia* to "turn her attention to the harbor of Yorktown, if it is considered safe for her, under cover of night, to pass Fort Monroe."[196] Mallory finally agreed to such an evening sortie in late April. Tattnall made one attempt to strike at Yorktown under cover of darkness but was ordered back to Norfolk by Major General Benjamin Huger. There could be no attack against Yorktown until the newest Confederate ironclad, the CSS *Richmond*, also called the *Merrimac II*, was complete. Naval constructor John Luke Porter modified the design of the *Virginia* to make a smaller ironclad with less draft. The Confederates worked day and night to complete the second ironclad, believing that the *Richmond* could handle the *Monitor*, as the *Virginia* wreaked havoc among the Union fleet in the York River. The *Virginia* was in real need of additional support. On 18 April, the *Jamestown*, the *Raleigh* and the *Teaser* were sent up the James River to support Magruder's right flank. Two days later, the *Patrick Henry* and the *Beaufort* were assigned to Commander John Randolph Tucker's squadron, stationed off the mouth of the Warwick River. This left the *Virginia* as Norfolk's only defense against an attack by the U.S. Navy.

The Federal noose around Norfolk was tightening more every day. Major General Ambrose Everett Burnside was threatening Norfolk by way of the canals linking the port city with North Carolina. Brigadier General Jesse Lee Reno had already attempted to march his division through the weak Confederate forces at South Mills, North Carolina, defending the entrance to the Chesapeake & Albemarle Canal. While Reno was repulsed at South Mills on 19 April 1862, it was clear that Burnside's command was in an excellent position to move against Norfolk whenever the Union general wished to make a concentrated effort.

The Union navy added a third ironclad on 27 April 1862, when the *Galena* arrived in Hampton Roads. The *Galena*, built by Cornelius Bushnell at the Maxson Ship Yard in Mystic, Connecticut, was one of the three ironclads selected by the Ironclad Board. This experimental warship was designed by Samuel Pook and featured $3\frac{1}{2}$-inch clapboard iron plating in a "tumble-home" design to help deflect shot, but the deck was unarmored. She was armed with two one-hundred-pounder Parrott rifles and four IX-inch Dahlgrens. The *Galena* had a length of 210 feet, a 13-foot draft and could make eight knots. When the men of the *Monitor* reviewed the *Galena*, George Geer wrote his wife, "The wonderfull [sic] *Galena* has arrived, but she is not much account. The *Merrimack* would make very short work of her; her own men have very little confidence in her, and would be very glad of the chance to get on here."[197] William Keeler agreed with this assessment; however, he also noted that Captain Jeffers believed that the "moral effect of having them here is great." Keeler concluded his thought, stating that the "*Galena* and the *Naugatuck*…would be totally unable to cope with the *Merrimack*… They could be used to good effect against the *Yorktown* & *Jamestown*."[198]

Time was quickly running out for the Confederate navy in Hampton Roads. General Joseph Eggleston Johnston, whose army had reinforced Magruder's command during the Yorktown-Warwick Siege, ordered the evacuation of the Lower Peninsula. Johnston believed that "the fight for Yorktown must be one of artillery, in which we cannot win. The result is certain; the time only doubtful."[199] Thus, faced with defending both Norfolk and Richmond, Johnston chose to protect the Confederate capital. Johnston advised both Mallory and Tattnall that "the abandonment of the peninsula will, of course, involve the loss of all our batteries on the north shore of the James River. The effect of this upon our holding Norfolk and our ships you will readily perceive."[200]

The *Virginia* was Norfolk's primary defense, and as the Confederate army retreated toward Williamsburg, the ship's crew made her presence known in Hampton Roads. "The Big Thing," as the *Monitor*'s crew referred to the Confederate ironclad, came out from Sewell's Point "like the genius of evil men." Keeler watched as the *Merrimack* steamed to Newport News Point and then back into the middle of the harbor before returning into the Elizabeth River.

Keeler believed that the *Virginia* had but a couple of options following the Confederate retreat from Yorktown: "When they learn of the evacuation of Yorktown…they must soon decide upon the course they will pursue…I see but two things left them—to stay in their hole & be nabbed there as they surely will, or come out into the Roads & fight as a last desperate resort."[201]

The crew could hear the rear guard attack as McClellan struck at the retreating Confederates, and they were overjoyed to learn of the artillery and stores left behind by Johnston's army. This confidence prompted Jeffers to request permission for the *Monitor* to either shell Sewell's Point or to go up the river after the *Virginia*. Of course, Goldsborough refused, believing that his forces were just not strong enough to simultaneously take on Confederate water batteries as well as the *Virginia*. Goldsborough lamented that the *Galena* was "beneath naval criticism," and despite McClellan's pleas for Union gunboats to enter either river, Goldsborough appeared satisfied to remain on the defense.

Keeler was amazed at this attitude, as he wrote his wife:

> *Capt. Jeffers has sent down to the "Old Flag"* [Goldsborough] *for permission to shell out Sewall's* [sic] *Point tomorrow—of course we shan't*

Stephen Russell Mallory, circa 1855. *Courtesy of The Mariners' Museum.*

get it. The supineness & want of energy exhibited in keeping two hundred guns that are afloat here, silent & useless when they could render effective service, is disgraceful & shameful & it is no wonder that people are beginning to ask why something is not done. Day after day the emblem of rebellion flaunts in our very faces, & day after day we remain torpid & inactive.[202]

Meanwhile, Mallory recognized that Norfolk would soon be abandoned by the Confederate army, despite the presence of the *Virginia*. Even though the CSS *Richmond* was launched on 1 May 1862, the ironclad was far from being ready for combat. That same day, Mallory visited Portsmouth and directed Commandant Sidney Smith Lee to begin the evacuation of Gosport Navy Yard. All of the yard's salvageable material was dismantled and removed to Richmond, Virginia, and Charlotte, North Carolina. It was a herculean effort, made desperate by Joe Johnston's retreat from the Lower

The Great Monster *Merrimack* Is No More

Peninsula. Mallory, in an effort to save as much property as feasible, ordered Commander John Randolph Tucker's squadron to assist the evacuation. On the evening of 5 May, the *Patrick Henry* towed the incomplete ironclad the *Richmond* (called *Merrimac II* by most Union naval personnel) and the unfinished gunboat CSS *Hampton*. The *Jamestown* pulled a brig containing heavy guns and ordnance supplies intended for use aboard the *Richmond*.

The *Virginia* was once again the lone sentinel guarding the Elizabeth River. The Confederate ironclad maintained a daily station off Sewell's Point as a threat to any Union attempt to attack Norfolk. William Keeler remembered, in a 7 May letter, that "The Big Thing"

> *again made her appearance and another just after dinner while she was in status quo under Craney Island, apparently chewing the bitter end of reflection and ruminating sorrowfully upon the future. She remained there smoking, reflecting, and ruminating till nearly sunset, when she slowly crawled off nearly concealed in a huge murky cloud of her own emission, black, and repulsive as the perjured hearts of her traitorous crew. The water hisses and boils with indignation as like some huge shiny reptile she slowly emerges from her loathsome liar [sic] with the morning light, vainly seeking with glairing [sic] eyes some mode of escape through the meshes of the net which she feels is daily closing her in. Behind her she already hears the hounds of the hunter and before are the ever-watchful guards whom it is certain death to pass. We remain in the same position—a sort of advanced guard for the fleet.*[203]

The Confederate ironclad was trapped, and a change in command structure would serve to tip the balance in favor of the Union.

President Lincoln, dismayed by McClellan's slow progress up the Peninsula and the U.S. Navy's inaction, decided to go to Fort Monroe to prompt more resolute action. The president, accompanied by Brigadier General Egbert L. Viele, Secretary of the Treasury Salmon P. Chase and Secretary of War Edwin Stanton, arrived at Fort Monroe aboard the USRMS *Miami* during the evening of 6 May 1862. Lincoln had been invited to Fort Monroe by General Wool in conjunction with Wool's plans to strike against Norfolk.

On 7 May, President Lincoln held a council of war with Flag Officer Goldsborough and Major General Wool. While Wool reiterated his willingness to organize an amphibious operation to capture Norfolk, Goldsborough was reluctant to send even the *Galena* up the James until the *Virginia* was neutralized. No decision was made. The president then went on a tour of the fleet. Keeler recorded Lincoln's visit:

We received a visit today from President Lincoln, in company with Secretaries Chase & Staunton [Stanton] & other dignitaries, attended by Gen. Wool & staff in full uniform. Mr. Lincoln had a sad, careworn & anxious look in strong contrast with the gay cortege by which he was surrounded.

As the boat which brought the party came alongside every eye sought the Monitor *but his own. He stood with his face averted as if to hide some disagreeable sight. When he turned to us I could see his lip quiver & his frame tremble with strong emotion & imagined that the terrible drama in these waters of the ninth [eighth] and tenth [ninth] of March was passing in review before him.*

As the officers were introduced I was presented as being from his own state. He was very happy to find one from Illinois on board the Monitor. *He examined everything about the vessel with care, manifesting great interest, his remarks evidently shewing [sic] that he had carefully studied what he thought to be our weak points & that he was well acquainted with all of the mechanical details of our construction.*

Most of our visitors come on board filled with enthusiasm & patriotism ready, like a bottle of soda water, to effervesce the instant the cork is withdrawn, but with Mr. Lincoln it was different. His few remarks as he accompanied us around the vessel were sound, simple, & practical, the points of admiration & exclamation he left to his suite.

Before he left he had the crew mustered on the open deck & passed slowly before them hat in hand.[204]

Lincoln then visited the *Galena*, where he met, for a second time, Commander John Rodgers. Rodgers, a descendant of one of the nation's most distinguished naval families, had learned of the president's arrival the night before and had gone that evening to see him. Lincoln was in bed, but Rodgers insisted on an interview. After being introduced as the brother-in-law of Quartermaster General Montgomery Meigs, Rodgers urged that the *Galena* and several gunboats be sent up the James River. Lincoln agreed with Rodgers; however, when the president later discussed the concept with Goldsborough, the flag officer was unconvinced. Upon seeing the president again aboard the *Galena*, Rodgers reiterated his plan to take the *Galena* into the James River at low tide so that the *Virginia* could not follow because of her deep draft. Rodgers noted that, once in the James River, the *Galena* and other Union gunboats could disrupt the Confederate retreat. He added that the Union ships might even reach Richmond before the Confederate army could arrive to defend their capital.

The Great Monster *Merrimack* Is No More

When the *Virginia* made her appearance at the mouth of the Elizabeth River, the politicians scurried back to Fort Monroe. There, President Lincoln received a telegram from General McClellan, informing the president that the Confederate fortifications on Jamestown Island were abandoned and McClellan wished gunboats to come up the James. President Lincoln was now determined to press Goldsborough to enact Rodgers's plan. Goldsborough conceded and added a naval assault against Sewell's Point.

Commander John Rodgers assumed command of the James River Task Force. The *Galena* was supported by two other wooden gunboats, the USS *Aroostook* and the USS *Port Royal*. At daybreak on 8 May, Rodgers's command entered the James River and began shelling Fort Boykin on Burwell's Bay. Simultaneously, Goldsborough directed another force to attack Sewell's Point. The *Monitor* and the *Naugatuck*, supported by several wooden warships—including the twenty-four-gun *San Jacinto*, the sixteen-gun USS *Susquehanna*, the six-gun *Dacotah* and the six-gun USS *Seminole*—moved past the Rip Raps and began their cannonade of the Sewell's Point battery. The Union warships were circling "in front of Sewell's Point and throwing their broadsides into our works as they passed."[205]

Jeffers moved the *Monitor* to within three hundred yards of the Confederate batteries. The *Monitor*'s commander then positioned himself in one of the turret hatches, with his head exposed, to view the effects of the *Monitor*'s shot. The *Monitor*'s incendiary shells caused the Confederate barracks to burn, and the battery's flagstaff was shot down twice, but somehow, the Confederates managed to raise their flag for a third time. The batteries slowly returned fire, but none of the Union vessels was struck. Based on information gleaned later from deserters, the Federals had apparently caught the Confederates dismantling their guns.

When the Federal ships began their bombardment of Sewell's Point, the *Virginia* was at Gosport Navy Yard undergoing some repairs and loading supplies. Upon hearing the shelling, Tattnall immediately steamed his ironclad down the Elizabeth River at full speed, ready to contest the Union advance. As the *Virginia* neared Sewell's Point, Tattnall realized that he faced a difficult decision. He could either send his ironclad to block the *Galena*'s advance up the James River or engage the *Monitor* to protect his base. Tattnall knew that he could not reach the *Galena* because of his vessel's draft, and he had a peremptory need to protect his base. Accordingly, Tattnall headed his ironclad toward the *Monitor* and the other Union warships. While it appeared that a second contest between ironclads might occur, Goldsborough ordered the Union warships to their anchorage beyond Fort Monroe. While the

CSS Virginia *Forces the Retreat of the USS* Monitor, *circa 1907. Courtesy of John Moran Quarstein.*

Virginia had forced the Federals to end their bombardment of Sewell's Point, the Confederates were disappointed that their ironclad could not reengage the *Monitor*. Tattnall continued to steam around in Hampton Roads for the next two hours, hoping that he might induce the *Monitor* to attack but still refusing to be baited into the channel near Fort Monroe. "Finally the commodore," John Taylor Wood recounted, "in a tone of deepest disgust, gave the order: 'Mr. Jones, fire a gun to windward, and take the ship back to her buoy.'" This act of disdain and defiance was considered most appropriate by the Confederates, as Wood noted: "It was the most cowardly exhibition I have ever seen…Goldsborough and Jeffers are two cowards."[206]

Goldsborough's actions on 8 May even surprised many of the Union officers. While Goldsborough knew that the Confederate ironclad was now isolated in Norfolk and would eventually have to engage the Federal fleet to survive, his orders to the commanders of the ships shelling Sewell's Point were for them to remain on the defense. Lieutenant David C. Constable, commander of the *Naugatuck*, noted that his instructions were to "engage the battery, and if the *Merrimack* made her appearance to fall back out of the way to induce her to come out into the roads, so that she could be attacked by the larger steamers which were then at anchor below the fortress."[207]

The Great Monster *Merrimack* Is No More

Members of the *Monitor*'s crew were not only critical of Goldsborough's tactics but also of their own commander. Keeler noted:

> *A good deal of fault has been found with Captain Jeffers by the officers on board for not attacking the* **Merrimack** *as we had her in a very favourable [sic] spot that would have given us every advantage we desired. He has always complained that he could not get permission to attack her from the Flag-Officer, but we have reason to think that he had the consent [of] the President to "pitch into her" if a favourable [sic] opportunity offered. Still if his orders were simply to make a reconnaissance to discover if the batteries had been strengthened or re-enforced with men or guns, he accomplished his object.*[208]

Despite Lincoln's presence, Goldsborough had managed to let slip another opportunity to engage the Confederate ironclad. Goldsborough must have been content to let events unfold, as Norfolk would surely soon be captured, leaving the *Virginia* with few options.

President Lincoln was dismayed with the U.S. Navy's failure to reduce the Sewell's Point fortifications. The president was even more disappointed with the *Monitor*'s failure to confront the Confederate ironclad, even though Goldsborough was merely following Lincoln's own orders. President Lincoln viewed the entire action from the Rip Raps battery, renamed Fort Wool in March 1862 in honor of Major General John Ellis Wool. Secretary of War Edwin Stanton telegraphed Washington, D.C.:

> *President is at this moment (2 o'clock PM) at Fort Wool witnessing our gunboats—three of them besides the* **Monitor** *and Stevens—shelling the Rebel batteries at Sewell's Point. At the same time heavy firing up the James River indicates that Rodgers and Morris are fighting the* **Jamestown** *and* **Yorktown**... *The Sawyer gun at Fort Wool has silenced one battery on Sewell's Point. The James rifle mounted on Fort Wool also does good work... The troops will be ready in an hour to move.*[209]

Though the *Virginia* stopped any landing at Sewell's Point, President Lincoln, according to William Keeler, "seems to have infused new life into everything, even the superannuated old fogies begin to shew [sic] some signs of life & animation."[210] The president's two-pronged naval assault worked. Rodgers was now up the James River, and Norfolk appeared ready for conquest.

When President Lincoln realized that Norfolk could not be captured by naval attack, he began a personal reconnaissance of the coastline east of Willoughby Spit, identifying the Ocean View area as perfect for amphibious assault. The president toured the *Monitor* again on the morning of 9 May and instructed Jeffers to steam to Sewell's Point and, if the *Virginia* emerged, to engage her. The *Monitor* found Sewell's Point deserted and no sign of the *Virginia*. The Union ironclad's crew was anxious to fight the Confederate ironclad, as William Keeler noted: "Anything that will give us the *Merrimac*—she is the summit of our wishes, the height of our ambition."[211] By late afternoon on 9 May, under cover of a second naval bombardment of Sewell's Point, over six thousand troops had been ferried across the Chesapeake Bay from Fort Monroe to Ocean View in canalboats. The first wave was commanded by Brigadier General Max Weber and the second by Brigadier General Joseph K.F. Mansfield. Major General J.E. Wool retained command of the entire operation.

The Federal troops marched on Norfolk with little opposition. When Wool's command reached the outskirts of Norfolk by 5:00 p.m., they were met by Mayor William Lamb and members of the Norfolk municipal council. Lamb welcomed the Union army with a well-planned ceremony, designed, according to General Viele, as "a most skillful ruse for the Confederates to save their retreat from the city." The ceremony lasted until dark. Viele noted:

> *In the meanwhile the Confederates were hurrying with their artillery and stores over the ferry to Portsmouth, cutting the water-pipes and flooding the public buildings, setting fire to the navy yard, and having their own way generally, while our General was listening in the most innocent and complacent manner to the long rigmarole so ingeniously prepared by the mayor.*[212]

This oratorical delay enabled the destruction of Gosport Navy Yard and other military equipment being left behind by the Confederate army.

The Confederate commander of Norfolk, Major General Benjamin Huger, had ordered Norfolk's evacuation the day before. When Huger, a fifty-seven-year-old graduate of West Point, learned that Union gunboats had reached Jamestown Island, he feared that his ten-thousand-man command would be cut off from its retreat, via Suffolk, by the U.S. Navy. Huger abandoned Norfolk with such haste that he neglected to inform Tattnall of the evacuation.

Tattnall was furious when he learned, on 10 May, that Norfolk was abandoned. The Confederate ironclad's commander was faced with a difficult decision. He could take his ship out and attack the Union fleet, perhaps destroying several enemy vessels before sinking in a blaze of glory. Neither this course of action nor any effort to take the *Virginia* out to sea en route to another Southern port was advisable. The ironclad needed to be taken up the James River to serve as a floating battery defending Richmond. The pilots advised that this could only be achieved if the huge ironclad could reduce her draft from twenty-three feet to eighteen feet so she could cross Harrison's Bar. The crew immediately went to work throwing coal, ballast and everything else overboard except the ironclad's guns and ammunition. The *Virginia* had been lightened to twenty feet by 1:00 a.m. on 11 May. The pilots, however, informed Catesby Jones that the wind was from the west rather than the east, blowing water away from the bar and making it even shallower. Not only could the *Virginia* not be taken to Richmond, but she was now no longer an ironclad and therefore unable to engage the Federal fleet.

The *Virginia* had to be destroyed to prevent her capture. The ironclad was run aground at 2:00 a.m., and once the crew was rowed ashore, the ship was set afire at 4:58 a.m. The Federals were overjoyed as they witnessed the *Merrimack*'s demise. One young officer, S.R. Franklin, observed:

> *It was a beautiful sight to us in more senses than one. She had been a thorn in our side for a long time, and we were glad to have her well out of the way. I remained on deck for the rest of the night watching her burning. Gradually the casemate grew hotter and hotter, until finally it became red hot, so that we could distinctly mark its outlines, and remained in this condition for fully half an hour, when, with a tremendous explosion, the* Merrimac *went into the air and was seen no more.*[213]

George Geer gleefully wrote his wife: "The great Monster *Merrimack* is no more."[214] While satisfied that the Confederate ironclad had been destroyed, William Keeler was dismayed by the event, as he and others aboard the *Monitor* felt that the *Merrimack* was their "exclusive game." They had hoped to capture and inspect the ironclad, as Keeler wrote: "Her career was short & infamous as her end was sudden & unexpected. We knew her days were numbered but felt confident that she would die game rather than fall by her own hand."[215]

Sunrise brought confirmation that the *Virginia* had indeed been scuttled, and orders were sent that the *Monitor* was to immediately proceed to Norfolk. The *Monitor* moved up the Elizabeth River, passing by the formidable Confederate fortifications, which were "much larger, stronger & more extensive than we had supposed." Lieutenant Thomas O. Selfridge was the first Federal officer to land at Sewell's Point aboard a tug and ceremoniously raised the Stars and

Destruction of the Rebel Ram Merrimac. Currier & Ives, circa 1862. *Courtesy of The Mariners' Museum.*

Stripes above the battery. When the *Monitor* noticed "two rebel rags" flying over the Craney Island batteries, crew members went ashore and had those flags "taken down & presented to Mr. Lincoln & Our Flag now occupied their place."[216] The *Monitor* continued her journey eight miles up the river, stopping to take a few souvenirs from the blackened remains of the *Virginia* and then anchoring at the *Virginia*'s old mooring. Several other Union vessels came upriver to Norfolk. The last vessel was the steamer the USS *Baltimore* with President Lincoln, Chase, Stanton and Goldsborough aboard. The president took his hat off and bowed as he passed the *Monitor*.

Abraham Lincoln was obviously pleased with this course of events. He lauded General Wool for his role in the campaign, "which resulted in the surrender of Norfolk and the evacuation of strong batteries erected by the rebels on Sewell's Point and Craney Island and the destruction of the rebel iron-clad steamer *Merrimack*."[217]

President Lincoln believed that the capture of Norfolk and the destruction of the Confederate ironclad were "among the most important successes of the present war."[218] It was a huge defeat for the Confederacy. McClellan was ecstatic and telegraphed Secretary of War Edwin Stanton at Fort Monroe: "I congratulate you from the bottom of my heart upon the destruction of the *Merrimack*."[219]

Chapter 8
The *Monitor* Was Astonished

The *Monitor* soon returned to Hampton Roads, as the ironclad had unfinished business. President Lincoln recognized that Richmond was now within the grasp of the Union navy. The president, as well as Gideon Welles and George McClellan, pressed Louis Goldsborough to strike quickly against the Confederate capital. Goldsborough ordered the *Monitor* and the *Naugatuck* to reinforce Rodgers and "to reduce all the works of the enemy as they go along, spike all their guns, blow up their magazines, and then get up to Richmond." He noted to Rodgers that he must "push on up to Richmond if possible, without any unnecessary delay, and shell the place into a surrender…Should Richmond fall into our possession, inform me of the fact at the earliest possible moment."[220]

On the morning of 12 May, the *Monitor* and the *Naugatuck* began their mission to Richmond. As the *Monitor* passed Fort Huger, the Confederates sent several shells at the ship, one of which passed three feet above the ironclad's deck. The *Monitor* did not stop, as it appeared that the fort was only making a token resistance.

The *Galena* had already engaged the Confederates south of the James water batteries on 8 May. Fort Boykin was actually a colonial-era fortification, while Fort Huger was a modern work designed by Colonel Andrew Talcott. Unfortunately, despite these forts mounting modern naval ordnance, they were designed to combat wooden ships, not ironclads. When the *Galena* engaged Fort Boykin, the fort was quickly evacuated by the Confederates. Fort Huger, however, was a more extensive work, mounting twelve heavy cannon. As Rodgers's squadron approached the fort, the Confederates opened fire, and several shells struck the water near the Union gunboats. The *Aroostook* and the *Port Royal* counter-fired, using explosive and incendiary shells. The wooden buildings within the fort were soon ablaze. One shot from the *Port Royal* dismounted one of the Confederate rifled guns, and, in turn, one shot from Fort Huger struck the *Aroostook*, causing little damage. While the *Port Royal* and the *Aroostook* dueled with the fort from a distance, the *Galena* stood in as close as the pilots would take the ironclad. The *Galena* ran past the fort seven times. After dismounting all but one gun, Rodgers then stopped his ironclad in front of the fort and raked the fortification at point-blank range as the *Aroostook* and the *Port Royal* steamed past the fort. The *Galena* then proceeded up the river.

The *Galena* grounded twice as she steamed toward Hog Island. This situation caused some significant mechanical difficulties. Like most oceangoing steamers, the *Galena* took in water for her boilers through fittings in her bottom. River mud, sand and debris had entered her machinery when the *Galena* grounded, and the engine would not restart. Consequently, the engine room crew was forced to disassemble the engine to clean valves and other equipment fouled by the mud. When this work was completed, the engine was restarted, and *Galena* proceeded upriver. By 11 May, the squadron was anchored off Jamestown Island. Rodgers telegraphed Goldsborough that he needed additional warships, particularly ironclads, to make his thrust against Richmond.

The *Monitor* and the *Naugatuck* arrived the next day. While now only consisting of five ships, Rodgers's flotilla was a powerful force, and the Confederates no longer had a warship to counter Union ironclads. In addition to his flagship, the six-gun ironclad *Galena*, Rodgers had two additional ironclads, as well as the ninety-day gunboat the *Aroostook*, commanded by Lieutenant John C. Beaumont, and the side-wheel double-ender the *Port Royal*. The *Port Royal* was captained by Lieutenant George Upham Morris, who had survived the sinking of the *Cumberland* on 8 March 1862. The *Aroostook*'s armament consisted of two twenty-four-pounder smoothbores, one XI-inch Dahlgren and one twenty-pounder Parrott rifle, while the *Port Royal* mounted one hundred-pounder Parrott, one X-inch Dahlgren smoothbore and six twenty-four-pounder howitzers.

Rodgers's progress up the river was slow. The squadron, due to Goldsborough's orders, had to stop at each Confederate fort. The Confederates, however, had abandoned these works and spiked their guns as Johnston's army withdrew up the Peninsula. By 13 May, the Union ships had reached the confluence of the James and Appomattox Rivers, where City

The *Monitor* Was Astonished

Point was a small port with a railroad link to Petersburg. The Confederates had just left after setting ablaze the railroad depot and warehouses. Keeler was unsure about the white Southerners he encountered at City Point, but he noted that the African Americans were happy because "Massa Lincoln's ships be comin'."[221] A Unionist woman, according to George Geer, insisted on seeing Commander Rodgers. She gave him a newspaper telling about the destruction of the *Virginia* and how bitter the public was about the war. She asked Rodgers not to shell the town. Rodgers advised the lady, "We carry on no war with peaceable people—our arms are for those in arms."[222]

Rodgers and his warships left City Point at 3:20 p.m. on 13 May. The ships had not steamed far upriver when they encountered a runaway slave who warned that there was a huge Confederate fort on Drewry's Bluff, just eight miles below Richmond. By nightfall, the squadron had anchored in the river several miles above City Point.

The Union squadron resumed its way upriver by 8:00 a.m. on 14 May. By 2:00 p.m., the ships reached Dutch Gap, where the *Galena* ran aground. The Federal seamen continued to interact with African Americans, many of whom wished to gain their freedom by getting aboard the Union ships. One runaway told the men on the *Monitor* that Confederate soldiers were waiting around the next bend. Some Confederate cavalry appeared on a knoll before the *Galena* was freed, and the *Naugatuck* sent a shell that ended their observations. According to William Keeler, "They left with more haste than I suppose Southern cavalry would use."[223] Rodgers then took the *Galena* and the *Naugatuck*, leaving the other three vessels behind at Dutch Gap. When the two ironclads reached Chaffin's Bluff, they spied Confederates lurking in ambush. Two shells from the *Galena* dispersed them. The *Naugatuck* was sent downriver to bring the other vessels up to join the *Galena*. When these ships got underway, the *Port Royal* sent shells onto the riverbank to scatter Confederates intent on another ambush. Rodgers anchored his command two miles below Drewry's Bluff at Kingsland Creek, based on the news he had gleaned of the fort at Drewry's Bluff and encounters with sharpshooters along the shoreline. The *Naugatuck* was positioned ahead of the rest of the squadron with chains and grappling hooks to protect against fire rafts. The *Galena* fired one of her one-hundred-pounder Parrott rifles toward Drewry's Bluff. Then Rodgers had his ships prepare for the action that was sure to come in the morning.

There were no Confederate batteries guarding the James River just below Richmond until March 1862. Captain Alfred Rives and Lieutenant Charles Mason of the CS Engineer Corps surveyed the river with Captain Augustus

Herman Drewry, commander of the Southside Heavy Artillery, to find a suitable location to fortify and obstruct the river. Property owned by Drewry himself, known as Drewry's Bluff, was selected as the best site. The bluff, rising almost one hundred feet above the river, commanded a sharp bend in the James River and was the last place available to effectively mount a defense before reaching Richmond.

Lieutenant Mason designed the fort, and Captain Drewry's command was assigned to build and defend it. As the earthwork began to take shape, three heavy guns, two eight-inch Columbiads and one ten-inch Columbiad, were placed in battery. While this work continued, Thomas Wynne organized the construction of stone cribs in order to obstruct the river. Both projects progressed slowly.

The Union navy's entry into the James River on 8 May sent shock waves reaching all the way to Richmond. Commander John Randolph Tucker's squadron was forced to retreat up the river. When the little squadron passed Drewry's Bluff, Tucker noticed the unpreparedness of the defenses. He immediately wrote Commander Ebenezer Farrand, the recently appointed commander of Drewry's Bluff: "I feel very anxious for the fate of Richmond and would be happy to see you about the obstruction placed here—I think no time should be lost in making this point impassable."[224]

The Confederate capital itself was in an uproar over the approach of the Union fleet. Preparations were begun by the Confederate administration to abandon Richmond, and the city's government vowed to burn the city rather than see it fall to the Union. Richmond's weak river defenses had been an issue for several months. General Robert E. Lee, however, was determined that Richmond "shall not be given up" and ordered his son, Colonel George Washington Custis Lee, to help coordinate efforts to improve Drewry's Bluff. Colonel Lee feverishly worked with Captain Drewry and Commander Farrand, CSN, to improve the gun emplacements known as Fort Darling.

Farrand was ordered to "lose not a moment in adopting and perfecting measures to prevent the enemy's vessels from passing the river."[225] Rodgers's delays, investigating each Confederate fort below City Point, gave the Confederates precious time to enhance their defenses. The CS Navy began mounting additional guns on the bluff, as well as improving the obstructions. Reinforcements sent to Drewry's Bluff included the Bedford Artillery, infantry from Brigadier General William Mahone's brigade and the crew of the *Virginia*.

Catesby Jones and his *Virginia* shipmates arrived at the bluff on the morning of 13 May 1862. He understood that "the enemy is in the river, and extraordinary exertions must be made to repel him."[226] Even though Josiah Tattnall's decision to scuttle the *Virginia* saved the crew from death or capture, their morale was very low over the loss of their ship. Corporal Thomas Mason, of the Southside Artillery, remembered one sailor from the *Virginia* lamented "that to attempt to defend the place would only make it a slaughter pen."[227] Although John Taylor Wood thought that the Confederate navy "for the time had been destroyed," they "must seek other ways of rendering ourselves useful."[228]

Jones organized his crew members into work parties, assisting Commander John Randolph Tucker's sailors in constructing new gun emplacements. The men worked diligently, in the rain and mud, for the next two days. By the

Drewry's Bluff Fortifications, circa 1865. *Courtesy of John Moran Quarstein.*

morning of 15 May, the seamen had mounted five guns: three thirty-two-pounders and two sixty-four-pounders taken from the *Patrick Henry* and the *Jamestown*. The gun positions were dug into the brow of the bluff and revetted with logs. Tucker's men mounted a seven-inch rifle from the *Patrick Henry* in an earth-covered log casemate located near the entrance to Fort Darling.

The Confederates made other arrangements to combat the Federals during those frantic two days. The *Jamestown* was sunk, along with several other stone-laden vessels, approximately three hundred yards in front of Drewry's Bluff to enhance the obstructions. This effectively closed the passageway to Union ships. Tucker held above the obstructions the remaining gunboats of the James River Squadron: the *Patrick Henry*, the *Teaser*, and the CSS *Nansemond*. These warships were ready to engage any Union vessel that might make its way past the defenses. A detachment of CS Marines, commanded by Captain John D. Simms and containing many crewmen from the *Virginia*, dug rifle pits below the bluff. Lieutenant Wood deployed sailors as sharpshooters on the opposite bank of the river to harass the Union ships as they neared Drewry's Bluff. Additional CS Army units, including the Washington Artillery, Dabney's Battery and the Fifty-sixth Virginia Infantry, were rushed to Chaffin's Bluff, two miles downstream from Drewry's Bluff, to impede the Federals' progress upriver.

About 6:30 a.m. on 15 May, Rodgers's command got underway from its anchorage and steamed around the bend at Chaffin's Bluff, with Rodgers in the *Galena* taking the lead. The ironclad's commander decided that the *Galena*'s thin armor should be tested under fire. "I was convinced as soon as I came onboard that she would be riddled with shot," Rodgers later wrote, "but the public thought differently, and I resolved to give the matter a fair trial."[229] The *Galena* had an experimental hull design, utilizing overlapping four-inch armor-clapboard strips. The ironclad's sides curved from the waterline to the top deck to give protection from shell fire from opposing ships. However, this armor-clad design would prove inadequate protection against plunging shot, since the deck was built solely of wood.

Despite these weaknesses, the *Galena* headed the squadron, followed by the *Monitor*, the *Aroostook*, the *Port Royal* and the *Naugatuck*. As the flotilla passed Chaffin's Bluff, Wood's sharpshooters began to fire upon the Union vessels. John Taylor Wood was furious that the batteries on Chaffin's Bluff did not shell the gunboats and ordered his riflemen to open fire on the passing Union warships. The Confederate riflemen began to pepper the Union ships. One sharpshooter picked off the leadsman on the *Galena*, and his replacement was also wounded. Not a man could be on an open deck without minié balls whizzing past him. William Keeler noted that one ball "passed between my legs

The *Monitor* Was Astonished

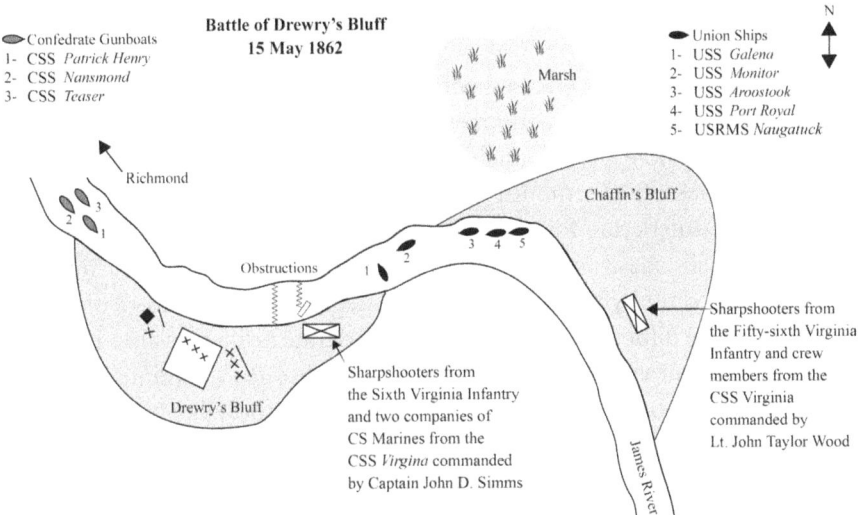

Battle of Drewry's Bluff on 15 May 1862. Illustration by Sara Johnston.

& another just over Lieutenant Greene's head."[230] The *Monitor*'s crew stayed within their ironclad throughout the engagement, while the wooden gunboats had to contend with the constant sniping. In response, the *Port Royal* or *Naugatuck* often raked the shoreline with their twenty-four-pounder howitzers.

By 7:45 a.m., Rodgers's squadron had neared the obstructions. He knew that he must reduce the batteries upon the bluff and then disperse the shoreline sharpshooters before he could open a door to Richmond through the obstructions. Therefore, the *Galena* steamed to within six hundred yards of Drewry's Bluff and dropped anchor. Former *Virginia* crew member Boatswain Charles H. Hasker marveled at the "neatness and precision" by which Rodgers maneuvered his ironclad into position in such a narrow river. He noted that the *Galena* "let go her starboard anchor, ran out the chains, put her head inshore, backed astern, let go her stream anchor from the starboard quarter, hove ahead, and made ready for action." Hasker concluded that it was "one of the most masterly pieces of seamanship of the whole war."[231] The *Galena* received two hits while completing the maneuver and quickly became the primary target of the Confederate cannon fire. Once in position, however, Rodgers ordered his ironclad's port broadside battery, consisting of two IX-inch Dahlgrens and one hundred-pounder Parrott, to open fire on the Confederate works atop the bluff.

The Federal fleet was at a distinct disadvantage. The obstructions blocked any opportunity to run past the batteries toward Richmond. The *Monitor* was anchored astern of the *Galena*. Sniper fire did not affect the *Monitor*'s crew, as they were safely encased within their iron vessel. However, the rifle fire from along the riverbank peppered the crews of the *Port Royal*, *Naugatuck* and *Aroostook*. These vessels anchored about a half mile downriver from the obstructions off the mouth of Cornelius Creek and faced their bows toward the Confederate fort. The *Port Royal* and *Naugatuck* then shelled the bluff with their one-hundred-pounder pivot rifles as the *Monitor* and *Port Royal* did the same with their XI-inch Dahlgrens. While this deployment presented more difficult targets for the Confederate artillerists atop Drewry's Bluff, Lieutenant Morris of the *Port Royal* was forced to continuously train his twenty-four-pounder howitzers on the nearby shoreline to disrupt the accurate fire of the Confederate sharpshooters in the woods. Morris would be the only casualty aboard the *Port Royal* when he was shot in the right leg. Lieutenant Beaumont of the *Aroostook* gave his men small arms to return the Confederate fire. These vessels encountered other, more serious, problems during the engagement. After seventeen rounds, the *Naugatuck*'s Parrott burst. One seaman, Peter Dixon, was severely injured by this incident. Nevertheless, the *Naugatuck* remained in action, using her howitzers against the Confederates along the shore. Another crew member, James Walton, was wounded in the arm by a Confederate minié ball. The *Port Royal* was also disabled during the fight. One shot crashed through the port bow of the gunboat just below the waterline, and she had to back downriver a short way to pump the water out and make temporary repairs. Once she came back into action, the *Port Royal* was hit again amidships on the port side, but the damage was minimal and the gunboat remained in action.

The *Galena* was the primary target for the Confederate gunners. The ironclad had anchored "within point-blank range" of the Confederate batteries, and the cannonade began to take effect. Ebenezer Farrand noted: "Nearly every one of our shots telling upon her iron surfaces."[232] Seeing that the Confederates were concentrating their fire on the *Galena*, Lieutenant Jeffers moved the *Monitor* virtually abreast of the *Galena* in an effort to draw some of the Confederate shot away from the larger ironclad. The *Monitor*'s new position and the dimensions of the turret's gun ports did not allow the ironclad to elevate its two powerful XI-inch Dahlgrens enough to hit the Confederate batteries. The Drewry's Bluff cannoneers, knowing of the *Monitor*'s shot-proof qualities, wasted little ammunition on the ironclad. Nevertheless, the *Monitor* was hit three times during the engagement. One eight-inch solid shot struck the turret; two others banged into the armor forward of the pilothouse. "Three shot struck us," William Keeler noted, "making deep indentations but doing no real harm."²³³ Eventually, the *Monitor* backed downstream and continued a deliberate fire from her final position below the *Galena*.

The two ironclads were also constantly shot at by the CS Marines who had dug rifle pits at the base of the bluff. The Confederates there often waited to fire their rifled muskets until a *Galena* gun crew member became exposed in a gun port when reloading a gun. Petty Officer Thomas Ready was shot in this manner. The *Monitor* was not impacted by these sharpshooters. George Geer wrote his wife that "the Rebel sharpshooters are firing at us all the time. None of us are allowed [*sic*] to go on deck, and I can assure you that I am not risk [*sic*] myself where there is any danger."²³⁴ While safe from rifle fire, the *Monitor*'s crew suffered in another way, as they were trapped within their iron machine. William Keeler recounted:

> *No one on board was hurt but all suffered terribly for the want of fresh air. It was one of those warm, muggy days with a very rare atmosphere which, shut up closely like we were, made ventilation very difficult. At times we were filled with powder smoke below threatening suffocation to us all. Some of the hardest men dropped fainting at the guns.*²³⁵

The outside humidity, coupled with the heat from the engines, led Jeffers to estimate that it was 140 degrees within the turret. The hell-like qualities of a mixture of coal fumes and smoke, as well as powder smoke and sulfuric smells, made it unbearable within the *Monitor*. Some men had to be taken out of the turret or engine room in order to breathe less-infected air.

When the Union ships first steamed into sight on 15 May, Farrand readied his command. The *Galena* maneuvered into position, and the three guns in the main battery opened fire. Captain Drewry fired the first shot, which passed over the *Galena* into the east side of the river. Most of the remaining shots, however, severely punished the *Galena*. Even though the Confederates appeared to have an advantage with their defensive position atop the bluff and the use of plunging shot, the Confederates still encountered numerous problems. The ten-inch Columbiad manned by the Bedford Artillery was loaded with a double charge of powder. Consequently, when the Columbiad was fired, it recoiled off its platform. This heavy gun was not brought back into action until near the end of the engagement. Also, the recent heavy rains caused the casemate protecting Tucker's seven-inch Brooke gun to collapse after the sailors from the *Patrick Henry* had fired just a few shots. The crew escaped and somehow was able to bring the Brooke gun back into action just before the battle concluded. Tucker commanded the five naval guns manned by sailors from the *Patrick Henry* and *Virginia*, while Catesby Jones was stationed at the Southside Artillery's position to assist the volunteer artillerists in managing their heavy guns. He was so exhausted by the efforts of the past five days that he actually dozed off on a shell box during the engagement. The naval battery, despite mounting very powerful ordnance, did not play as major a role in the engagement as the seamen wished. Positioned to the left of the fort on the bluff's brow, the sailors continued to bang away at the Union vessels throughout the engagement and needlessly exposed themselves to Federal fire.

Union cannon fire was rather effective early in the battle and inflicted thirteen casualties. Several men were killed by fragments from a one-hundred-pounder shell from the *Galena*. Thomas Mann of the Southside Artillery remembered:

> *Shells from the* Galena *passed just over the crest of our parapets and exploded in our rear, scattering their fragments in every direction, together with the sounds of the shells from the others, which flew wide of the mark, mingled with the roar of our guns, was the most startling, terrifying and diabolical sound which I had ever heard or ever expected to hear again.*[236]

"We could see large clouds of dirt & sand fly as shell after shell from our vessels exploded in the rebel works," Keeler remembered, "& no sooner was a gun silenced apparently in one portion of the batteries than they opened from some other part or from some new & heretofore unseen battery."[237] One way the Confederates avoided casualties is that the men retreated into bombproofs during periods of heavy Union shelling. When Federal fire slackened, the defenders would rush back to their guns to return fire. This action also helped to preserve the Confederates' limited supply of ammunition. Despite the Union ships sending "their iron messengers with remarkable accuracy," the "batteries on the Rebel side were beautifully served," noted John Rodgers, "and put shot through our sides with great precision."[238]

As the duel between the batteries and Union warships continued through the morning, Farrand noticed he was running low on ammunition, so he let the Southside Heavy Artillery gun crews take a half-hour break. When the Confederate gunfire slackened, Rodgers thought his squadron was about to silence the Confederate guns; however, he also learned that the *Galena* was running out of ammunition for her IX-inch Dahlgrens. Rodgers had to switch from explosive shells to solid shot, which had little effect on

The *Monitor* Was Astonished

the Confederate earthworks. Farrand ordered Drewry's men back to their Columbiads, and they reopened their cannonade of the *Galena* with terrible effect. Every shot from the Confederate cannons seemed to strike the *Galena*. The ironclad's armor was broken, bent and pierced in many places.

Despite the advantages of the Confederate position, some of the defenders, like Thomas Mann, thought that the Federals "would finally overcome us."[239] Nevertheless, the battle was finally decided in the Confederates' favor after almost four hours of combat. Throughout the morning the Confederate gunners did splendid execution with their rifled ordnance, constantly sending shot and shell into the Union vessels. They almost never missed their targets. The *Galena* was struck by forty-three projectiles. Her weak iron plating was penetrated thirteen times during the engagement. About 11:05 a.m., the telling shot struck the *Galena*. An eight-inch shell from one of the *Patrick Henry*'s guns, manned by former *Virginia* crew members, crashed through the *Galena*'s bow port and exploded. The shell ignited a cartridge then being handled by a powder monkey, killing three men and wounding several others. The explosion sent "volumes of smoke," according to William Keeler, "…issuing from the *Galena*'s ports

USS *Galena*, circa 1862. *Courtesy of John Moran Quarstein.*

& hatches & the cry went through us that she was on fire, or a shot had penetrated her boilers—her men poured out of her open ports on the side opposite the batteries, clinging to the anchor, to loose ropes…We at once raised our anchor to go to her assistance but found she did not need it."[240]

When the Confederates saw the smoke and flames rise from the *Galena*, the gunners on the bluff "gave her three hearty cheers as she slipped her cables and moved downriver."[241] It appeared to the Confederates that the *Galena* was retreating because of the explosion. Rodgers, who had barely escaped injury when the final shell exploded on the *Galena*'s gun deck, had already decided to hoist the signal to break off action. Rodgers knew that his ironclad was running low on ammunition and had barely survived a damaging hail of Confederate fire. The entire squadron retreated down the river. As he watched the Union vessels steam away from Drewry's Bluff, Lieutenant John Taylor Wood, formerly of the *Virginia*, hailed the *Monitor*'s pilothouse from the riverbank shouting, "Tell Captain Jeffers that is not the way to Richmond."[242]

Rodgers's squadron endured "a perfect tempest of iron raining down upon and around us." The *Galena* suffered the greatest damage: her railings were shot away, her smokestack was riddled and she suffered twenty-four casualties. The fight demonstrated that the ironclad was "not shot-proof."[243] William Keeler of the *Monitor* commented that the *Galena*'s "iron sides were pierced through and through by the heavy shot, apparently offering no more resistance than an eggshell," verifying the opinion that "she was beneath naval criticism." When Keeler went aboard the *Galena*, he thought that the ship

> looked like a slaughterhouse…of human beings. Here was a body with the hand, one arm & part of the breast torn off by a bursting shell—another with the top of his head taken off the brains still steaming on the deck, partly across him lay one with both legs taken off at the hips & at a little distance was another completely disembodied. The sides & ceiling overhead, the ropes & guns were spattered with blood & brains & lumps of flesh while the decks were covered with large parts of half coagulated blood & strewn with portions of skulls, fragments of shells, arms, legs, hands, pieces of flesh & iron, splinters of wood & broken weapons were mixed in one confused, horrible mass.[244]

The *Monitor* Was Astonished

The Battle of Drewry's Bluff was a dramatic Confederate victory. As one Southerner wrote:

> *The* Monitor *was astonished,*
> *And the* Galena *admonished,*
> *And their efforts to ascend the stream*
> *Were mocked at.*
> *While the dreaded* Naugatuck,
> *With the hardest kind of luck,*
> Was very nearly knocked
> Into a cocked-hat.[245]

While the battle saved Richmond from capture by the U.S. Navy, Keeler considered:

> *We do not regard the matter in the light of a defeat as we accomplished our purpose, which was to make a reconnoisance* [sic], *ascertain the nature & extent of the obstructions, the position & strength of the batteries. We found them of such a nature that it was an impossibility to force them with the means at our command & the river so narrow it is equally impossible to bring a much larger force to bear.*[246]

It was indeed a Union defeat, as George Geer wrote his wife: "We have been fighting all day and have come off 2nd best."[247]

Chapter 9
Long, Hot Summer

The Union flotilla steamed downriver to City Point. Commander John Rodgers recognized that his ships were needed to stop any Confederate effort to refortify the riverbanks below Drewry's Bluff in order to support Major General McClellan's operations against Richmond. Squadron Commander Louis Goldsborough sent supplies to City Point and additional gunboats, including the USS *Maratanza*, *Wachusetts*, *Island Belle*, *Stepping Stones* and *Coeur de Lion*. Because Commander William Smith was senior to Rodgers, Smith assumed command of the flotilla at City Point. He was destined to retain the command for only a few days.

William Keeler called the *Monitor*'s new anchorage "out of humanity's reach," and it was there that he was soon to witness new facets of the war. The Union soldiers were operating "in an enemy's country," and consequently, armed guards were posted every evening in expectation of sharpshooters or a raiding party. During the night of 18 May, an alert was called, "Boat ahoy!" and a shot fired on an approaching boat. Captain Jeffers exclaimed "Boarders!" and all available crewmen rushed to the deck. Once on deck, Keeler "found the vast array of 'Monitors' armed to the teeth drawn up confronting the enemy—a poor trembling contraband—begging not to be shot." The "contraband," Siah Hulett Carter, confessed that he was the first slave to escape from Shirley Plantation. Carter had been warned by his master, Colonel Hill Carter, not to go on board any of the "Yankee ships" because "the Yankees would carry them out to sea…& throw them overboard."[248] Siah enlisted as a first-class boy, ship's number fifty-three, on 19 May 1862.

On that very same day, several officers and sailors went ashore at City Point on a humanitarian mission, only to be ambushed and nearly captured by Confederates. Keeler was amongst those who had gone ashore, but before the Confederate trap was sprung, the *Monitor*'s paymaster was warned by an African American that the Confederates were waiting on the bluff. As Keeler remembered the escape:

> *I started for my boat…& ordered the men to shove off, shouting to the crew of the other boat to look out for their officers as I believed the "Secesh" were coming…they waited a few moments & shoved off also but had got but a short distance when a smart fire of rifles was opened up on us…Half a dozen or so were all that were directed at us, but a shower of them fell around the other boat & I saw two men fall.*
>
> *By this time I had got pretty much out of range, one ball going through the boat about 18 inches from me.*[249]

Once Keeler was back on board, the *Monitor* immediately got underway. Her two Dahlgrens were charged with cannister and raked a warehouse where the Confederates had previously been deployed; unfortunately, they had already withdrawn. The Union ironclads then commenced shelling City Point. The *Monitor* was ordered to fire two shells and a round of cannister into the Eppeses' family home, Appomattox Manor. The shells passed through the beautiful, century-old house before exploding, and the residents pleaded for the Union ships not to destroy the place. Jeffers noted, according to William Keeler, that "he did not come to make war on unarmed men or women & children."[250] The residents, who seemed more afraid of their own soldiers than the Federals, still abandoned their home.

As the *Monitor* awaited further action at City Point, many crew members began to recognize various problems with living in an ironclad during the heat of the summer season. Keeler noted that the "warm weather is making it very uncomfortable on board our vessel—there is not sufficient ventilation." A few days later, he reiterated this message to his wife: "Yesterday was a hot uncomfortable day & we lay broiling on or in our iron box or Cage as it has now become."[251]

A large number of crewmen sought relief from the heat by swimming. Under the supervision of an officer, and with a boat crew standing by ready to throw a line, any man with the ability and desire to swim could take off his clothes and jump in the river. George Geer was able to see various forms of artwork in the form of tattoos that many a seaman had placed upon his body:

Long, Hot Summer

> *I wish you could see the body's* [sic] *of some of these old saylors* [sic]*: they are regular Picture Books.* [They] *have India Ink pricked all over their body. One has a Snake coiled around his leg, some have splendid done pieces of Coats of Arms of states, American Flags, and most of all have the Crusifiction* [sic] *of Christ on some part of their Body.*[252]

Food was another source of complaint at City Point. George Geer often wrote his wife about his various meals. He noted that for breakfast, the sailors would have coffee and crackers cooked in pork fat. The men had two gills of grog each day. Other meals would include roast beef and potatoes. Geer described them: "Potatoes taste like I don't know what—anything that no taste at all—and the Beef is all parts of the Cow cooked together until it is next to a Jelly and will drop to pieces. It is good when there is none bettor [*sic*]."[253]

Keeler also complained to his wife about the food. He noted emphatically that there were no fresh fruits or vegetables, nor any fresh meat. It bothered Keeler that there seemed to be plenty of Confederate cattle, and other types of food, available, but the crew was not allowed to forage for it. The *Monitor* men were, however, able to supplement their meals with items purchased from the many "contrabands" who visited the Union ships anchored off City Point. Keeler remembered purchasing fresh shad at ten cents cash.

George Geer would often share additional details about his daily routine with his wife. He noted that on Sundays, after "breakfast everything is cleaned up about the ship, which takes about one hour, and after that there is nothing to do but keep watch, which amounts to laying around the deck for the saylors [*sic*] and laying around the engine room for the firemen."[254] Keeler also noted that Sunday was "quiet" and that the "usual routine of daily work, men drilling & at quarters, of painting & scraping &c is not carried on."[255] Sunday was muster day, the day the captain performed an inspection of the crew. The boatswain's "shrill whistle" would begin the day at 6:00 a.m., and "everybody must turn out and lash their hammock up and stow them away. All hands make their way on deck, get a pail when their turn comes, and have a good wash." William Keeler recalled the captain's review once all the men were "mustered for inspection":

> *Each one is expected to be dressed in his Sunday best…The seamen & petty officers are drawn upon one side of the deck, the firemen & coal heavers on the other. Each man answers to his name as the Lieut. calls the roll.*
>
> *The Capt. is then informed that the men are ready for inspection. He passes slowly along in front of the lines of men looking closely at their dress,*

> *appearance &c—"Jones why are your shoes not blacked?" Jones having no good excuse, the Paymaster's steward is ordered to stop his grog for a day or two.*
> *"Lieut. What is this man's name?"—"Smith, sir."—"Well have his grog stopped for a week for coming to inspection without a cravat." "Do you belong to the ship?"—"Yes sir."—"Well you are a filthy beast, a disgrace to your ship mates, the dirt on you is absolutely frightful. If I see you so again I will have the Master at Arms strip you & scour you with sand & canvas." & so the inspection goes on.*[256]

Once the muster was over, the men enjoyed their free-time pursuits of reading, writing, mending, fishing and gambling. Geer noted that the "gambling still keeps going, through some of the worst ones have lost all he had, and everybody is glad. One boy who only had 50¢ when he commenced has won $40. He is the youngest boy (Thomas Carroll #2) in the ship, and it is fun to see him take the money from some of the old gamblers."[257] Geer also noted there often were fights between the enlisted men. "The English and Irish portions of the crew are out with each other," Geer wrote his wife, "and have several times come very near having a general fight, and have had several small fights[.] I wish that they would have a big fight and eat each other up, so we could get clear of some of them and get some deasant [*sic*] men here—if there is any in the navy, which I begin to doubt."[258]

The *Monitor* spent little time in actual combat, so consequently, when the ironclad was anchored on guard duty in the James River, there was much idle time. William Keeler noted that time was usually spent in reading and conversation, and that there wasn't much else to do "to kill time." He wrote his wife, "Here we lie, day after day & week after week, prisoners to all purposes, no going ashore—no nothing, but eat, drink & sleep, & while away the tedious hours as best we may."[259]

While there were a few military diversions, excitement was more often caused by the unexpected. During the early hours of 22 June, a fire broke out in the *Monitor*'s galley. According to William Keeler, it was "a narrow escape last night from destruction."[260] The wooden backing, where the galley's stovepipe passed through the deck, caught fire and had burned to within six feet of one of the shell rooms before it was discovered. It appeared that embers from the stack caused the blaze, but it was quickly extinguished. George Geer commented:

> *I suppose nothing will be said about it, as people would smile at the idea of an Iron Ship getting on Fire, but then they do not know we have wood*

Long, Hot Summer

View of USS *Monitor*. *Left to right:* Robinson W. Hands, Louis Stodder, Albert B. Campbell and William Flye. Photograph by J. Gibson, 4 July 1862. *Courtesy of The Mariners' Museum.*

> *inside. But there would be very little chance for a fire to get under much head way, as there is always some 10 or 12 on watch at a time. In all Men of War there is an arrangement to Drownd* [sic] *the Powder Magazine and Shell Room. Here we can by simply turning a Lock, full* [sic] *them with Water in five moments, so there would be no chance of us Blowing up.*[261]

However, the stove was damaged beyond repair, so that morning, Keeler reported that "our breakfast was a scanty one, crackers & coffee, the latter being made in the fire room, in one of the furnaces."[262] Shortly thereafter, a fireplace was rigged on deck using bricks and iron plates to enable the cook to produce meals.

Following the Drewry's Bluff fight, Lieutenant Jeffers decided to write a detailed report to Flag Officer Goldsborough regarding the *Monitor*'s fighting capabilities. Jeffers was considered an ordnance expert and was assigned to the *Monitor* to survey the experimental ironclad's strengths and weaknesses.

Interior Views of the USS Monitor, *1862. Courtesy of John Moran Quarstein.*

He did not like the ship and was actually quite contemptuous of its design. His report to Goldsborough highlighted several significant problems:

- The captain could not control the ship from the pilothouse. The pilothouse should have been placed atop the turret.
- The turret should have a protective shield to allow riflemen to be deployed.
- The turret did not have a full range of fire due to the location of the pilothouse and air intakes. Jeffers thought the range of fire to be 200 degrees rather than 360 degrees.
- Only one gun could fire at a time. The port stoppers were too heavy to operate in combat. The gun commander had limited visibility, that being over the guns out the gun ports. The ironclad had limited fire control.
- The gun ports did not allow for adequate elevation of the guns.
- Ventilation is intolerable. Jeffers recorded on 16 June 1862 the following temperatures: Galley—164 degrees, Engine Room—128 degrees and Berth Deck—120 degrees.

Long, Hot Summer

Jeffers summarized his report:

> *Notwithstanding the recent battle in Hampton Roads and the exploits of the plated gunboats in the Western rivers, I am of the opinion that protecting the guns and gunners does not, except in special cases, compensate for the greatly diminished quantity of artillery, slow speed, and inferior accuracy of fire; and that for general purposes wooden ships, shell guns and forts, whether for offense or defense, have not yet been superseded.*[263]

John Ericsson responded to Jeffers's report with a fury. He felt that the U.S. Navy did not understand how to operate his vessel and arrogantly answered each one of Jeffers's complaints. Ericsson noted that the port stoppers were not to be opened and closed after each shot. The turret, he also noted, should merely be turned away from the enemy's fire when reloading, as was done during the engagement with the *Virginia*. He agreed that the pilothouse placement should be atop the turret. Ericsson wrote that this was an obvious solution to fire control and communication and that he had considered doing this while the *Monitor* was under construction. The pilothouse was not moved on top of the turret because it would have delayed completion of the ship for over a month. The inventor concluded that the "damage to the national course which might have resulted from the delay is beyond computation."[264] Even though Ericsson was correct about the turret's operation and pilothouse location, he admitted that the poor ventilation system was a significant problem when he wrote Isaac Newton: "You have had a very severe trial and cannot imagine anything more monotonous and disagreeable than life onboard the *Monitor* at anchor in the James River, during the hot season." George Geer was well aware that the *Monitor* "was not properly ventilated for men to live in in hot weather, and I do not think she will ever go in another action until she has some alteration made, as the men would drop at the Guns before they fought half [an] hour."[265]

Geer noted that the temperatures throughout the ship were just intolerable: 110 degrees in the storeroom, 127 degrees in the engine room, 155 degrees in the galley and 85 degrees on the berth deck. He concluded his letter to his wife stating "so you can see what a hell we have."

First Assistant Engineer Isaac Newton also corresponded with Ericsson on several occasions in May and June concerning various improvements and enhancements to the *Monitor*. Newton commented that a separate blower was needed for venting the vessel. He also believed that the pipes that sent steam to the blowers' engines and the auxiliary pumps should be altered to allow them to operate when the engine was stopped. The engineer also noted that the pumps required a larger main injection pipe on the ship's side. Newton believed that the pipe was not large enough to simultaneously handle the water from both pumps. Furthermore, he noted that the wooden backing and deck were beginning to rot.

Newton reported to Ericsson about many other minor problems he discovered with the *Monitor*'s systems. The engineer noted that "nothing in the world but my great interest in the *Monitor*"[266] prompted him to suggest the various changes. Newton feared that the vessel could be rammed—or "punched"—and so believed additional pumps were required to keep the vessel afloat if such a "catastrophe" should occur. He installed an Andrew's centrifugal pump and wanted to acquire a Woodward crank pump to replace one of the Worthington pumps already installed. Engineer Newton rationalized: "For the reason that, notwithstanding the good workmanship on Worthington pumps, they are not as reliable as one which employs a crank."

The engineer informed Ericsson about the ironclad's engine system encountering problems on 2 June 1862. While Newton was able to organize temporary repairs, he noted that new parts were needed to ensure the ship's proper operation. He later advised Ericsson:

> *On the supposition that you would like to know the whereabouts and condition of* Monitor *I have determined to write you. There is still no prospect whatever for us to be engaged in any active duty or to be allowed time to overhaul the machinery, we are doing nothing here but swelter in this river. No opportunity has been given me since the 26th of February to haul the fires from the furnaces & the blowers of course have been in operation <u>unremittingly</u> during that time.*
>
> *The trouble about the excessive heat in the gallery* [sic] *could not have been foreseen, hot weather brought it out, but as I have already written you, it can <u>easily be remedied</u>, when the authorities give us time to do so—It appears to me that for some reason or other Comdr Goldsborough has determined to keep us in this river for some time yet…We will be hurried off without a moment's notice and without a chance to do any overhauling.*[267]

The *Monitor* men became rather lackluster as they waited at City Point for Major General McClellan to make his final push on Richmond. Throughout the last week of May, and for most of June, the ship remained at City Point, with occasional forays up the James or Appomattox Rivers. Many crew

Long, Hot Summer

members complained about the lack of activity; they gambled, fought one another or drank their spare time away. Officers and crew members alike recognized that their ship was not in good condition and she should be sent to the Washington Navy Yard for a refit because the ship's bottom was fouled with marine growth and the engines required some repairs. Jeffers requested the ironclad be sent to Washington; however, Goldsborough refused to move the *Monitor* out of the James River. Consequently, Jeffers asked to be reassigned. His request was ignored. Boredom and living conditions prompted Acting Master Edwin Gager to send in his resignation, but the U.S. Navy refused to accept it.

On 25 June, crew members "could hear the almost constant roar of artillery in the direction of Richmond."[268] The Battle of Oak Grove was McClellan's first attack against the Richmond defenses. His eight divisions were unable to advance due to the maneuvering of Major General John Bankhead Magruder. The next day, General Robert E. Lee seized the initiative with his assault against McClellan's exposed right flank at Mechanicsville. While Lee was initially unable to dislodge the Federals, the next day, he hurled his troops against Major General Fitz John Porter's command at Gaines Mill. Porter was defeated, and McClellan ordered his army to change its base of operations to the James River.

USS Monitor *in the James River*, circa 1862. *Courtesy of The Mariners' Museum.*

The men of the *Monitor*'s crew did not hear the sounds of either of these battles, as they were deployed on a special mission to steam up the Appomattox River to destroy the Richmond & Petersburg Railroad Bridge. This task force included the gunboats USS *Port Royal, Mahaska, Southfield, Jacob Bell, Stepping Stones, Satelitte, Maratanza* and *Island Belle*, as well as the hand-propelled submarine USS *Alligator*. This unusual submersible craft was ordered to the James River by Gideon Welles. She was conceived by Brutus de Villeroi of Philadelphia. The *Alligator* carried two spar torpedoes and was propelled by a hand-cranked screw propeller. Keeler described the vessel:

> *A submarine battery intended to work beneath the water. It resembles in appearance & is about the size of a large steam boiler pointed at both ends, with a row of small glass lights along the top & a manhole for entrance. It is sunk by admitting water into one of the water-tight compartments with which it is furnished & is propelled by means of a screw turned by the men inside.*[269]

The expedition's purpose was to use the *Alligator* to blow up the Appomattox River railroad bridge. If that was a success, the submarine would then be used to clear a channel through the Drewry's Bluff obstructions or to torpedo the ironclad CSS *Richmond*. However, both Goldsborough and Rodgers questioned the *Alligator*'s value.

The task force encountered constant sharpshooting as the ships worked their way up the narrow channel. The *Maratanza* ran aground, and the expedition was subsequently abandoned. As the *Maratanza* was freed, the *Island Belle* then became stuck in the shallows. News reached the Federal ships of McClellan's defeat at Gaines Mill and the subsequent need to bring all of the gunboats to Harrison's Landing to help protect the retreating Union army. The *Island Belle* was abandoned and set afire. All the Union gunboats were ordered into the James River. There, the *Galena* and *Maratanza* were able to shell the advancing Confederates during the 1 July 1862 Battle of Malvern Hill. The *Monitor* did not participate, as she could not effectively elevate her Dahlgrens to provide ground support fire and was therefore ordered back to City Point.

Once McClellan's army was safely encamped at Harrison's Landing, the U.S. Navy sent the *Monitor* and *Maratanza* up the James River to investigate Confederate naval activity. On 4 July 1862, the two Union ships encountered the CSS *Teaser* when they rounded Turkey Point. The *Teaser*, formerly an eighty-foot-long tug, was now armed with one thirty-two-pounder rifle and

Long, Hot Summer

CSS *Teaser* following her capture by USS *Maratanza* and USS *Monitor*. Photograph by J. Gibson, 4 July 1862. *Courtesy of The Mariners' Museum.*

one twelve-pounder pivot gun. This gunboat had served with distinction during the Battle of Hampton Roads as part of the James River Squadron, under the command of Lieutenant William Webb. She was, however, no match for either of the Federal warships. The first shot from the *Maratanza* was high, yet the next shot of cannister prompted the Confederates to quickly abandon their gunboat. As soon as the Southern sailors were off the *Teaser*, a second shell struck her boiler and left her dead in the water. When Federals inspected their prize, they discovered that the *Teaser* was commanded by Lieutenant Hunter Davidson. The lieutenant was working with Matthew Fontaine Maury on developing electric battery torpedoes (mines). Several torpedoes were found on board the *Teaser*, as well as various documents detailing where the torpedoes were to be placed in the river.

Of particular note was the Confederate hot air balloon. It was the same device used by the Confederates along the Warwick River during the early stages of the Peninsula Campaign. What William Keeler found most interesting

> *was the private memorandum book of Hunter Davidson who was in command. He was one of the officers of the* Merrimac *& in this book was* [sic] *drafts of the* Monitor *& sketches of the mode of our capture, as they intended to attempt it. It was minute in all its details. We were to be boarded from four tugs at the same time (one of them the* Teaser*) by men*

The *Monitor* Boys

carrying turpentine, ladders, fire balls, wedges, sheets of metal, chloroform &c. The names of the men were given, just what article each one was to carry, to what part of the Monitor *he was to go &c, it even gave the men who were to carry matches & sand paper to rub them on.*[270]

Keeler realized that the capture of the *Teaser* would bring the *Monitor*'s officers and crew some prize money. He also reflected that had the *Monitor* captured the *Virginia*, they would have shared $1 million in prize money.

The *Monitor* continued her service in the James River throughout July. On 9 July 1862, the ironclad was visited once again by President Lincoln. Lincoln was joined by Assistant Secretary of War Franklin Blair and Flag Officer Louis Goldsborough. Also touring the *Monitor* that day was photographer James F. Gibson. Gibson, who had studied under Mathew Brady, was at Harrison's Landing, Virginia, photographing McClellan's army. While there,

Crew of the USS *Monitor*. Photograph by J. Gibson, 4 July 1862. *Courtesy of The Mariners' Museum.*

the photographer also took several images of the ironclad's officers and crew. The views of the crew captured the *Monitor* Boys reading newspapers, playing checkers, smoking and cooking on a stove. Gibson's work created the only photographic images to be made of the *Monitor*. That same day, the James River Flotilla was made a separate command by Gideon Welles. Goldsborough considered this an insult and requested reassignment. Rear Admiral Samuel Phillips Lee replaced him on 2 September 1862. The new flotilla was placed under Captain Charles Wilkes, famed Antarctic explorer and captor of Confederate ministers James Mason and John Slidell during the infamous Trent Affair.

While Wilkes agreed that the *Monitor* was defective and required repairs, the continued fear of the emergence of the *Richmond* kept the *Monitor* stationed in the James River between City Point and Harrison's Landing "in constant preparation for the *Merrimac*…Some of us will die off one of these days with *Merrimac*-on-the-brain," William Keeler believed. "The disease is raging furiously, especially among those inclined to old fogyism." While Keeler acknowledged that the *Monitor* was considered the savior of the U.S. Navy, since only she could contest the advance of the newest Confederate ironclad, it meant that the officers and crew were left "close prisoners in the bowels of our iron monster, not a very enviable situation I assure you in the present hot weather."[271]

The heat and inactivity had a definite negative impact on the crew's morale. The soaring temperatures were only worsened by the flies and mosquitoes. Keeler wrote his wife, "You may imagine me writing this with a towel in one hand brushing off mosquitoes & wiping off perspiration, my brains in a sort of mix with the buzzing of the mosquitoes, the hum of flies, the trickling of sweat." A few days later, he continued his lament with "Hot, hotter, hottest—could stand it no longer, so last night I wrapped my blanket 'round me & took to our iron deck—if the bed was not soft it was not as insufferably hot as my pen…What with heat, mosquitoes & a gouty Captain have really gone distracted."[272]

By mid-August, the situation would completely change for the *Monitor*'s men. The food situation had improved since May. The Federal sailors were now allowed to forage, and the *Monitor* men were able to go on pig hunts (while still being targeted by Confederate sharpshooters) or to acquire cattle, sheep, fish and vegetables from slaves. The U.S. Navy also made a major effort to provide the James River Flotilla with fresh provisions, including vegetables and fruit, to improve the health of the men.

The *Monitor*'s purpose as the guardian of McClellan's army at Harrison's Landing, Virginia, also ended in August. Even though McClellan had hoped to resume operations against Richmond by way of Petersburg, the Army of the Potomac was instead shifted to Aquia Landing to operate with Major General John Pope's Army of Virginia.

Several crew changes would occur while the *Monitor* was at anchorage in the hot Virginia sun on the James River. A few crew members were discharged, transferred or sent on sick leave to Hampton Roads. Landsman George W. Burrows was transferred from the *Port Royal* to the *Monitor* and was transferred back to the *Port Royal* after brief service. Seaman John Hardy was transferred from the *Wachusetts* as ship's number fifty-four.

There were also major changes to the officer complement. On 14 August 1862, First Assistant Engineer Isaac Newton Jr. was detailed as superintendent of construction for the Office of the General Inspector of Ironclads in New York. The next day, Lieutenant William Jeffers was reassigned, to be replaced by Commander Thomas Holdup Stevens Jr. on the sixteenth.

Stevens, who had previously served as captain of the *Maratanza*, was the son of one of the heroes of the Battle of Lake Erie. He was born in Middletown, Connecticut, on 27 May 1819. He entered the U.S. Navy as a midshipman on 14 December 1836 and cruised on the *Independence*. Stevens studied at the Philadelphia Naval School, graduated third in his class and became a passed midshipman on 1 July 1842. Following duty as naval aide to President John Tyler, he was eventually assigned to the *Michigan* on Lake Erie. Stevens was then detailed as a naval storekeeper in Honolulu, Hawaii, from 1845 to 1848. While returning home with his wife and daughter aboard the Chilean ship *Maria Helena*, they were shipwrecked on Christmas Island. Rescued three months later, in April 1848, he was promoted master and detailed to Sacketts Harbor Naval Station, New York.

Promoted to lieutenant on 10 May 1849, Stevens returned to serve on the *Michigan*. He was named commander of the USS *Ewing* in 1852 and completed West Coast coastal survey work for three years. Assigned to the USS *Colorado* of the Home Squadron, he was later detailed as the commander of the USS *Ottawa* in the South Atlantic Blockading Squadron. Stevens participated in the capture of Fort Walker and Fort Beauregard and several other engagements in Port Royal Sound, as well as the capture of Fort Clinch, Florida, and the occupation of Jacksonville, Florida. His first prize during the war was the CSS *America*. Stevens was reassigned to the North Atlantic Blockading Squadron and given command of the *Maratanza*. He guided the *Maratanza* during the early stages of McClellan's Peninsula Campaign, including the Battle of Eltham's Landing. He also provided support for the Union positions at Cumberland and White House Landings.

Long, Hot Summer

Commander Thomas Holdup Stevens Jr., circa 1870. *Courtesy of The Mariners' Museum.*

Keeler was pleased with this change of command.

> *Yesterday our "most noble Captain" bade us adieu & left for the east where he is ordered to superintend the building of "ironclads." I can assure you we parted from him without many regrets. He is a person of a good deal of scientific attainment, but brutal, selfish & ambitious. Commander Stevens, who takes his place, has the appearance of a quiet modest man, so far I like him, but I find that first impressions are not always to be trusted. He has the reputation of taking a glass too much occasionally, the curse of the navy.*[273]

While the change of command was well received, the news that the *Monitor* was ordered back to Hampton Roads was met with overwhelming approval. On Saturday, 30 August 1862, the *Monitor* dropped her anchor off Newport News Point, "but a few rods from the sunken *Cumberland*." The *Monitor*'s sole purpose was to "blockade James River against the egress of the new *Merrimac*."[274]

The *Monitor*'s month-long assignment in Hampton Roads was a positive change for all. Once again, the crew was feted by the army, navy and Unionists. The food and temperatures were much improved. Stevens proved to be a much more pleasant commanding officer, but he was not destined to stay with the ship for long, as he was an "intemperate man." It appears that Stevens may have run afoul of a new act of Congress that ended the twice daily grog ration. The act stated that no liquor would be allowed on U.S. Navy vessels except for medicinal purposes. George Geer had already told his wife with disdain about how Captain Jeffers strived to circumvent the law:

> [He] *is going to do something he thinks smart. The new law about liquor says that NO liquor shall be brot* [sic] *on any government vessel after 1st of September, but says nothing about what is on hand at the time, so our Captain has sent for three 40-gallon barrels to have on hand. That is enough to last us one year. I hope the government will catch him at it someway, and make him trouble. I hope that he will get his deserts* [sic] *yet, before the war is over.*[275]

Drinking was considered a curse by many of the *Monitor* Boys. Keeler often wrote his wife about incidents such as the expedition up the Appomattox River when the ship commanders became too drunk to direct their vessels forward. Another story he shared was particularly sad. Lawrence Murray, the wardroom steward, had become drunk on shore leave, and when he returned to the *Monitor*, he took an axe and tried to split Keeler's servant's head open. Murray was placed in double-irons but somehow managed to get to the ship's side and jumped overboard. The heavy irons doomed him, and his drowned body was discovered three days later, on 5 September 1862. Geer noted that no one would lament his loss, as all Lawrence Murray did was drink and gamble his time and money away without sending support to his wife and child in California.

In the U.S. Navy, drinking stimulants was a serious problem shared by both enlisted personnel and officers. However, while enlisted men could only drink on duty under supervision—as when their grog ration was issued—officers drank without care and often were punished only when their intoxication became debilitating. Rumors persisted about Commander Stevens's drinking habits, and it was certainly the cause of his reassignment.

Commander John Pyne Bankhead was named Stevens's replacement on 10 September 1862. Bankhead arrived on 15 September and advised Stevens

Long, Hot Summer

Commander John Pyne Bankhead, circa 1865. *Courtesy of The Mariners' Museum.*

that he was being replaced because of a drinking incident a few days before, when he was too drunk to entertain several visitors, including the captain of the USS *New Ironsides*. When the crew learned about the circumstances surrounding Stevens's removal as the *Monitor*'s commander, the officers and crew wrote many positive letters on his behalf. When presented with the letters, Stevens gave a brief speech to the crew, and as he got on the tug to leave, the men gave him "9 as hearty chears [*sic*] as ever a man had."[276]

The new commander, John Pyne Bankhead, was a well-respected veteran naval officer, born at Fort Johnston, Charleston Harbor, South Carolina, on 3 August 1821. His father was Brigadier General James Bankhead, a close friend of Lieutenant General Winfield Scott and a Mexican-American War hero; his cousin was Confederate General John Bankhead Magruder.

John Pyne Bankhead entered the U.S. Navy as a midshipman on 10 August 1838 and was detailed to the USS *Macedonian*. His next assignment was the USS *Concord*, which ran aground, and her commander drowned. Bankhead was reassigned to the flagship of the Home Squadron, the *Independence*. He graduated second in his class from the Philadelphia Naval School and was promoted a passed midshipman on 20 May 1844. Bankhead served as chief of staff for his father's brigade at Vera Cruz, Mexico, during Lieutenant General Winfield Scott's march on Mexico City. He then was assigned to the USS *Truxtun*, which ran aground and was captured by the Mexicans. Following the war, he was on an extended sick leave. When he returned to duty, Bankhead was assigned to the USS *Vandalia* in the Pacific Squadron. Promoted to master in 1851 and lieutenant in 1852, Bankhead went on a two-year sick leave in 1853. Prior to the war, he served aboard the USS *Columbia* and the USS *Constellation* in the Mediterranean Squadron. At the outbreak of the war, he commanded the USS *Cranford*. He then was assigned to the *Susquehanna* and participated in the bombardment of the forts guarding Hatteras Inlet, North Carolina. Bankhead took command of the USS *Pembina* in the South Atlantic Blockading Squadron on 8 October 1861 and participated in the capture of Port Royal Sound on 5 November. Flag Officer Samuel Francis DuPont called Bankhead a "superior officer" and secured his promotion to commander on 16 July 1862. Bankhead was assigned to the *Monitor* on 10 September 1862.

The need for the *Monitor* to remain in Hampton Roads ended with the arrival of the most powerful ironclad built by the Union during the war, the *New Ironsides*. This ironclad was 230 feet in length and very similar to the European casemated armor-clads. The *New Ironsides* mounted twenty cannon, including two 150-pounder Parrotts, fourteen XI-inch Dahlgrens and two 50-pounder Dahlgren rifles. She was commanded by Captain Thomas Turner.

Awaiting a new assignment, the *Monitor*'s crew remained in Hampton Roads, and the men feasted on fish, crabs, oysters, peaches, watermelons and more. They enjoyed the fresh ocean breezes and were able to visit Old Point Comfort or Norfolk on leave. Furthermore, the men enjoyed the new command atmosphere since Jeffers and Newton had left the ironclad. Many men took to making trinkets to sell from pieces of the wreckage. The enterprising George Geer continued his business of selling knives, silk threads and other items, fulfilling the needs of sailors.

On 30 September 1862, Bankhead received orders to take the *Monitor* to the Washington Navy Yard for repairs. The ironclad's bottom had become so fouled that she had to be towed by a small tug. The *Monitor* arrived at the yard by 9:00 a.m. on the morning of 3 October. The officers and men longed for leave, and the *Monitor* was long overdue for an overhaul.

Chapter 10
Respite and Refit

As soon as the *Monitor* arrived at the Washington Navy Yard, the officers and men were met by crowds "huzzahing their service and marveling at the ship that saved the nation." The *Monitor* became Washington's premier tourist attraction. The crowds were so large that the Marine guards at the yard's entrance just opened the gates to allow the thousands and thousands of curiosity seekers in to see the *Monitor*. Many sought, and took, a souvenir from the famous warship. "When we came up to clean that night," Louis Stodder remembered, "there was not a key, doorknob, escutcheon—there wasn't a thing that hadn't been carried away."[277] While the officers were expecting a large public reception to honor their service, instead, the public flocked to them, and the ship was overwhelmed by soldiers, politicians and civilians. The crew had to post Marine guards around the ship so that the crew could have an undisturbed dinner. William Keeler described the chaotic scene to his wife:

> *Our decks were covered & our wardroom filled with ladies & on going into my stateroom I found a party of the "dear delightful creatures" making their toilet before my glass, using my combs & brushes. We couldn't go to any part of the vessel without coming in contact with petticoats. There appeared to be a general turn out of the sex in the city, these were women with children & women without children & women—hem—expecting, an extensive display of lower extremities was made going up & down our steep ladders.*
>
> *The decks were lined with carriages—& it was in fact a perfect jam—no caravan or circus ever collected such a crowd, not only in*

numbers but respectability. I made a large number of what would no doubt be very pleasant acquaintances if I had the time & disposition to follow them up—as all that I showed over the vessel gave me their addresses with an invitation.[278]

However, the *Monitor* was not in the Washington Navy Yard to welcome visitors. Consequently, the officers and men were sent aboard the steamer the USS *King Phillip* when workers took over the ironclad to initiate repairs.

The *Monitor* Boys were not destined to remain on the *King Phillip* for long. Captain Bankhead immediately went on leave and left behind authorization that officers and crew were to be allowed two to four weeks of leave. Keeler would be one of the last officers to depart, as he had the responsibility to pay off the men with several months' worth of pay. Most of these men had been away from home for over eight months and were overjoyed that their furlough freed them from the cramped quarters on the *Monitor*.

When the officers and crew returned, they did not board the ironclad immediately upon their arrival but were again temporarily berthed on the *King Phillip*. Once aboard the *Monitor*, they found many changes, including new faces and an improved warship. The engines had been completely overhauled and the bottom thoroughly cleaned. In an effort to make the ironclad more seaworthy, a large thirty-foot-tall telescopic funnel was placed over the low smokestack boxes, and taller fresh-air funnels were installed. The improvements were an effort to ensure that water did not enter the vessel from heavy seas. In addition, davits and cranes to help transfer the new cutters in and out of the water were also installed.

Several interior improvements were also completed in order to make the ship more livable. The fire damage to the galley was repaired, and the berth deck was enlarged. This was accomplished by removing the side storerooms, which gave eight feet more width in which to work. The floor of the berth deck was raised, and below it were placed two storerooms for provisions and a shell room. This improvement reduced the room's height "so that we can barely stand erect under the deck above," William Keeler reflected, "however is what we want." A large blower, with its own engine, was installed partly above and below the deck. The blower was designed to draw air down through the pilothouse "& through the deck lights (when open) in the wardroom & our staterooms & forces it into the engine room to aid the draft of the furnaces," Keeler explained.[279] George Geer thought that the blower "next summer will be very nice…and blows such a blast it would nock [*sic*] one down."[280] Keeler also felt that the additions, such as "new awnings have been furnished us, ventilators for our deck lights & many other little conveniences which would have added greatly to our comfort last summer could we have had them then."[281]

Geer was very impressed by the engine overhaul, noting that "it works first-rate now." In turn, Keeler was most impressed with how the Dahlgrens were engraved with large letters:[282]

MONITOR & MERRIMAC
WORDEN

MONITOR & MERRIMAC
ERICSSON

He also noted that the ironclad was covered with iron patches where shot had made indentations. Each place was marked: "'Merrimac,' 'Merrimac's Prow,' 'Minnesota,' and 'Fort Darling,' to indicate the source from where the blow was received."[283]

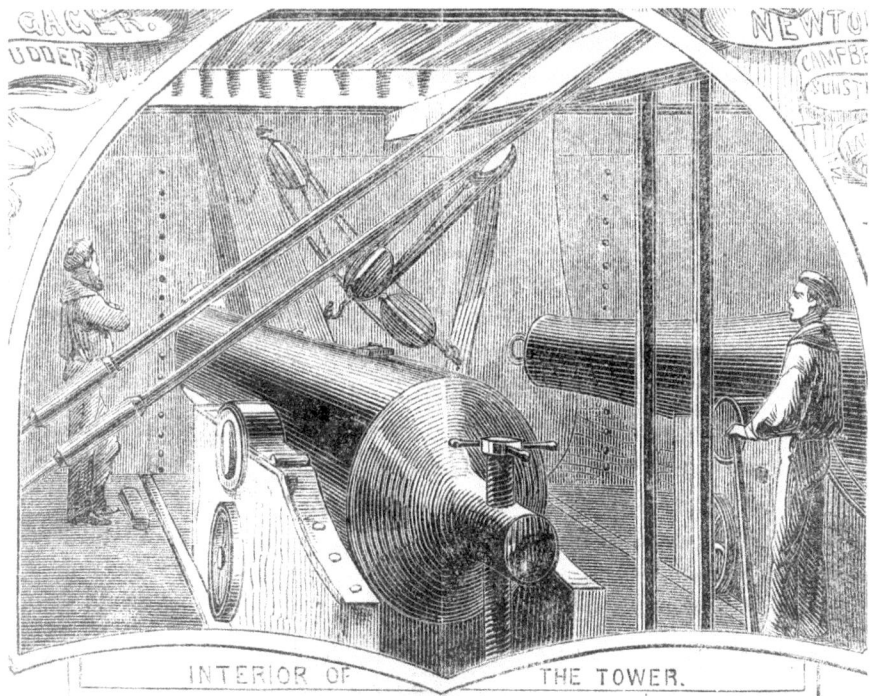

Monitor *Turret. Courtesy of John Moran Quarstein.*

157

Besides improvements to the *Monitor*, the officer cadre and crew went through some major changes. A new recruit transferred from the Brooklyn Navy Yard to the Washington Navy Yard, Ordinary Seaman Jacob Nicklis, noted:

> *I do not know what they intend doing with us[.] some say they are going to drill us and place us in Forts others say in Gun Boats but I care not where. The* Monitor *lies in the Yard at present for repairs & she will probably take some of us, as her crew ran away when they landed in the Yard on account of her not being sea worthy. But since then they have altered her so I think that there will be no danger.*[284]

In fact, there were numerous transfers, resignations and outright desertions during this time period. Acting Volunteer Lieutenant William Flye was transferred off the *Monitor* to become commander of the *Underwriter* on 30 October 1862. Dr. Daniel Logue resigned from the U.S. Navy shortly after arriving at the Washington Navy Yard. Dr. Thomas W. Meckly was initially named as Logue's replacement to serve as the *Monitor*'s assistant surgeon. Meckly, whose father was also a doctor, was born in Milton, Pennsylvania, on 27 August 1840. He graduated with honors from the Pennsylvania College Medical Department in 1861. Meckly was commissioned as surgeon on the USS *Tuscarora* but then resigned from the U.S. Navy in order to accept an appointment, in July 1862, as a U.S. Army assistant surgeon. The surgeon served in the Sixth Corps, Army of the Potomac, from the Peninsula Campaign through to the Battle of Antietam. He then resigned from the army, this time to accept an appointment as acting assistant surgeon in the U.S. Navy. Meckly was assigned to the *Monitor* on 30 October but was soon transferred to the USS *Lodona* on 8 November 1862. He was replaced that very day by Dr. Grenville Mellen Weeks.

Grenville M. Weeks, also the son of a doctor, was born in New York City on 22 November 1837. He attended the College of Physicians and Surgeons (now Columbia University) and University Medical College (now New York University). Weeks volunteered for service following the Battle of Bull Run, eventually being assigned to the USS *Valley City* as an assistant surgeon. Transferred to the *Brandywine* on 31 October 1862, Weeks was assigned to the *Monitor* on 8 November.

Two other officers were assigned to the *Monitor* when she was in the Washington Navy Yard. When First Assistant Engineer Isaac Newton was transferred off the *Monitor*, Second Assistant Engineer Albert Campbell

Respite and Refit

Second Assistant Engineer Albert Campbell, circa 1862.
Courtesy of The Mariners' Museum.

assumed the role as the *Monitor*'s chief engineer. Acting Third Assistant Engineer George H. White had been transferred to the *Monitor* on 18 August 1862, after Newton left the ironclad. Two months later, White was assigned to the *Passaic* class monitor the USS *Nantucket* on 29 October. This action left a vacancy in the engineering department. Even though George Geer was striving to receive an appointment as a third assistant engineer, he was unable to secure this rank, despite all of his lobbying efforts. Instead, Third Assistant Engineer Samuel Augee Lewis of Baltimore, Maryland, was assigned to the engine room. Keeler referred to Lewis as "a mere boy, merely a cypher [*sic*] in our little society."[285] The other new officer was Norman Knox Atwater of New Haven, Connecticut, who had been in the U.S. Navy for less than two months.

Several former slaves also mustered on the ironclad while she was at the yard, such as the "contrabands" Robert Howard and Daniel Moore. Many "contrabands" were already serving in the U.S. Navy. The *Minnesota* had several African Americans onboard. The frigate's aft pivot gun was manned by an all–African American gun crew, and these men bravely fought during the Battle of Hampton Roads. Captain Van Brunt of the *Minnesota* lauded their service during the engagement, noting that the "Negroes fought energetically and bravely—none more so. They evidently felt that they were thus working at the deliverance of their race."[286] Accordingly, the U.S. Navy's long-standing tradition to recruit African Americans enabled runaways like Robert Cook and Edward Cann to join the crew of the *Monitor* as first-class boys.

George Geer noted that over two hundred new recruits arrived at the Washington Navy Yard on 2 November, and he speculated that "some of them I suppose will go on us, as ten of our men have not come back yet." Jacob Nicklis was one of the new recruits for the *Monitor*. While Geer was happy, thinking that the new crewmen were all American-born (they actually were not, as one Canadian and one native of Ireland were among the new recruits), he felt they were all "green hands as I was eight months ago."[287]

Many *Monitor* Boys simply did not return from their furloughs. Several of these men were unable to get back to the *Monitor* due to illness or misfortune. Many, like Hospital Steward Jessie M. Jones and Seaman Thomas Loughran, just disappeared from the records. Coal Heaver William Durst was given two weeks liberty and spent his leave first in Philadelphia and then in New York City, where he became ill. He claimed his illness to be typhoid fever, but he also believed he was "sick for some trouble to my heart and stomach, contracted from the bad air on the *Monitor*." Regardless, he was unable to return to duty and apparently did not know what to do. Durst went out drinking one day, shortly thereafter: "I was taken in charge of by a 'runner' and the next thing I knew I was on board the *North Carolina*. I asked for an explanation, and was informed that I had been shipped under the name of Walter David, that there was nothing for me to do but to serve out my enlistment for one year."[288] William Durst's "WD" tattoo had earned him the alias of "Walter David."

Another tale of apparent desertion was Seaman Hans A. Anderson's story. Just before he was due to return to Washington, D.C., from New York City, Anderson "fell in with some men and went into a rum hole, and took a glass of beer which was evidently drugged."[289] He soon found himself shanghaied on an unknown barque sailing for London. Anderson eventually made his way to a U.S. consul, returned to the United States and surrendered himself on 2 February 1863. He was allowed to return to duty and, except for loss of pay while absent from duty, was not penalized for his misadventures.

Respite and Refit

Actual deserters included Anton Basting, Derick Bringman and John Conklin, none of whom appeared on the *Monitor*'s 7 November 1862 muster roll. John Driscoll also did not appear at this muster; however, on 15 December 1862, he resurfaced to enlist on the USS *Connecticut*.

Two other original *Monitor* Boys also did not appear on the 7 November muster: Edward Moore and Moses Stearns. Ship's Cook Edward Moore suffered an unknown injury prior to the *Monitor*'s arrival at the Washington Navy Yard. Although Moore's injury does not appear in any surgeon's report, he was discharged on 28 November 1862. Likewise discharged that day was Moses Stearns. Quartermaster Stearns suffered, according to John Conklin and Thomas Carroll, "while engaged in passing shot from berth deck received a strain which caused hernia of the left side." Even though Stearns was seen by Dr. Logue and excused from duty, his condition was not included in the surgeon's report after the 9 March engagement. When the *Monitor* arrived at the Washington Navy Yard, Stearns was examined and discharged as unfit for service.

All of these men needed to be replaced, and Captain Bankhead called for volunteers from the unassigned seamen recently transferred to the Washington Navy Yard. Jacob Nicklis did not wish to ship onboard the *Monitor* "on account of her accommodations they are very poor."[290] He volunteered anyway because several of his friends from Buffalo, New York, like Isaac H. Scott, did so. Scott, at five feet, eleven inches tall, was the tallest of the new crew members. Originally from Quebec, Canada, he was a cabinetmaker who lived and enlisted in Buffalo and had been in the U.S. Navy only two weeks when he was transferred to the *Monitor* to become ship's number sixty. The shortest new sailor on the *Monitor* was a native Irishman, James Malone. He was a chandler in New York before the war and was listed as being just four feet, four and a half inches tall. William Henry Remington was a prewar farmer from Syracuse, New York, who became a coal heaver, ship's number sixty-two, on 7 November 1862. Other new coal heavers detailed to the *Monitor* included William Morrison, ship's number sixty-four, from England; Charles Smith, ship's number sixty-seven, from Rome, New York; George Littlefield, also given as ship's number sixty-seven,[291] from Saco, Maine; and James Smith from Haverstraw, New York.

Francis Banister Butts, commonly called Frank, was a farmer from Cranston, Rhode Island. He had originally mustered, at the age of seventeen, as a corporal in Battery E, First Rhode Island Light Artillery, on 16 September 1861. When his army enlistment expired, Butts enlisted as a landsman on 3 October 1862 at the Brooklyn Navy Yard. Butts was eventually transferred to the Washington Navy Yard and volunteered for service aboard the *Monitor* as ship's number sixty-three. Several other landsmen transferred to the *Monitor* on 7 November, including Irish immigrants James Coleman, ship's number sixty-five, and William Egan, ship's number seventy-three; John Monaghan, ship's number sixty-nine, from New York City; and William Allen, ship's number seventy, originally from England. Seaman John Brown, from Germany, with less than a month in the navy, also joined as ship's number seventy-two.

George Geer was not pleased with most of the new recruits once he got to know them. "These new men," he wrote his wife, "are [the] damdest [sic] lazyest [sic] set of Hoggs [sic] I ever saw, and out of 13 there is [sic] but two saylors [sic]. The rest of them never were on salt water."[292]

Now with the officers and crew at full complement, the *Monitor* was once again ready for active duty, and she left the Washington Navy Yard at noon on 8 November 1862. The ironclad moved down the Potomac River en route to Hampton Roads. She had an uneventful voyage down the Chesapeake Bay, and by noon on 10 November 1862, the *Monitor* was back at her old anchorage off Newport News Point.

The ironclad remained at this mooring for the next six weeks. Fears that the newest Confederate ironclad would soon steam down the James River were rampant. The *Richmond* never appeared. Keeler knew that the Confederates would never risk their only James River ironclad, as an attack on Hampton Roads "would only result in her destruction. They know by keeping her up the river she neutralizes three or four of our heavy ironclads which we kept blockading the mouth to prevent her exit."[293] As the crew waited for any action in the James River, they speculated where the *Monitor* might be sent to operate against a major Southern port. In actuality, the U.S. Navy was considering two different operations: one against the blockade runners' haven of Wilmington, North Carolina, or another to recapture Fort Sumter in the hated home of secessionism, Charleston Harbor, South Carolina.

The monotony wore on as the *Monitor* awaited new orders. Several tasks were completed to improve the ironclad. Both George Geer and Jacob Nicklis wrote home that they had been detailed to paint the turret. While walking on the deck with Captain Bankhead, Keeler suggested that an iron breastwork should be built atop the turret to protect men positioned there from musket fire. A few days later, Bankhead told Keeler that the breastwork would be installed, complete with loopholes so the crew could fire back at sharpshooters. Additional repairs were made to the turret engines, as they had not been effectively repaired before leaving the Washington Navy Yard.

Respite and Refit

Officers of the USS *Monitor*. *Back row, left to right*: Albert Campbell, Mark Sunstrom, William Keeler and L.H. Newman (USS *Galena*). *Second row, seated, left to right*: Louis Stodder, George Frederickson, William Flye, Daniel Logue and Samuel Dana Greene. *Front row, left to right*: Robinson Hands and Edwin Gager. Photograph by J. Gibson, 4 July 1862. *Courtesy of The Mariners' Museum.*

USS Passaic, circa 1862. *Courtesy of The Mariners' Museum.*

Some excitement was generated by the arrival of one of the new class of monitor, the USS *Passaic*. The *Passaic* was commanded by Captain Percival Drayton. Like Bankhead, Drayton was a native of South Carolina. The new ironclad was the first of the *Passaic* class, which Ericsson designed as an improvement over the original *Monitor*. These monitors were twenty-one feet longer than the original *Monitor* and displaced 1,875 tons, as compared to the *Monitor*'s 987 tons. The *Passaic* class featured a pilothouse above the turret. Furthermore, they packed a bigger punch, as the turret contained one XV-inch and one XI-inch Dahlgren. When the *Passaic* arrived on 30 November, she joined the *Monitor*, *New Ironsides* and *Galena*, making Hampton Roads temporarily home to the most powerful ironclads afloat in the world. Unfortunately, the *Passaic* encountered problems with her boilers and was sent to the Washington Navy Yard for repairs. However, the sailors aboard the *Monitor* were excited about the arrival of several other *Passaic* monitors, especially the USS *Montauk*. John Worden had partially recovered from his wounds received on 9 March and had been named commander of the *Montauk*. It was well known that she would soon arrive in Hampton Roads. The men had been happy to learn that Worden received "the thanks of Congress upon the recommendation of the President, which will promote him one grade making him a captain," and all of the *Monitor*'s officers and men were anxious to see their "old captain" again.[294]

USS New Ironsides, circa 1862. *Courtesy of The Mariners' Museum.*

Respite and Refit

Captain John Bankhead had proven to be a well-organized and respected commanding officer. "Everything goes along quiet & smoothly on board," Keeler recounted, "—very different from what it was under our former captain."²⁹⁵ Bankhead's past service afloat and as chief of staff to his father during the Mexican-American War enabled him to give valued opinions about ongoing military affairs. The officers recognized that his calm demeanor disguised his seriousness of purpose, duty and honor. Geer thought that the captain was stricter than he was before the *Monitor*'s refit because "he has all hands Muster every morning at 9 Oclock [*sic*] in clean blue working clothes."²⁹⁶ It seems he was instilling a sense of discipline in preparation for future operations.

The fall and early winter spent by the *Monitor* was far more acceptable than their days spent on the James River the previous summer. The steam-heating system worked very well, and the enlisted men had more room on the berth deck for their hammocks. Keeler commented how comfortable and cozy the wardroom appeared. "One would hardly suppose from the quiet stillness that pervades our submarine abode that a gale was raging over us." Keeler also shared this with his wife:

> *I have spent the day in my little snuggery reading & writing by the aid of the few rays of light which straggled down from a bright unclouded sun through the little circular opening over my head, closed by the thick plate of glass & covered by some six (or eight) inches of water. I went to sleep last night to the swish, swash, of the waves as they rolled over my head & the same monotonous sound still continues & will be my lullaby tonight.*²⁹⁷

Besides the better, more comfortable accommodations, the food had vastly improved. Geer noted that the crew members went "oystering" often:

> *Been out Oystering once sence* [sic] *I wrote you last…We have a way of cooking diferant* [sic] *from any you ever done. We take and put them in a box with holes bored in the bottom and put the box over a steam pipe and let them Steam until they open, and I can tell you they are lucious* [sic] *to put them in Butter and gullup* [sic] *them down. I can put away about two dozen any time.*²⁹⁸

Jacob Nicklis wrote home that "I am getting fat and look tough & hardy."²⁹⁹

The officers also had a bountiful mess. William Keeler remembered all the provisions acquired before the *Monitor* left the Washington Navy Yard:

I doubt if you could find better at few hotels on shore. Mr. Stodder is Caterer, he bought in Baltimore just before we left Washington a bill of provisions for mess amounting to over $700 intending it for two months…

We breakfast at 8, lunch at 12 & dine at 5. Our breakfast is usually fried oysters, beefsteak, fish balls, mutton chops, with an ambundance [sic] *of vegetables, sweet & common potatoes &c. For lunch we usually have oysters or cold tongues, lobster (in cans), cold roast or corned beef, sometimes cold boiled ham, sardines, crackers, cheese, etc., etc.*

Dinner is <u>the</u> meal. Soups, stew'd oysters, boiled Salmon, Roast beef, Mutton or Turkey, or boiled ham, & so on through a whole hotel bill of fare with all the sauces, condiments & fancy pickels [sic]*…Don't imagine we are going to Starve.*[300]

On 23 December 1862, all hands were mustered, and an order from the War Department was read forbidding any army or navy personnel from communicating any news about military operations. This dictum sent a clear message to the crew that they would soon be headed south. Many thought it would be Charleston.

Christmas Day was celebrated in Hampton Roads in a most merry fashion. Visiting English and French warships traded salutes with Fort Monroe. The HMS *Ariadne* and USS *Colorado* commenced target practice with their heavy guns. Soon, according to William Keeler, the "powder smoke hung like a thick fog over the water & was so dense that at one time it was impossible to see vessels lying a few yards distant."[301] There were celebrations everywhere along the shoreline.

The shipboard meals were fabulous. The officers supplemented their Christmas Dinner with specialties sent from home, including raisins, oranges, pies, cakes and nuts, which enhanced the variety of fish, fowl and meats that were served. George Geer and many other crew members did not enjoy the day. Geer noted that he had paid one dollar for the cook to create a splendid meal; unfortunately, it was poorly cooked. Jacob Nicklis thought otherwise and described the meal: "We had chicken stew & then stuffed Turkey mashed potatoes & soft bread after this we had a plum pudding & some nice fruitcake with apples for desert [sic]."[302] Despite these differing opinions about their Christmas meal, the enlisted men still had to work most of the day preparing the ironclad for an ocean voyage. Bankhead had received orders on that festive day to take the *Monitor* out to sea.

Chapter 11
Hatteras

The *Monitor*'s Christmas Day orders detailed the ironclad to Beaufort, North Carolina. There she was to join two of the new *Passaic* class monitors, the *Passaic* and the *Montauk*, for a joint army-navy expedition against Wilmington, North Carolina. Eventually two other monitors, the USS *Weehawken* and the USS *Patapsco*, would join this squadron in January 1863.

Union Secretary of the Navy Gideon Welles had sought to strike at Wilmington since the summer of 1862. Located on the Cape Fear River, Wilmington was an excellent blockade runner's haven. Just 670 miles from Bermuda, the Cape Fear River had two entrances—New Inlet and Old Inlet—into the river from the Atlantic Ocean. This situation made it difficult for the North Atlantic Blockading Squadron to stem the flow of material in and out of Wilmington.

Welles's concept was to move the monitors into the Cape Fear River via the Old Inlet and to bombard Fort Fisher into submission. Fort Fisher was the largest earthen coastal fortification in the South, and it guarded the entrance to the New Inlet. Both entrances to the Cape Fear River were rather shallow. Commander John Bankhead, who had worked on the coastal survey of this section of North Carolina's coast, believed that the *Monitor*, with her draft of ten feet, six inches, could only navigate into the Old Inlet. The other monitors drew too much water to enter even the Old Inlet's depth of eleven feet, six inches.

Despite these problems, Welles wanted the operation to move forward. While the monitors were to subdue the Cape Fear forts, Major General

John Gray Foster's army at New Bern, North Carolina, would move south and besiege Wilmington. Once the monitors had passed the forts, the port city would be shelled into submission. The plan had many merits, but unfortunately, the monitors had to steam from Hampton Roads down the coast to Beaufort to implement it.

News of the impending voyage south was not well received by either the officers or enlisted men, particularly those crew members who had experienced the ironclad's harrowing voyage from New York to Hampton Roads in March. Lieutenant Samuel Dana Greene warned, "I do not consider this steamer a seagoing vessel. She has not the steam power to go against a headwind or sea, and…would not steer even in smooth weather, and going slow she does not mind her helm readily."[303] Rumors of the *Monitor*'s last sea voyage were heard by many of the new crew members, prompting Jacob Nicklis to write his father, "They say we will have a pretty rough time going around Hatteras, but I hope that it will not be the case."[304]

Despite the fears associated with the voyage, Bankhead prepared his ship for sea. The new ship's surgeon, Dr. Grenville Weeks, noted that the "turret and sight holes were caulked and every possible entrance for water made secure, only the smallest opening being left in the turret top."[305] George Geer worked on securing the hatches "with Red lead putty, and the Port Holes I made Rubber Gaskets one inch thick and in fact had everything about the ship in the way of an opening water tight."[306] Bankhead thoroughly followed the Navy Department's instructions to guard the vessel against leaks. Unfortunately, once again, the navy ignored the fact that Ericsson had designed the turret to fit snugly onto a brass ring set into the deck. This design feature was rejected as not watertight, and instead, the *Monitor*'s turret was jacked out of its ring and oakum was packed around its base. Geer noted that the men working on this task "did not put any Pitch over it and the sea soon washed the oakum out."[307]

The powerful 236-foot-long side-wheeler the USS *Rhode Island*, which had recently arrived in Hampton Roads following a refitting in Boston's Charlestown Navy Yard, was detailed to tow the *Monitor* to Beaufort. The *Rhode Island* was captained by Commander Stephen Decauter Trenchard. Trenchard's ship had run aground entering the Chesapeake Bay on 19 December but was declared undamaged. The *Rhode Island* and *Monitor* were to be accompanied on their voyage south by the USS *State of Georgia* towing the *Passaic*.

The expedition was delayed by a heavy storm that struck Hampton Roads on 27 and 28 December. William Keeler thought the storm was a good omen and noted that the *Monitor* "shall hold on here till the storm is over & take advantage of the calm that follows for our trip down the coast."[308] As the crew waited for the storm to abate, Second Assistant Engineer Albert Campbell was testing the pumps, and he "got caught in one of the small Pumping engines and got his leg so badly brused [sic] he will be confined to his bed some weeks. It is the greatest wonder he has not broken a leg, but it was his carelessness and nobody is to blame."[309] Campbell was so badly injured that he was transferred to a hospital ashore and would miss the forthcoming trip south. Campbell was replaced by Third Assistant Engineer Joseph Watters from Bordentown, New Jersey. Watters transferred from the USS *Chattanooga* to the *Monitor* on 28 December 1862.

The day of 29 December began "clear and pleasant, and every prospect of its continuation," prompting Commander Trenchard to prepare the *Rhode Island* for the voyage. At 2:30 p.m., the *Rhode Island* took up the two towlines attaching her to the *Monitor* and steamed past the Virginia Capes south to North Carolina. The only disappointment felt by the crew is that the long-anticipated arrival in Hampton Roads of the *Passaic*-class ironclad *Montauck*, captained by John Worden, occurred just after the *Monitor* began to work her way down the coast.

The first evening at sea was pleasant. Keeler noted that "a smooth sea & clear skies seemed to promise a successful termination of our trip & an opportunity of once more trying our metal [sic] against rebel works & making the 'Little *Monitor*' once again a household name."[310]

The next morning, clouds were seen off to the south and west. Commander Bankhead reported:

Third Assistant Engineer Joseph Watters, circa 1862. *Courtesy of The Mariners' Museum.*

We began to experience a swell from the southward with a slight increase of wind from the southwest, the sea breaking over the pilothouse forward and striking the base of the tower, but not with sufficient force to break over it. Found that the packing of oakum under and around the base of the tower had loosened somewhat from the working of the tower as the vessel pitched and rolled. Speed at this time was about five knots, ascertaining from the engineer of the watch that the bilge pumps kept her perfectly free, occasionally sucking. Felt no apprehension at this time.[311]

George Geer remembered the waves washing the oakum out from beneath the turret, where the water "came down under and down on the Berth Deck in Torents [sic]. But our pumps were sufficient."[312]

Shortly thereafter, the *Monitor* made signal to the *Rhode Island* to stop. Bankhead wanted to make adjustments to the tow line. This action enabled the work to be accomplished, and the two ships proceeded on their course.

The winds increased as the *Monitor* steamed southward. By midday, George Geer remembered that "soon the sea commenced to break over us and wash up against the tower with a fearful rush, and the sea was white with foam." By 1:00 p.m., the *Rhode Island* and *Monitor* passed the Cape Hatteras Lighthouse and began to work their way around the Cape itself. The winds

USS Rhode Island *towing USS* Monitor, *1863. Courtesy of John Moran Quarstein.*

continued to increase as the two vessels made little headway against the storm. Geer hoped that "as soon as we got pass [sic] the cape it would clear up."[313] Instead, the storm increased. Winds reached gale force, and the sea had grown very rough. Bilge pumps were started to remove the small amounts of water coming into the ironclad.

Despite the storm, the officers sat down to dinner, as Keeler remembered, "everyone cheerful and happy & though the sea was rolling and foaming over our heads the laugh & jest passed freely 'round; all rejoicing that at last our monotonous, inactive life had ended & the 'gallant little Monitor' would soon add fresh laurels to her name."[314]

Meanwhile, Bankhead, using chalk messages written on a blackboard, informed the *Rhode Island* that if the *Monitor* needed help during the evening, a red lantern would be displayed next to the ironclad's white running light.

Bankhead recorded: "Toward evening the swell somewhat decreased, the bilge pumps being found amply sufficient to keep her clear of the water that penetrated through the sight holes of the pilothouse, hawse hole, and the base of the tower (all of which had been well caulked previous to leaving)."[315]

As darkness came, the *Rhode Island* and *Monitor* were separated from the *Passaic* and *State of Georgia*.

The storm soon increased in its ferocity. Waves dashed across the deck and broke with tremendous force. Bankhead recalled that he "found the vessel towed badly, yawing very much, and with the increased motion making somewhat more water around the base of the tower. Ordered engineers to put on the Worthington Pump and bilge injection and get the centrifugal pump ready and report to me immediately if he perceived any increase of water."[316]

While the Worthington steam pump temporarily stemmed the flow of water, the *Monitor* was suddenly hit by a series of fierce squalls. The ironclad was now in "very heavy weather, riding one huge wave, plunging through the next, as if shooting straight to the bottom of the ocean." The *Monitor*'s helmsman, Francis Butts, continued his description of the effects of the heavy gale on the *Monitor*, stating that the ironclad would drop into a wave "with such force that her hull would tremble, and with a shock that would sometimes take us off our feet."[317]

When the Worthington Pump failed to stem the rising water, the call came from the engine room, "the water is gaining on us, sir." This report "sounded ominously,"[318] according to William Keeler; however, Bankhead ordered the large Adams centrifugal steam pump into action. This pump was capable of removing three thousand gallons a minute in a stream "as large as your body,"[319] according to George Geer. Unfortunately, this pump also proved inadequate to stop the flow of water, which had by 9:00 p.m. risen over a foot deep in the engine room. As the storm increased in its fury, Bankhead put the crew to work on the hand pumps and organized a bucket brigade. The bailing served little purpose other than to lessen the panic among the crew. "But our brave little craft struggled long and well," wrote William Keeler. "Now her bow would rise on a huge billow and before she could sink into the intervening hollow, the succeeding wave would strike her under her heavy armor with a report like thunder and violence that threatened to tear apart her thin sheet iron bottom and the heavy armor which it supported."[320]

Unfortunately, the water was rising rapidly. When it reached above the engine room floor, Keeler noted that it "was the death knell of the *Monitor*."[321] Bankhead later wrote:

> *The sea about this time commenced to rise very rapidly causing the vessel to plunge heavily, completely submerging the pilothouse and washing over and into the turret and at times into the blower pipes. Observed that when she rose to the swell, the flat under surface of the projecting armor would come down with great force, causing considerable shock to the vessel and turret, thereby loosening still more packing around its base.*[322]

Leaks appeared everywhere. The *Monitor* was going "head on" into the storm, and the constant pounding of the waves forced the upper deck to begin separating from the lower hull. Louis Stodder believed that the situation was exacerbated by "the *Rhode Island*, being a powerful steam ship, towed us faster than our engines could keep up with, and the sea beating under our 15-foot overhang at the bow ripped us apart."[323] Every time the ironclad crashed down into the sea, more water was forced into every hole, gap, crevice and crack.

The situation had become desperate aboard the ironclad. Nothing seemed to arrest the influx of water. By 10:00 p.m., furnace fires were extinguished by the ever-rising seawater, rendering the *Monitor* helpless. Joseph Watters reported that when the boilers lost steam, "the main engines stopped, the Worthington and centrifugal pumps still working slowly, but finally stopped."[324] The foundering ironclad appeared isolated in a sea of hissing, seething foam. Bankhead ordered the red lantern displayed and thus tried to signal the *Rhode Island* for assistance. Signal flares were launched, yet the *Rhode Island* still did not notice the *Monitor*'s dilemma. "Send your boats immediately, we are sinking," Bankhead shouted at the side-wheeler.

Hatteras

Fighting the Incoming Seas—USS Monitor, *1863. Courtesy of John Moran Quarstein.*

"Words cannot depict the agony of those moments as our little company gathered at the top of the turret, stood with a mass of sinking iron beneath them, gazing through the dim light, over the raging waters and an anxiety amounting almost to agony for some evidence of succor from the only source to which we could look for relief,"[325] Keeler painfully remembered. Even though the *Monitor* had been fitted with two new ship's boats while being overhauled at the Washington Navy Yard, those "lifeboats" had been stored aboard the *Rhode Island* for the voyage south. The crew was helpless to save themselves without assistance from the *Rhode Island*. When all seemed lost, "the clouds now began to separate, a moon of about half-size beamed out upon the sea," Francis Butts remembered, "and the *Rhode Island*, now a half mile away, became visible. Signals were exchanged, and I felt that the *Monitor* would be saved."[326]

When Bankhead realized that the *Rhode Island* was finally responding to his pleas for assistance, he also immediately recognized that the towline to the *Rhode Island* could be the end for both vessels. He feared that when the ships hove together, the *Monitor* could surge forward and pierce the *Rhode Island*'s wooden hull with her iron bow. Accordingly, Bankhead ordered several men forward to cut the cables connecting the *Monitor* and the *Rhode Island*. Quarter Gunner James Fenwick attempted it but was swept overboard. Then Boatswain's Mate John Stocking took an axe and frantically chopped at the cables until he, too, was washed into the sea by a giant wave. Finally, Acting Master Louis Stodder rushed forward to cut the thirteen-inch line. "It was not an easy job," Stodder recalled, "and while I was hacking at it a big sea came over the bow."[327] Somehow, Stodder held on and finished the job. Once the line was parted, Bankhead ordered the anchor dropped in order to stop the ironclad's pitching. While this action was somewhat successful, dropping the anchor would also loosen the watertight packing around the anchor well, allowing still more seawater to flow into the wallowing ironclad.

As the *Rhode Island* backed toward the *Monitor*, one of the towlines fouled the port paddlewheel. This caused the side-wheeler to temporarily lose control, and she almost collided with the sinking ironclad. The *Rhode Island*'s crew worked feverishly to stabilize the steamer and then launched lifeboats to retrieve the *Monitor*'s crew.

The *Monitor*'s crewmen anxiously waited for the *Rhode Island*'s boats to arrive. Keeler was placed in charge of a bailing party to endeavor to staunch the flow of water. Francis Butts was in the turret to pass bails up and down through the hatch of the turret. Butts claimed that he became so annoyed by the wailing of a cat that he placed the feline into a barrel of one of the XI-inch Dahlgrens and stuffed a wad in after it. Unfortunately, this action did not stop the cat's mournful howling.

Finally, the *Rhode Island*'s rescue boats were seen nearing the ironclad, with the *Rhode Island* following close behind them. The rough seas forced the side-wheeler toward the *Monitor* and nearly crushed the lifeboats betwixt the two larger vessels. The *Rhode Island* managed to pull away and stood off the *Monitor*, a quarter mile away.

In the meantime, Bankhead readied the crew to abandon the ironclad. The *Monitor*'s captain, and several other officers, went through the ship to make sure all were ready to enter the lifeboats. According to Francis Butts, Third Assistant Engineer Samuel Augee Lewis was so seasick that he was unable to get out of his bunk. Lewis had been aboard the *Monitor* only sixty days, and this was his first sea voyage; it is no wonder he had become so ill. William Keeler went to his stateroom to recover his books and papers and found "the water nearly to my waist & swashing from side to side with the roll of the ship." Keeler groped his way through the "thick darkness rendered more dense if possible by the steam, heat & gas which was finding its way from the half-extinguished fires of the engine room."[328] The paymaster quickly realized that he could not retrieve his files without endangering his own life. Therefore, Keeler started to make his perilous way back to the turret's top.

> *Everything was enveloped in a thick murky darkness, the waves dashing violently across the deck over my head; through the wardroom where the chairs & tables were surging violently from side to side, threatening severe bruises if not broken limbs; then up a ladder to the berth deck; across that & up another ladder into the turret; around the guns & over gun tackle, shot, sponges & rammers which had broken loose from their fastenings, & up the last ladder to the top of the turret.*[329]

George Geer had finally abandoned his pump when it stopped working and waded through knee-deep water toward the turret. Francis Butts realized that everyone was ready to leave the ship when there was no one else to pass a bail to.

It seemed like an eternity to the *Monitor*'s crew, as they waited for the *Rhode Island*'s boats to come alongside the ironclad. "It was a scene well calculated to appall the boldest heart," Keeler wrote, adding:

> *Mountains of water were rushing across our decks & foaming along our sides; the small boats were pitching & tossing about on them or crashing against our sides, mere playthings on the billows; the howling of the tempest, the row & dash of waters; the hoarse orders through the speaking trumpets of the officers; the response of the men; the shouts of encouragement & words of caution:*
>
> *"the bubbling cry*
> *of some strong swimmer in his agony."*
>
> *& the whole scene lit up by the ghastly glare of the blue lights burning on our consort, formed a panorama of horror which time will never efface from my memory.*[330]

Many crew members weakened as they waited. George Frederickson returned a watch he had borrowed from Peter Williams, stating, "Here, this is yours; I may be lost."[331] A statement made in simple fear proved to be uncannily prophetic.

Finally, Bankhead ordered Keeler to lead the first party to the boats. This meant the men had to descend a ladder down the side of the turret and

USS Rhode Island *rescuing USS* Monitor *crew*, circa 1863. *Courtesy of John Moran Quarstein.*

then move across the wind- and wave-swept deck to the waiting rescue craft. Keeler was hit by a strong wave and tossed into the sea as he crossed the deck. Somehow, the waves threw him back against the wallowing ironclad, which enabled him to grab a line from one of the deck stanchions. He held on and successfully made his way safely into a boat.

Once the two craft were filled, they rowed toward the *Rhode Island*. Keeler later noted that "our dangers were not yet over. We were in a leaky, overloaded boat, through whose crushed sides the water was rushing in streams & had really a half a mile to row over the storm tossed sea before we could reach the *Rhode Island*." When the boats finally reached the pitching ship, the men had to climb or be pulled up by ropes onto the steamer's deck. Keeler, who had hurt his hand, was hauled up by a loop of rope, called a bight. As Keeler reached the deck, he and the others received "the congratulations & hospitalities of her officers, & I assure you they were not deficient in either."[332]

When the unloading was accomplished, the boats made ready for a return to the *Monitor*. One vessel was crushed by the *Rhode Island*, and only a single remaining craft, commanded by Acting Master's Mate D. Rodney Browne, made its way back to the *Monitor*.

Bankhead stood on the deck and held the painter of the craft close to the *Monitor* as others struggled toward the *Rhode Island*'s rescue boat. Butts watched from the turret as the men before him made their way across the deck and into the boat. Many misjudged the rise and fall of the waves as they jumped toward the lifeboat and fell, lost into the sea. Butts then crossed the deck to help Bankhead, leaving behind him men too terrified to take their chances. These men appeared resigned, hoping they may survive with the *Monitor* if she were to last out the storm. Peter Truscott was the last to leave the turret. His friend ahead of him was washed overboard as he jumped toward the lifeboat and then disappeared beneath the waves, exclaiming, "Oh, God!" George Geer dashed across the deck and watched his shipmate, Daniel Moore, being forced into the sea and then disappearing. Geer was washed across the deck, but he waited for another wave and "this time reached the boat and was saved."[333] Bankhead and Butts were then carried into the ocean by a wave; however, each of them were pulled into the boat by Samuel Dana Greene. Greene himself had been saved in a similar fashion by Dr. Weeks, who had thrown the executive officer a line and saved his life.

As Browne ordered his loaded craft back to the *Rhode Island*, Bankhead pleaded with those who remained to come with them, but they refused. Browne promised to return for them. "As we pulled away," Peter Truscott remembered, "I saw in the darkness some black forms clinging to the top of the turret."[334]

When they reached the *Rhode Island*, the rescue boat was thrown up against the towering steamer by a wave. Dr. Grenville Weeks instinctively used his right arm in an attempt to keep the lifeboat from crashing into the larger ship. The arm was caught between the ship and the lifeboat and was dislocated, three of his fingers crushed by the collision. He never regained the use of his arm, a dreadful injury for a practicing surgeon. However, he remained stoic about his destiny. "An arm," Weeks would later write, "was a small price to pay for life."[335]

The rescued men were taken aboard the *Rhode Island* and given blankets, coffee, food and other necessary comforts. Joseph Watters remembered that he "was about played out" by the experience and that he "was taken by the Chief Engineer of the *Rhode Island* who gave me a dry set of clothes, he also gave me his room and treated me very kindly indeed."[336]

Weeks immediately received medical attention. The *Rhode Island*'s surgeon quickly amputated three of Weeks's fingers and reset his arm. Weeks then

Sinking of USS Monitor, *1863. Courtesy of John Moran Quarstein.*

rejoined his shipmates on the *Rhode Island*'s deck to watch the *Monitor*'s light shine above the raging sea. "It was half past twelve, the night of the 31st of December, 1862," remembered Francis Butts, "when I stood on the forecastle of the *Rhode Island*, watching the red and white lights that hung from the pennant staff above the turret, and which now and then were seen as we would perhaps both rise on the sea together, until at last, just as the moon had passed below the horizon, they were lost, and the *Monitor*…was seen no more."[337]

Chapter 12
Aftermath

Rodney Browne had steered the *Rhode Island*'s boat toward the *Monitor*'s lights. They had disappeared several times, hidden by the huge waves, only to reappear. Browne had gotten close to the *Monitor* during his third rescue effort, only to be forced away by the heavy seas. When the rescuers finally reached where they had thought the *Monitor* once was, the brave *Rhode Island* men found nothing but an eddy "apparently produced by a sinking vessel." The *Monitor* had sunk in 220 feet of water fifteen miles south of Cape Hatteras, with the loss of four officers and twelve men.

Unfortunately, Browne and his shipmates were, by then, over two miles from the *Rhode Island*, and they could not get back to her in the storm. Browne kept his little craft afloat until the gale lessened and, in the morning, was rescued by the schooner *A. Colby* out of Bucksport, Maine.

Aboard the *Rhode Island*, the *Monitor* survivors were thankful they had made it through the storm. Many quickly wrote letters home to their loved ones to assure them that they had not gone down with the ironclad. "I am sorry to have to write you that we have lost the *Monitor*," George Geer wrote his wife, "but do not worry. I am safe and well."[338] William Keeler wrote his wife several times about the events off Cape Hatteras:

> *The telegraph has properly informed you before this of the loss of the* Monitor *& also of my safety. My escape was a very narrow one…*
>
> *I have been through a night of horrors that would have appalled the stoutest heart.*[339]

Of course, it fell on the shoulders of Bankhead and other officers to write the relatives of those who had perished that evening. Despite his injury, Dr. Grenville Weeks wrote to the sister of crew member Jacob Nicklis:

> *I am too unwell to dictate more than a short sad answer to your note. Your brother went down with other brave souls, and only a good providence prevented my accompanying him. You have my warm sympathies, and the assurance that your brother did his duty well, and has I believe gone to a brighter world, where storms do not come.*[340]

The *Rhode Island* steamed back to Hampton Roads, where the crew disembarked. Secretary of the Navy Gideon Welles ordered on 4 January 1863 that all of the survivors who had served on the ironclad from its first muster through to its sinking were granted "a two weeks leave, with 20% of all they may have been due them," and they were given the opportunity to "return to the receiving ships nearest their residence."[341] Weather and wartime conditions limited travel, and most of the crew were initially detailed to the *Brandywine*. The *Brandywine* had been a forty-fun-gun frigate launched on 16 June 1825. The old frigate had been redesignated as a storeship for the North Atlantic Blockading Squadron, Hampton Roads Station. Conditions aboard the *Brandywine* were difficult for the former *Monitor* Boys. The men were not immediately rationed new clothing or blankets. They were cold, depressed and demoralized.

George Geer even thought about deserting while at Gosport Navy Yard and noted in a letter to his wife that he hoped he would be given liberty and a reassignment to a receiving ship in New York. He sadly commented, "I very much doubt their ever seeing any of us again if we once get outside of the Navy Yard Gate." Geer thought most of the sailors and soldiers were against the war and "hardly one but would desert the first chance."[342] Many did. Christy Price, Robert Quinn, William Remington, Henry Harrison, Anthony Connoly, William Marion, Daniel Walsh and William Scott all deserted within a year of the *Monitor*'s demise. John Rooney was assigned to the *North Carolina*; however, due to his injuries sustained during the sinking, he was sent to the Brooklyn Naval Hospital and eventually discharged on 29 May 1863. The "mulatto" William H. Nichols was also assigned to the *North Carolina*. Nichols never appeared for duty; he later claimed that he was injured and was subsequently discharged. Since there are no medical records to support this claim, Nichols probably deserted. Most of the surviving crew members were not taken off the muster roll of the *Monitor* until 31 January 1863.

Aftermath

Several of the survivors were rewarded for their service before and during the ironclad's sinking. Quartermasters Richard Anjier and Peter Williams both displayed, according to Commander Bankhead, "the highest quality of men and seamen" during the emergency. Bankhead further added that Anjier "remained at his post at the wheel when the vessel was sinking, and when told by me to get into the boat replied, 'No, sir, not until you go.'"[343] Anjier was promoted to the rank of acting master's mate and assigned to the North Atlantic Blockading Squadron. Peter Williams received the highest accolades of any of the *Monitor*'s crew. He was awarded the U.S. Medal of Honor for his service during the *Monitor*'s engagement with the *Virginia*. John Driscoll remembered that "Peter saw more of her [the *Virginia*] than anyone else. He say right into the bore of the gun…Pete says, 'Captain, that is for us,' and rip! she came."[344] Williams is often credited with steering the *Monitor* away from the *Virginia* when the Confederate ironclad attempted to ram the *Monitor*, as well as when Worden was wounded. He was promoted to acting master's mate on 28 March 1862 for his service during the battle and again, to the rank of acting ensign, on 10 January 1863. Williams was then assigned to the USS *Florida*. Ensign Williams ended the war as commander of the steamer USS *Clematis* as part of the West Gulf Blockading Squadron.

George Geer attained his dream to be promoted to third assistant engineer on 19 January 1863 and was detailed to the *Galena*. Geer would eventually be promoted to second assistant engineer on 1 June 1864 and end the war aboard the side-wheeler *Philadelphia*.

Despite their experience with the poor seagoing qualities of the *Monitor*, many men accepted reassignment to various improved monitors. Second-Class Fireman Michael Mooney, Chief Boatswain Hans Anderson, Quartermaster Peter Truscott, Gunner's Mate Joseph Crown and Second-Class Fireman William Durst, a deserter before the sinking, all later served on the *Passaic*-class monitor the USS *Catskill*. First-Class Firemen John Garety, William Richardson and Patrick Hannan served on the tower ironclad the USS *Keokuk*. The unusual, poorly armored ironclad mounted two XI-inch Dahlgrens and was struck ninety times by Confederate shot during Rear Admiral Samuel Francis DuPont's failed 7 April 1863 ironclad attack upon Charleston. The *Keokuk* foundered the next day off Morris Island, and all three *Monitor* veterans survived. Acting Ensign Robert Hubbell went on to serve aboard another ironclad, the USS *Osage*, a single-turret monitor designed by James Eads. This type of monitor had a wooden hull with a "turtleback" design and very shallow draft. The *Osage* and her sister ship, the USS *Neosho*, were the only stern-wheel monitors.

The *Monitor* Boys

Another *Monitor* Boy who served on an ironclad was First-Class Fireman John Mason. Mason was detailed to the *Passaic*-class monitor the *Patapsco*. While this monitor class had been proven virtually shot-proof, their thinly armored hulls were still vulnerable to Confederate torpedoes. On 16 January 1865, the *Patapsco*, commanded by Lieutenant Commander Stephen P. Quackenbush, struck a torpedo in Charleston Harbor and sank immediately with the loss of sixty-four officers and crew. John Mason was listed as lost at sea.

Mathew Leonard would also perish during his naval service. After the *Monitor*'s sinking, First-Class Fireman Leonard was detailed to the USS *Granite City*. The *Granite City* was formerly a blockade runner and was pressed into service with the West Gulf Blockading Squadron. This iron-hulled side-wheeler participated in the unsuccessful Union operations against Sabine Pass (8 September 1863) and Matagorda Peninsula (31 December 1863) in Texas. The *Granite City* and the USS *Wave* were captured by Confederate batteries at Calcasieu Pass, Louisiana, on 28 April 1864. Leonard became a POW and died of disease at Camp Groce, Texas, on 15 September 1864.

The remainder of the *Monitor*'s crew went on to serve aboard various warships until their discharge. Coal Heaver Charles Smith was assigned to the *Stepping Stones* until his discharge on 30 September 1863. Four other *Monitor* Boys were also attached to this former New York side-wheel

Interior of USS Montauk, *circa 1862. Courtesy of The Mariners' Museum.*

Aftermath

ferryboat. Landsmen Francis Butts and James Coleman and Coal Heavers James Smith and William Morrison also served aboard this Potomac Flotilla gunboat and participated in actions up the Nansemond River during the 11 April–4 May 1863 siege of Suffolk, Virginia. Butts was later promoted and served as paymaster's steward on the USS *Flag.* Abraham Tester, who was a first-class fireman on the *Florida*, was promoted as an acting third assistant engineer and reassigned to the USS *Montgomery.*

While the *Monitor*'s enlisted men served with various levels of success during the rest of the war, many of the former officers went on to distinguished careers. John L. Worden received a gold sword made by Charles Tiffany from the State of New York and, after recovering from his wound, was detailed as commander of the *Passaic*-class monitor the *Montauk*. Worden was assigned the task of testing this new class of monitor against fixed fortifications by Admiral Samuel DuPont. On 27 January 1863, the *Montauk* engaged Fort McAllister, located on the Ogeechee River below Savannah, Georgia. DuPont noted:

> *The monitor was struck some thirteen or fourteen times, which would have sunk a gunboat easily, but did no injury whatever to the* Montauk— *speaking well for the impenetrability of those vessels—though the distance was greater than what would constitute a fair test. But the slow firing, the inaccuracy of aim, for you can't see to aim properly from the turret…I asked myself this morning while quietly dressing, if one ironclad cannot take eight guns—how are five to take 147 guns in Charleston Harbor.*[345]

Worden steamed the *Montauk*, along with three other wooden gunboats and the mortar schooner USS *C.P. Williams*, back into the Ogeechee River. The squadron stopped within six hundred yards of Fort McAllister and, at 7:45 a.m., began to shell the fort. During the next four hours, the *Montauk* was struck forty-eight times by the guns of Fort McAllister. While the *Montauk* did not incur any significant damage, neither did the fort. The fort's earthen exterior walls absorbed much of the impact of Federal shot and shell and could easily be repaired between bombardments.

Worden returned once again to the Ogeechee River on 28 February 1863. The *Montauk*, accompanied by the wooden gunboats the USS *Wissahickon*, the *Seneca* and the *Dawn*, commenced a bombardment of Fort McAllister. The *Montauk* concentrated on the blockade runner the CSS *Rattlesnake*. The *Rattlesnake* was formerly the commerce raider CSS *Nashville*, which had a brief career in 1861, sinking two Northern merchant vessels. The *Rattlesnake* had now run aground beyond Fort McAllister, and only her superstructure was visible past the fort. The *Montauk* closed to within fifteen hundred yards of the fort and opened fire. Despite the heavy fire directed at the *Montauk* from Fort McAllister, Worden moved his ironclad three hundred yards closer. Using seven-second fuses, the *Montauk*'s gunner quickly found the range. After a twenty-minute bombardment, the *Rattlesnake* was ablaze from stem to stern. When the fires reached the Confederate commerce raider's magazine, she exploded with a tremendous roar. Worden noted that the moment was the "final disposition of a vessel which had so long been in the minds of the public as a troublesome pest."[346]

The Union cannon, however, were unable to silence Fort McAllister. Since the tide was falling, Worden headed his ironclad back down the river. Unfortunately, the *Montauk* struck a torpedo, and the explosion fractured the ironclad's iron hull. She was only saved thanks to Worden's quick thinking: he ordered the *Montauk* run aground, and mud sealed the hole. She underwent temporary repairs and then returned to Port Royal Sound for permanent repairs.

Worden commanded the *Montauk* during DuPont's ironclad attack on Charleston Harbor on 7 April 1863. Then, in May 1863, he was assigned to

John L. Worden, Commander of the Montauk, *circa 1863. Courtesy of The Mariners' Museum.*

Aftermath

the Navy Department in Washington, advising on the development of new monitors until the war's conclusion. Worden briefly commanded the USS *Idaho* and USS *Pensacola* until promoted commodore on 27 May 1868. After a one-year leave of absence in Europe, Worden was named superintendent of the U.S. Naval Academy and served in that capacity from 1869 to 1874. During his tenure, the first minority midshipman was admitted to the academy. Worden was promoted rear admiral in 1872 and served as commander in chief of the European Squadron from 1874 to 1877. He was responsible for concentrating American forces in Turkish waters after Russia declared war on that nation on 24 April 1877.

His final assignment was as a member of the Naval Retiring & Examining Board from 1877 to 1886. Rear Admiral John L. Worden retired with full sea pay and died of pneumonia on 18 October 1897. President William McKinley Jr. led the funeral procession at St. John's Episcopal Church in Washington, D.C. The *Monitor*'s first commander has had four naval vessels named in his honor, as well as the parade field at the U.S. Naval Academy and Fort Worden in Port Townsend, Washington.

Other *Monitor* commanders also achieved great distinction and rank following their service with the *Monitor*. Thomas Oliver Selfridge Jr. served as Flag Officer Goldsborough's flag lieutenant during the initial stages of the Peninsula Campaign and participated in the recapture of Norfolk on 10 May 1862. He briefly commanded the submarine the *Alligator* during her operations in the James River. Selfridge then assumed command of the *City*-class river ironclad the USS *Cairo* on 16 August 1862. With this ironclad as his flag ship, he took a flotilla of gunboats on an expedition up the Yazoo River, beginning 21 November 1862. On 12 December, the *Cairo* struck a torpedo and sank. It was the first warship to be sunk by a torpedo during the war, and Selfridge was able to save his entire crew. He then assumed command of the gunboats the USS *Conestoga* and the USS *Manitou*. During the siege of Vicksburg, he commanded a land battery using men and cannons from the *Manitou*. Following the fall of Vicksburg, he captured the steamers the CSS *Louisville* and the CSS *Elmira*. Unfortunately, the *Conestoga* was later rammed and sunk, on 8 March 1864, by the Confederate ram CSS *General Price*. Selfridge then was detailed as commander of the *Osage*, a *Neosho*-class monitor.

While in command of the *Osage*, he participated in the Red River Campaign, a series of battles fought along the Red River in Louisiana from 10 March to 22 May 1864. When the *Osage* ran aground near Blair's Landing, Louisiana, on 12 April 1864, she was attacked by Confederate dismounted cavalry and artillery commanded by Major General Thomas Green. Green, well fortified with rum, led an impetuous charge on the Union ironclad. Selfridge reported, "I waited till they got into easy shelling range, and opened upon them a heavy fire of shrapnel and cannister. The rebels fought with unusual pertinacity for over an hour, delivering the heaviest and most concentrated fire of musketry that I have ever witnessed." The devastating Union cannon fire decided "this curious affair...a fight between infantry and gunboats."[347] General Green was killed during the attack.

When Rear Admiral David Dixon Porter went east to assume command of the North Atlantic Blockading Squadron, he brought Selfridge with him and placed him in command of the gunboat USS *Huron*. Selfridge participated in the capture of Fort Fisher and other Confederate forts defending the entrance to the Cape Fear River, North Carolina, commanding the third division of the assaulting columns of sailors and Marines. The battle resulted in the capture of Wilmington, North Carolina. He was on duty on the James River in Virginia when the war ended. Fearing that Confederate President Jefferson Davis would try to flee the country, Selfridge was ordered to Key West, Florida, to keep a lookout for him.

After the war's conclusion, Selfridge was appointed an instructor in the Department of Seamanship at the U.S. Naval Academy and commander of the *Macedonian*. Following a brief tour as commander of the USS *Nipsic*, he was selected to command an expedition to survey the Isthmus of Darien (now Panama) for an inter-oceanic canal. After he took a two-year leave of absence, Selfridge was named commander of the USS *Enterprise* in the European Squadron in 1878 and selected to complete a survey of the Amazon and Madeira Rivers in South America. Accordingly, he was invited as a special delegate by Ferdinand de Lesseps to attend the May 1879 International Canal Congress in Paris, France. There, Selfridge was presented with the Legion of Honor by the French government and made an honorary member of the Royal Geographic Society of Belgium—all for his isthmus-surveying work. Selfridge returned to the United States and served as commander of the Newport, Rhode Island Torpedo Station from 1880 through 1884.

He then was given command of the USS *Omaha* in the Asiatic Squadron. Afterward, Selfridge returned to the United States and served on various boards, finally being named commander of the European Squadron, with the rank of acting rear admiral, on 12 November 1895. While in European waters, he attended the coronation of Tsar Nicholas II and received a Gold Coronation Medal in May 1896. He was later received by Pope Leo XIII in March 1897. Selfridge retired on 6 February 1898 and

Aftermath

resided in Washington, D.C., until his death on 4 February 1924. A *Porter*-class destroyer, the USS *Selfridge*, was named for both him and his father, Rear Admiral Thomas O. Selfridge Sr.

Another *Monitor* commander to have a ship named in his honor (the destroyer the USS *Jeffers*) was William Nicholson Jeffers III. Following his service on the *Monitor*, Jeffers was assigned as inspector of ordnance, Washington, D.C., and remained in this post for the rest of the war. While in Washington, Jeffers was one of three officers assigned to prepare the powder-boat the USS *Louisiana* to support Major General Benjamin Franklin Baker's attack on Fort Fisher, North Carolina. Jeffers developed the clock-fuse system used to ignite the

Thomas O. Selfridge Jr., circa 1880. *Courtesy of The Mariners' Museum.*

explosives on the *Louisiana*. However, the system did not operate properly, and the powder-boat's partial explosion failed to damage the fort.

In March 1865, Jeffers was promoted commander and given command of the *Resca*-class steam-screw wooden gunboat the USS *Swatara*. He helped prepare this vessel for service, and the *Swatara* was commissioned on 15 November 1865 and assigned to the European Squadron. One of Jeffers's major assignments while in European waters was to return Lincoln assassination conspirator John Harrison Surratt to America. Following Lincoln's assassination, Surratt had first escaped to Canada and from there to Europe, where he was finally discovered serving in a Zouave regiment in the Papal Army. Although the suspected conspirator managed to elude his original captors, he was later re-apprehended in Alexandria, Egypt. There, Jeffers placed him on the *Swatara*, "heavily chained and handcuffed" and under constant Marine guard. Jeffers delivered Surratt to the Washington Navy Yard on 18 February 1866, and the *Swatara* returned to European waters.

In 1868, Jeffers was detailed to duty at the Naval Observatory. He was then assigned to the Navy Department as an assistant to Commodore A. Ludlow Case, chief of the Bureau of Ordnance. After the war, Jeffers continued his work on ordnance, investigating the cause of rupture of 150-pounder rifles and improving the navy's gunpowder program. Jeffers was promoted captain in 1870, and the next year, he assumed command of the venerable frigate the *Constellation* after convincing the navy to make the vessel a gunnery practice ship for training midshipmen and seamen. In 1873, Jeffers was named chief of the Bureau of Ordnance, and in this position, he sought to modernize U.S. naval armament, in keeping with the European advances in naval gunnery. His most notable achievement was doubling the power of the XI-inch Dahlgren by converting it into an eight-inch rifle. Jeffers was elevated to the rank of commodore in 1878. Jeffers published several manuals, including *Inspection and Proof of Cannon* (1864), *Nautical Surveying* (1871), *Care and Preservation of Ammunition* (1874) and *Ordnance Instruction for U.S. Navy* (1866 and 1880). He died in Washington, D.C., on 23 July 1883 and was buried in the Naval Cemetery at Annapolis.

Thomas Holdup Stevens Jr. did not have a ship named in his honor; nevertheless, he did have a distinguished post-*Monitor* career. He was considered a "valuable officer" by John Rodgers and eventually took command of another monitor, the *Patapsco*. He commanded that ship during DuPont's 7 April 1863 attack on Charleston Harbor and fought at Mobile Bay on 2–5 August 1864 as commander of the USS *Winnebago*. He later assumed command of the Texas division of the Gulf Blockading Squadron, and in June 1865, he witnessed Confederate Generals Kirby Smith and John Bankhead Magruder surrendering the last Southern armies at Galveston, Texas. Admiral S.F. DuPont advised him that he was "skillful and brave in all of the duties pertaining to your command."

During the postwar era, Stevens served as lighthouse inspector and then as commander of the flagship of the European Squadron, the USS *Guerriere*, in 1870 and 1871. He was promoted commodore in 1872 and became commandant of the Norfolk Navy Yard from 1873 to 1876. Stevens then became a member of the U.S. Advisory Board to the Harbor Commissioners of Norfolk and Portsmouth, Virginia, in 1876 and was assigned special duty at Norfolk Harbor from 1878 to 1880. Promoted to rear admiral on 27 October

Aftermath

1879, he was named the next year as commander of the Pacific Squadron and served as such until 1881. His final duty took him to Guatemala, aboard the *Pensacola*, to meet with President Justo Rufino Barrios. He retired on 27 May 1881. Stevens was president of the U.S. Naval Academy Board of Visitors and worked as an author until his death on 15 May 1896. He was laid to rest at Arlington National Cemetery in Washington, D.C.

Though John Pyne Bankhead survived the *Monitor*'s sinking, he remained on sick leave, due to exposure, until he assumed command of the blockader USS *Florida* on 7 March 1863. The *Florida* was a side-wheel steamer originally built for the New York & Savannah Steam Navigation Company. The steamer was acquired in 1861 and, after service with the South Atlantic Blockading Squadron, was sent to the Brooklyn Navy Yard for a refit. She was armed with four IX-inch Dahlgrens, one one-hundred-pounder rifle, one fifty-pounder rifle and one twelve-pounder rifle. Several of Bankhead's former *Monitor* Boys were also assigned to the *Florida*, including William Keeler, Samuel Dana Greene, David Ellis, Abraham Tester, Peter Williams and Siah Carter.

Bankhead left New York on 9 March 1863, with the *Passaic*-class monitor the *Nantucket* under tow. His clerk, Charles Post, wrote: "It is strange that [Bankhead's]...next job should be to tow a monitor over the old ground. I fancy he prefers the role of tower to that of towee." The two vessels survived several storms and finally arrived at Port Royal, South Carolina, on 15 March 1863. Bankhead then steamed the *Florida* to Hampton Roads, where he met with Flag Officer Samuel P. Lee and was assigned to the Wilmington Blockade. Even though this was monotonous service, Bankhead always kept his ship prepared for the chase. Charles Post noted, "At night the ship is required to be kept in perfect silence. No lights of any kind are shown. It remains one of a great hunt, the way we lie, waiting for our game." The *Florida* discovered its game in June 1863, when the steamer was able to capture two runners, the *Calypso* and *Hattie*. The 11 June 1863 capture of the *Calypso* was an exciting three-hour chase. As the runner prepared to surrender, the *Calypso*'s crew began dumping items overboard. According to Post, Bankhead reacted swiftly, shouting, "Train the two pivots on her." He then ran out the guard and, hailing them, said: "Throw one more thing of any kind overboard and I'll send a broadside into you...if you attempt to scuttle the ship, or blow her up, you can take care of yourselves. I shan't pick up a single man."

The *Florida*'s success was envied by the other blockaders serving on the Wilmington Station. After the capture of the richly laden schooner *Hattie*, Bankhead's clerk noted:

We feel quite excited at this last capture...How sore [the other blockaders]...*will feel when they hear of it. Some of them have been down here over eighteen months and have not made a capture for the want of judgment and energy on the part of the commanding officer. They will lie at anchor day after day without attempting to give chase to strange sails that are often reported. Not so Capt. Bankhead. No sooner comes our anchor (if it is down for we lie a good deal of time floating around, not coming to anchor), off we go flying the signal, "a strange sail in sight," not waiting for orders to give chase. If Capt. B. had the control of matters on the blockade, things would be hurried up some I assure* [you]. *There would be less writing done & but little regard paid to red tape but the blockade would be far more effective. He is untiring, up day & night. When he does sleep, it is usually during the day & in clothes. No one knows when he will be on deck during the night & woe betide the one found wanting when he does come.*

William Keeler agreed with Post's point of view, noting that Bankhead "was a strict disciplinarian exacting obedience and respect from both officers and men...His officers he always expected to do their duty and while they did, were always treated with the greatest courtesy and consideration."

Unfortunately, J.P. Bankhead became seriously ill with "inflammation of the bowels" and left the *Florida*. He was eventually transported to New York City and was not able to resume active command until 3 February 1864, when he was detailed as commander of the double-ender side-wheel gunboat *Otsego*. Bankhead took his vessel to Hampton Roads and then was assigned to the North Carolina Sounds as commander of the Union naval forces there on 21 June 1864. His primary duty, along with four other gunboats, was to guard the entrance to the Roanoke River to block any effort by the ironclad CSS *Albemarle* to gain the sounds. By October, he had left the *Otsego* due to ill health.

On 27 March 1865, Commander Bankhead was assigned to command the six-gun steam sloop of war USS *Wyoming*. Bankhead was sent to the Indian Ocean in search of the Confederate commerce raider the *Shenandoah*. When the *Shenandoah* had eluded the Federals by sailing to Liverpool, Great Britain, by November 1865, the *Wyoming* was assigned to the Asiatic Squadron. Bankhead was promoted to captain on 25 July 1866, but his health quickly began to fail, and he was forced to ask to be relieved of his command. Unfortunately, on 27 April 1867, Bankhead died aboard a steamer bringing him home, and he was buried near Aden, Arabia. His clerk, Charles Post, believed that Bankhead's continuing ailment was appendicitis.

Aftermath

John Pyne Bankhead's untimely death was lamented by many of those who had served with him during the war. William Keeler noted that Bankhead "while…a model officer…never forgot he was a gentleman." Indeed, Bankhead possessed many of the social graces and "courtly manners" displayed by his cousin, Confederate General John Bankhead Magruder. "Captain Bankhead," his clerk wrote, "is a splendid fellow and charming companion, though he never forgot that he is captain."

Several of the *Monitor*'s officers did not remain in service long after serving aboard the ironclad. Albert B. Campbell resigned, due to sickness, in May 1863. Darius Farrington Gallagher served as a third assistant engineer on the *Monitor* during her engagement with the *Virginia* but transferred off the ship just before her sinking and was later dismissed from the service on 15 July 1863. He worked in the postwar era as a leading man at the Brooklyn Navy Yard until he died on 15 June 1905. Edwin V. Gager resigned his commission on 8 July 1862 and lived in Newark, New Jersey, until his death in 1914. Robert K. Hubbell was promoted to acting ensign and served under Thomas Selfridge on the *Osage* during the Red River Campaign. He was dismissed from the service on 22 April 1864; however, four days later, he volunteered to retrieve the body of a fellow crewman who had been shot while on picket duty at Deloach's Bluff, Louisiana.

The *Monitor*'s first surgeon, Daniel C. Logue, resigned from the U.S. Navy on 7 October 1862. He worked as a surgeon until his death on Long Island, New York, in 1914. Dr. Thomas Meckly had a brief stint on the *Monitor* and served the rest of the war as the *Lodona*'s surgeon. Postwar, he worked as an ophthalmic surgeon in Jersey Shore, Pennsylvania, until his death in 1890. The *Monitor*'s other surgeon, Dr. Grenville M. Weeks, recovered from his severe injury that occurred during his rescue from the floundering *Monitor*. Despite the loss of three fingers and the use of his right arm, he was able to apply his surgical knowledge as a consulting surgeon, appointed with the rank of major by President Abraham Lincoln. His assignment for the rest of the war was as acting medical director of the Union Department of Florida. After the war, he served for six years as secretary of the Indian Commission for the Pacific Northwest. He then returned to private practice in New Jersey. Weeks is noted for being the author of the "Resolution Recognizing Cuban Independence by U.S. Congress" in 1898. He died twenty-one years later in the Home for Disabled Soldiers in Kearny, New Jersey.

Samuel Howard volunteered to serve on the *Monitor* to act as the ironclad's pilot during her engagement with the *Virginia*. Howard returned to his ship, the *Amanda*, and was named acting volunteer lieutenant for his efforts on

The *Monitor* Boys

Samuel Howard, circa 1880. *Courtesy of The Mariners' Museum.*

the *Monitor*. He served aboard three other vessels—the USS *Neosho*, the *Vermont* and the *Pensacola*—for the war's duration and received an honorable discharge in November 1868. Howard served in New Orleans, Savannah and Mobile for the rest of his career as a lieutenant in the U.S. Revenue Marine. He died on 14 January 1900 and was buried at Arlington National Cemetery in Washington, D.C.

William Dunlap Park was assigned to the *Monitor* on 7 February 1862, but he was dismissed only ten days later for unknown causes. Park later served aboard the side-wheel frigate USS *Mississippi*. The *Mississippi* went aground trying to pass the Confederate batteries at Port Hudson, Louisiana, and was destroyed by shell fire. Park died three months later, in July 1863, due to his injuries. He was buried on Profit Island in the Mississippi River.

Alban C. Stimers was never officially part of the *Monitor*'s crew, but he played an important role in the ironclad's construction and engagement with the *Virginia*. After the battle, Stimers joined in collaboration with John Ericsson to build the *Passaic* and *Canonicus* classes of monitors. He was on

Aftermath

board the *Passaic* during her 5 March 1863 attack on Fort McAllister, South Carolina. He observed, and reported on, the performance of ironclads during DuPont's 7 April 1863 attack on Charleston Harbor. His derogatory remarks infuriated DuPont to the point that the admiral filed charges against the engineer. Stimers was tried, and acquitted, in May through October 1863 by a court of inquiry for spreading falsehoods and conduct unbecoming an officer. Once he was exonerated, he then began work on the *Casco* class of light-draft monitors. The project's failure resulted in an investigating committee ruling that Stimers had committed a "professional error." He next served on a *Casco*-class monitor, the USS *Tunxis*, and then the steamers the USS *Wabash* and the USS *Powhattan* until resigning on 3 August 1865. Stimers worked as a civil engineer until his death from smallpox on Staten Island, New York, on 3 June 1876.

Louis Napoleon Stodder had a very distinguished post–Civil War career. He joined the U.S. Revenue Cutter Service as a third lieutenant and was honorably discharged from the U.S. Navy shortly thereafter. Stodder was promoted to captain in 1879 and was commander of the USS *Oliver Wolcott* when he quelled an Indian uprising at Fort Simpson, British Columbia, Canada, on 11 January 1883. Stodder was named supervisor of anchorages, Port of New York, in 1892, and retained that position until 1901. He retired in 1902. He died of cerebral apoplexy and pulmonary edema, following a nervous breakdown, in 1911 and was buried in Brooklyn, New York.

William Keeler spent about a month after the sinking settling the *Monitor*'s accounts. While he "still hoped to visit Charleston in an iron-clad," Keeler was instead assigned to the *Florida* and served aboard that blockader off Wilmington, North Carolina. He was wounded in action by a shell fragment in his back during an engagement with shore batteries on 10 February 1864. Keeler was honorably discharged on 26 April 1866 and received a pension due to his wound. He returned to LaSalle, Illinois, and lived there until 1869, when he moved his family to Mayport, Florida, and built a comfortable home called Thalassa. Unlike many of the other men who served on the *Monitor*, Keeler disguised his naval service and enjoyed being called "Major" Keeler. The former paymaster received the post of deputy collector of customs, as well as election inspector and auditor for the local post office. Keeler also maintained a shop making and repairing iron items, as well as cultivating pecans and oranges. In 1884, he produced 6,664 oranges. Although he lost the sight of his right eye in 1880, he continued working and doing some writing. Besides keeping a daily journal, he wrote a weekly letter to Jacksonville's *Florida Times Union*. He provided news about activities

William Keeler, circa 1865. *Courtesy of The Mariners' Museum.*

in Mayport and signed the letters with the appellation "Silex," except for one he wrote commemorating the Battle of the Ironclads in 1885, which he closed with the signature "Monitor." Keeler died of heart disease on 27 February 1886.

Unfortunately, two of the *Monitor* officers met untimely ends by the taking of their own lives. Isaac Newton Jr. worked on several ironclad projects following his service aboard the *Monitor*. When the war was over, he became chief engineer of the Croton Aqueduct for the City of New York Public Works Department. Newton was a member of the Society of Civil Engineers and the Society of Mechanical Engineers. He became depressed over his ill health and committed suicide by cutting his own throat on 25 September 1884. Samuel Dana Greene also suffered from depression at the end of his life. While he rose to the rank of commander during the postwar era and held several notable assignments, such as the head of the Department of Navigation and Astronomy at the U.S. Naval Academy and executive officer of the Portsmouth Navy Yard, still questions lingered over his actions following Worden's wounding on 9 March 1862. Before his article, "Inside the *Monitor*'s Turret," was published, his depression became so severe that he killed himself with a gunshot to the head on 11 December 1884. Despite his sad demise, a *Clemson*-class destroyer was later named in his honor.

As the years began to slip by after the war, many of the *Monitor* Boys quietly passed away. Those who survived the war, and of whom records are available, reflect a proud series of stories. All of these men who saw service on the *Monitor* recognized that their time on that famous ironclad was the

Aftermath

highpoint of their lives. When seeking an invalid pension (the precursor to disability pension), Hans Anderson's attorney wrote: "The claimant is to my certain knowledge totally disabled from performing any work. Were it otherwise, he could and would, without doubt, be given some light work by Naval authorities, for they would not see a man who served on the *Monitor* in her fight be in want of employment, were he not totally disabled."[348]

Anderson did receive a pension; however, it was based on his injuries incurred during his service as chief boatswain on the ironclad *Catskill*, not the *Monitor*. Although Anderson was severely injured falling down a hatch on the *Catskill*, it appears that his most significant injuries occurred during the 16 January 1865 attack on Fort Fisher, North Carolina. On that date, Admiral David Dixon Porter opened the morning with a point-blank, heavy bombardment of Fort Fisher. Then, in conjunction with Major General Alfred Terry's land assault, Porter sent a landing party of one thousand sailors and four hundred Marines to attack the sea face of the fort. Porter hoped that his men would "board the fort on the run in a seaman-like way." These volunteers were commanded by Lieutenant Commander K. Randolph Breese. The sailors, only armed with Colt revolvers and cutlasses, were decimated by grapeshot, cannister and rifle fire. Thomas O. Selfridge Jr. remembered the scene of his men "packed like sheep in a pen, while the enemy were crowding the ramparts not forty yards away and shooting into them as fast as they could fire." The attack was a dismal failure. Among the three hundred dead and wounded was Hans Anderson, shot in the hip and pelvis. He was discharged ten days later. During the postwar era, Anderson served on several merchant vessels, including the *Alene*. The *Alene* was a Hamburg-American Line steamship that accidentally rammed the New York pilot boat the *James Gordon Bennett*. Hans Anderson was at the wheel of the *Alene* when the collision occurred; however, he was absolved of any wrongdoing. Anderson later worked as a night watchman and owned a "milk, bread and wood small way" on Hall Street. Like so many others, Anderson, known as "the 'shot man' on the famous *Monitor*," was called by his physician as the "last survivor of the USS *Monitor* while fighting the *Merrimac*." The doctor, who wrote a letter supporting Anderson's request for an increase in his pension, also noted that the "US will certainly erect a monument sooner or later to this great episode in US history and it is surely not the wish of the people to let him the last survivor suffer during his last days from lack of good food and reasonable comfort." Anderson died one year later. His obituary noted that he was the "last surviving member of the crew that manned the *Monitor*…He ended his days clinging to the belief that the famous naval hope of the Confederacy could have been either captured or sunk if her Union foe had followed up on the assault."³⁴⁹

Many other crew members sought invalid pensions due to injuries sustained from their service aboard the *Monitor*. Edmund Brown applied for a pension based on his having "contracted rheumatism from exposure" during the *Monitor*'s engagement with the *Virginia*. He was discharged as "unfit for service" on 24 November 1862 at the Washington Navy Yard. He worked as a machinist for some time after the war but later became a watchman due to his disability. He died at the New York State Soldier's Home in 1892.

While Isaac Scott only served aboard the *Monitor* from 7 November to 31 December 1862, he claimed two sicknesses from his time on the ironclad. Scott noted that "the trouble with my hearing came from my service on the *Monitor*." He believed the problem stemmed from gun practice in the turret. Furthermore, Scott also stated:

> *I was aboard the* Monitor *when she foundered off Hatteras. I was in the water & wet for hours before the vessel went down. We were bailing after the pumps stopped. When I got aboard the convoy* Rhode Island *I had nothing on but my shirt & pants. I caught a terrible cold & was unfit for duty all of 2 weeks…I date my asthma from that cold.*³⁵⁰

Despite these ailments, Isaac Scott became a deputy sheriff in Buffalo, New York. He also served as a coroner and then as an inspector in the U.S. Customs Service. He died in 1927.

Patrick Hannan also applied for an invalid pension because he had "contracted chronic rheumatism and general disability by long exposure to wet and cold in the water, at the time of the sinking of the USS *Monitor*." When he died on 2 March 1892, his death certificate listed, as an indirect cause of death, "heart disease and chronic rheumatism contracted during several hours' exposure after sinking of USS *Monitor* in 1862." Hannan had previously testified on behalf of fellow crew member Thomas Carroll #2. Patrick Hannan, like several other of his *Monitor* shipmates, noted that Carroll was a

> *strong robust boy well built and developed for his age…and at the time of said wreck was acting as "Coal Passer" in the hold of said vessel. When the* Monitor *was wrecked this deponent with said Carroll and others was in the same hold bailing water and they were compelled to stand in water and were completely wet. Finally all were compelled to jump into the open*

Aftermath

Patrick Hannan, circa 1880.
Courtesy of The Mariners' Museum.

sea from whence they were rescued and taken aboard the US vessel Rhode Island. *Deponent estimates the time spent in the water of the hold and open sea by himself, Carroll and others to be between 4½ and six hours.*[351]

Hannan noted that the next time he saw Carroll in 1873, "he was pale, emaciated and showed effects of the disease."

Like so many other *Monitor* Boys, Hannan made little mention of his subsequent naval service, even though he survived the sinking of the tower ironclad the *Keokuk* following DuPont's attack on Charleston Harbor. He expanded these stories about his time aboard the famous ship. Hannan claimed to have carried Worden to his room following the *Monitor* captain's wounding. This claim is made by more of the crew than the number of people needed to achieve this task. Perhaps these statements were made because of Worden's heroic attributes and the desire to have a closer connection to him. Hannan's obituary noted that he was a pallbearer during Worden's funeral, reinforcing Hannan's devotion to Worden to the very end. Although John Brown was not transferred onto the *Monitor* until 7 November 1862, he claimed to have been standing next to Worden when the captain was wounded on 9 March 1862. Brown claimed to have a small hole caused by shell fragments that never closed up. It was said that when he was smoking he would "amaze children by holding his finger against his nose and blowing a horizontal stream of smoke out the hole."[352] Brown would go to the school in Rolfe, Iowa, where he resided after the war, and tell students all about the battle between the *Monitor* and *Merrimack*, even though he had not even been there.

As the men grew older, their tall tales associated with the *Monitor* continued to grow. Wilhelm (William) Durst served on the *Monitor* until he deserted on 6 November 1862. Durst claimed that his desertion was due to an illness he had "contracted from the bad air on the *Monitor*" and he could not return to the ironclad when his furlough expired. He feared punishment if he returned to duty, Durst stated, and he instead reenlisted under the name of Walter David (Durst had a "WD" tattoo). However, in another statement, he noted that he was drinking one night and on the next day he discovered himself on board the receiving ship *North Carolina*. Unwittingly, he had reenlisted under the name Walter Davis. He then shipped aboard the ironclad *Catskill*. In his quest for a pension, he had to secure witnesses to prove that Durst and Davis were one and the same. Fortunately, several men who had shipped aboard the *Monitor* with him also served on the *Catskill*. Hans Anderson was one of many who attested to this truth. Anderson noted:

> *I did not see again until I saw him on the* Kaatskill *[sic] at the Brooklyn Navy Yard. I knew him as soon as I saw him and called him Durst. He laughed and said that that was his name once, while on the* Little Monitor *but now…they call me Walter David…I called him Durst…and he was sometimes called the Dusty Jew. He was the only Jew on board either the* Monitor *and the only Jew I ever saw on any United States vessel.*[353]

Many claimed that Durst had a "Jewish nose." This feature enabled his former shipmates to identify him; however, the unusual bend of his nose was due to it being broken, not by his being a Jew. Although Durst said he broke it aboard the *North Carolina*, his brother Louis stated that Durst's nose had been broken as a child when he fell out of a tree. The charge of desertion was dropped, as the Navy Department noted that "Durst was of foreign birth and unaccustomed to the ways of his adopted country" and that he had "in fact faithfully served in the navy during almost the entire period of the Civil War." Durst maintained great pride in his service aboard the *Monitor* and began to embellish his record as the years went by. He stated that

Aftermath

William Durst, circa 1880. *Courtesy of The Mariners' Museum.*

he was aboard the *Monitor* the night she sank. On 15 May 1885, he wrote: "I take pride in speaking…of that noble vessel the 'Monitor' although on that night when sinking little did I ever expect to see of land again, but thanks to the Lord, I was spared to return again to my home and for this my Patriotic feeling is strong within me."[354]

Again on 25 May 1885, he wrote Franklin Pierce:

> *I am pleased to know that the press recognizes the* Monitor *Crew as heroes, our heroism was well tried on the aweful* [sic] *night when our noble vessel was sinking and the seas were rolling mountain high and not a spar of wood to grasp to save a single soul, no, nothing but our iron warrior of the seas sinking beneath us, only those who were there can realize such a scent.*[355]

Although Durst was not there, he continued to state that he was, because he realized the ironclad's sinking to be an even more heroic event than the battle itself. In a letter to David R. Ellis's widow, he said that Ellis had saved him from drowning on the night of the sinking. Durst continued to promote his service on the *Monitor* by attending events, often stating that he was the last survivor of the *Monitor*. During a GAR encampment in Washington, D.C., Durst can be seen in a photograph supposedly wearing the same uniform he wore on the *Monitor*, saluting President Thomas Woodrow Wilson.

Many others claimed to have served on the *Monitor*, but there is no record of their service. Peter Brodie told stories about the ironclad to his daughter that made her feel part of history, but there is no indication that he was ever on the *Monitor*. Brodie actually was a mate aboard the *Cumberland* on 8 March 1862 when that warship was sunk by the *Virginia*. He was wounded in action on that day. Apparently because of his participation during the Battle of Hampton Roads, he sought to be connected in some way to the *Monitor*.

Thomas Taylor, formerly a slave who escaped to become a "contraband," actually enlisted on the *Roanoke* as a powder monkey. Taylor purported to have been wounded in action while serving in the *Monitor*'s turret on 9 March 1862, but no record exists to support his claim. This naval veteran served throughout the war and received an honorable discharge on 13 July 1865. During the postwar era, Taylor worked first as a mariner and then as a janitor in Putnam, Connecticut, where he died of a heart attack caused by wounds sustained during a robbery on 7 March 1932.

Other African Americans served on board the *Monitor*. Two had mustered with the original crew; however, several "contrabands" joined the ironclad during her service in Virginia and while at the Washington Navy Yard. Enlistment in the U.S. Navy was the only way an African American could serve the Union until the Militia Act of 17 July 1862. Siah Hulett Carter was the first "contraband" to join the crew. Siah, who enlisted under his master's name, Carter, but whose family name was actually Hulett, continued to serve in the U.S. Navy following the *Monitor*'s sinking. He subsequently served on the *Brandywine*, the *Florida*, the USS *Belmont*, the *Wabash* and the USS *Commodore Barney* until honorably discharged on 19 May 1865. Carter briefly lived in Hampton, Virginia, and then moved with his wife, Eliza, another former slave from Shirley Plantation, to Bermuda Hundred, Virginia, for five years. Siah Carter and his family then moved to Philadelphia, where he resided until his death in 1892.

Several of the African Americans who mustered onto the *Monitor* went down with the ironclad on 31 December 1862. Daniel Moore, Robert Howard and Robert Cook all were lost that night. Of the surviving African Americans, only Carter served throughout the war. William Nichols and William Scott deserted in 1863, and Edward Cann just disappeared from naval records.

Eventually, the *Monitor* Boys perished, and their obituaries always mentioned their heroic deeds aboard the *Monitor*. When the *Jersey City Evening*

Aftermath

Daniel Toffey, circa 1880. *Courtesy of The Mariners' Museum.*

Journal of 10 February 1893 reported Daniel Toffey's passing, the newspaper contained information about his successful business career and his service as a member of the Jersey City Board of Aldermen, but the article's main thrust was Toffey's service on the *Monitor*:

> *The dead man was a nephew of the great Admiral John Worden, USN, who won honor and distinction during the war of the rebellion as commander of the* Monitor...*Mr. Toffey was an officer on board the* Monitor *during the memorable conflict and caught his uncle when the Lieutenant was thrown to the deck and severely injured by the concussion caused by the explosion of a shell against the pilothouse of the little war vessel.*

While not connected in any fashion with the *Monitor*'s engagement with the *Merrimack*, Francis Butts was able to tell his stories about the Union ironclad when he published an article in the *Battles and Leaders of the Civil War*. Butts only briefly served on the *Monitor*; however, he collected everything he could that related to the ship and heard every story there was to hear from those who had served during the 9 March battle. Butts died 8 September 1905 and

Francis Butts, circa 1890.
Courtesy of The Mariners' Museum.

shared all of his knowledge about the battle as best he could until his death. While some of his stories seem inflated, Butts was typical of many of the *Monitor* Boys, who embellished their experiences aboard the famous ironclad.

When several men died, they were often referred to in newsprint as "the last *Monitor* survivor." When Peter Omer died on 23 January 1897, he was called the oldest surviving member of the *Monitor*'s crew.

Thomas B. Viall received a pension based on the claim that he, on 14 June 1862, had contracted "chills & fever & dysentery which has resulted in kidney disease & rheumatism from which he had first began to suffer about July 1889." He was discharged at Harrison's Landing, Virginia, on 19 July 1862. Viall later served on the USS *Onward* and was eventually appointed acting master's mate on 8 February 1865. Viall died on 11 November 1905 and was also called "the last survivor of the crew of the Yankee 'cheesebox,' the first *Monitor*."

Even though Anton Basting deserted from the *Monitor* on 6 November 1862, fifty years later, he was fêted as one of the *Monitor* Boys. On Decoration Day 1913, Basting was the guest of honor of the Cleveland, Ohio Epworth League, and later, he was the guest of honor at the unveiling

Aftermath

of the Lafayette Park Tablet in Jersey City, New Jersey, on July 4, 1914. Basting died on 10 February 1915.

Actually, the last documented crew member to pass away was Isaac H. Scott. Scott was eighty-nine years old when he died at his home in Buffalo, New York, on 17 May 1927, truly the last of the *Monitor* Boys.

The men who served aboard the *Monitor* helped that warship forever change naval warfare. While her career was ever so brief, the *Monitor*'s legacy was far-reaching. The *Monitor*'s tactical victory on 9 March 1862 produced an intense appreciation in the North for Ericsson's design. Less than three weeks after the battle, Ericsson and his partners received a contract to build "six boats on the plan of the *Monitor*." Other builders constructed these warships because the demand was so great. Eventually, the Union would produce over sixty ironclads as part of the war effort. Many other types of "monitors" would be created, featuring twin turrets like the USS *Onondaga* to enhance firepower or even oceangoing versions like the *Miantonomoh* class. A few designers modified the concept and built variations of low-freeboard, turreted, armored warships. Even the steam-screw wooden frigate the *Roanoke*, originally a sister ship of the *Merrimack*, was converted into an ironclad. The hull was plated, and three turrets were installed on her deck. She was top-heavy and not a success.

The revolutionary *Monitor* made the entire world recognize that ironclads were the key to naval supremacy. While she spawned a new era in warship construction, the design had numerous flaws. Monitors were so unseaworthy that they had to be towed from port to port and were often in danger of sinking during a storm. This circumstance meant that they could not effectively serve as blockaders, but they did prove valuable on inland waterways. Although virtually shot-proof, monitors lacked sufficient armament (two guns per turret) and had a slow rate of fire. Despite all of these problems, the *Monitor* established a vision for future navies to follow. The revolving, armored turret, with its concentration of guns able to fire in any direction, would dominate naval shipbuilding for the next seventy-five years. Many warships still feature iron construction, low profile, speed and maneuverability. From her conception and construction to her combat and conclusion, the *Monitor* defined the modern warship.

One of the *Monitor*'s commanders, Thomas O. Selfridge Jr., truly witnessed and understood naval warfare's evolution in 1862. Selfridge was aboard the *Cumberland* when she was rammed, and sunk, by the *Virginia* on 8 March. The *Cumberland*'s shot and shell could not stop the Confederate ironclad's ability to prove the power of iron over wood. From shore, he watched the next day's

Admiral John L. Worden, USN, circa 1890. *Courtesy of The Mariners' Museum.*

epic battle between the *Monitor* and the *Virginia*. He, like so many others, then believed that only an ironclad could stop another ironclad. While that was the initial wisdom, Selfridge soon learned otherwise. Selfridge eventually commanded the *Alligator*. This hand-propelled submarine, featuring two spar torpedoes, served in the James River during the summer of 1862 to operate about the Confederate ironclad *Richmond*, then being completed at the Navy Yard in Richmond. The *Alligator* proved to be a failure, and Selfridge then was transferred to command the *City*-class ironclad the *Cairo*. He was aboard that ship when she struck a torpedo in the Yazoo River and sank. He then recognized that the Civil War was the beginning of a major evolution in naval warfare. Once one technology was proven superior, a new one would quickly be created as a counter. And like Selfridge, John Worden understood that the *Monitor* was only a beginning. His service aboard monitors witnessed him almost sinking twice, while simultaneously becoming the toast of a thankful nation. Worden and all of the *Monitor* Boys proved that they were iron men aboard an iron ship. Together they changed history.

Appendix I
USS *Monitor* Chronology

1803

31 July
USS *Monitor* designer John Ericsson was born in Langbanshyttan, in the province of Varmland, Sweden.

1814

25 October
Acting Volunteer Lieutenant William P. Flye was born in Newcastle, Maine.

1817

30 October
Landsman James T. Slover was born in St. Michaels, Maryland.

1819

28 May
Commander Thomas H. Stevens Jr. was born in Middletown, Connecticut.

Appendix I

1821

12 March
Acting Master and Pilot Samuel Howard was born near Dublin, Ireland.

9 June
Acting Assistant Paymaster William F. Keeler was born in Utica, New York.

3 August
Commander John P. Bankhead was born at Fort Johnston in Charleston Harbor, South Carolina.

1823

Gunner's Mate Joseph Crown was born in New York City. First-Class Fireman Hugh Fisher was born in Ireland.

1824

Seaman Henry Harrison was born in Sweden. Officer's Cook Robert Howard was born a slave in Howard County, Virginia.

13 July
Quartermaster Hans A. Anderson was born in Gothenberg, Sweden.

16 October
Lieutenant William N. Jeffers III was born in Swedesboro, New Jersey.

1825

Quartermaster Richard Anjier was born in England. Quartermaster Moses Stearns was born in New Hampshire.

1826

Seaman John Atkins was born in Baltimore, Maryland. First-Class Fireman Patrick Hannan and Quartermaster William Marion were born in Ireland.

1827

First-Class Fireman Mathew Leonard was born in Ireland.

5 June
Chief Engineer Alban C. Stimers was born in Smithfield, New York.

1828

Seaman Thomas Loughran and Wardroom Steward Lawrence Murray were born in New York City, New York.

1829

Coal Heaver James Smith was born in Haverstraw, New York.

3 August
Acting Master John J.N. Webber was born in Brooklyn, New York.

1830

Yeoman William Bryan was born in New York City. Second-Class Fireman Christy Price was born in Ireland. Seaman Francis Riddey was born in Philadelphia, Pennsylvania.

15 March
Acting Ensign Norman K. Atwater was born in New Haven, Connecticut.

18 August
Boatswain's Mate John Stocking was born in Binghamton, New York.

Appendix I

1831

Quartermaster Peter Williams was born in Norway.

1832

Quarter Gunner John P. Conklin was born in Watertown, New York. Master at Arms John Rooney was born in Brooklyn, New York.

27 September
Acting Assistant Surgeon Daniel C. Logue was born in Otisville, New York.

1833

Captain of the Hold Thomas Carroll #1 was born in Boston, Massachusetts.

14 October
Acting Master Edwin Gager was born in Dutchess County, New York.

1834

Carpenter's Mate Derick Bringman was born in Bremen, Germany. Acting Master's Mate George Frederickson was born on the island of Møn, Denmark. Landsman James Malone was born in Ireland. Coal Heaver Charles Smith was born in Rome, New York.

2 November
Quartermaster Charles Peterson was born in Christinia, Norway.

1835

Seaman Anthony Connoly and Coxswain Daniel Walsh were born in Ireland. Third Assistant Engineer Robinson Hands was born in Baltimore, Maryland.

2 March
Seaman Anton Basting was born in Germany.

1836

Francis Petit Smith and John Ericsson, working independently, took out patents for screw propellers.

First-Class Fireman John Driscoll was born in County Cork, Ireland. First-Class Fireman William Richardson was born in Philadelphia, Pennsylvania.

6 February
Lieutenant Thomas O. Selfridge Jr. was born in Charlestown, Massachusetts.

6 May
Coal Heaver William Durst was born in Tarnow, Austria.

17 May
First-Class Fireman George Geer was born in Troy, New York.

1837

First-Class Fireman John Garety was born in Ireland. Officer's Steward William Jeffrey was born in Philadelphia, Pennsylvania. Coal Heaver George Littlefield was born in Saco, Maine. Wardroom Cook Edward Moore was born in Scotland. Second Assistant Engineer Joseph Watters was born in Bordentown, New Jersey.

12 February
Acting Master Louis N. Stodder was born in Boston, Massachusetts.

April
Ericsson's screw propeller was fitted on the forty-foot launch *Francis B. Ogden*, which towed a barge containing several senior Royal Navy officials down the Thames at ten knots.

Appendix I

4 August
First Assistant Engineer Isaac Newton Jr. was born in New York City, New York.

25 September
Second-Class Fireman Michael Mooney was born in Ireland.

16 October
Ordinary Seaman Isaac Scott was born in Quebec, Canada.

22 November
Assistant Surgeon Grenville M. Weeks was born in New York City, New York.

22 December
Captain's Clerk Daniel Toffey was born in Pawling, New York.

1838

Landsman William Allen was born in England. Seaman James Fenwick was born in Scotland. Seaman John Hardy was born in Portland, Maine. Coal Heaver James Seery was born in Ireland.

2 January
Seaman Thomas Viall was born in Bristol, Rhode Island.

1839

First-Class Fireman Thomas Joice and Quartermaster Peter Truscott were born in Ireland. First-Class Fireman Abraham Tester was born in England.

4 October
First-Class Boy Siah Carter was born a slave on Shirley Plantation, Charles City County, Virginia.

USS *Monitor* Chronology

1840

Second-Class Fireman Robert Quinn was born in Ireland.

11 February
Lieutenant Samuel Dana Greene was born in Cumberland, Maryland.

28 March
Captain's Steward David Cuddeback was born in Port Jervis, New York.

27 August
Acting Assistant Surgeon Thomas W. Meckly was born in Milton, Pennsylvania.

12 November
Coal Heaver David R. Ellis was born in Carmarthen, Wales.

1841

Swedish engineer John Ericsson arrived in the United States. He brought with him a 12-inch wrought-iron gun manufactured at the Mersey Iron Works near Liverpool named the "Orator." Renamed "Oregon," the gun had 3.5-inch wrought-iron bands shrunk on the breech, and its shot could penetrate 4.5 inches of wrought iron.

Ordinary Seaman John Brown was born in Germany. Landsmen James Coleman and William Egan were born in Ireland. Coal Heaver Thomas Feeney was born in Brooklyn, New York. Landsman John Monaghan was born in New York City, New York. First-Class Fireman John Mason was born in Providence, Rhode Island. Ordinary Seaman Jacob Nicklis was born in Buffalo, New York. Seaman Charles Sylvester was born in Sweden.

1842

Coal Heaver William Morrison was born in England.

APPENDIX I

1843

First-Class Boy Thomas Carroll #2 was born in Ireland. Officer's Steward William Nichols was born in Brooklyn, New York.

9 SEPTEMBER
USS *Princeton* commissioned, the first steam screw warship constructed for the U.S. Navy. Captain Robert S. Stockton supervised the sloop's construction. John Ericsson designed two vibrating-lever engines, three tubular iron boilers and a six-bladed screw propeller (fourteen feet in diameter). The *Princeton* was the first screw propeller warship in any navy, the first warship with machinery entirely below the waterline, the first to burn anthracite coal and the first to use fan blowers for her furnace fires.

1844

First-Class Boy Robert Cook was born a slave in Gloucester County, Virginia. Third Assistant Engineer Mark Sunstrom was born in Baltimore, Maryland.

27 JANUARY
Landsman Francis Butts was born in Providence, Rhode Island.

29 FEBRUARY
During a pleasure cruise on the Potomac River attended by over four hundred people—including President John Tyler Jr., his cabinet, congressmen and their families—the twelve-inch shell gun named the Peacemaker exploded, killing eight attendees, including Secretary of State Abel Upshur and Secretary of the Navy Thomas Gilmer. Robert Stockton and several others were severely injured. This event would cause the U.S. Congress to limit funds for ordnance and ship design development while simultaneously souring the relationship between the U.S. Navy and John Ericsson.

7 JUNE
Coal Heaver William Remington was born in Syracuse, New York.

USS *Monitor* Chronology

1847

AUGUST
Thomas L. Taylor was born a slave at Cole's Point, North Carolina.

1854

SEPTEMBER
John Ericsson sent plans for a turreted, iron-cased ship to Napoleon III, Emperor of France.

1859

MARCH
Captain Cowper Coles patented the idea of turrets aboard ships. His concept of centerline turrets gave the guns a wide arc of fire.

NOVEMBER
The French navy launched the *La Gloire* as an armored frigate. The *La Gloire* was protected by a 4.5-inch belt of iron supported by 17 inches of wood. The ironclad mounted fourteen 8.8-inch and 6.4-inch rifled, breech-loading guns.

1861

6 APRIL
Lieutenant John Lorimer Worden, a twenty-eight-year U.S. Navy officer who served at the Naval Observatory as a midshipman and then saw extensive sea duty in the Pacific Squadron, departed Washington, D.C., en route to Pensacola, Florida, with orders for the USS *Sabine* to reinforce Fort Pickens on Santa Rosa Island.

13 APRIL
Lieutenant John L. Worden was seized near Montgomery, Alabama, following his mission to Fort Pickens. He was the U.S. Navy's first prisoner of war in the Civil War.

Appendix I

3 August

U.S. Congress appropriated $1.5 million for ironclad ship construction. Secretary of the Navy Gideon Welles was authorized to "appoint a board of three skillful officers to investigate the plans and specifications that may be submitted for the construction or completing of iron or steel-clad steamships or steam batteries." Flag Officer Joseph Smith, Flag Officer Hiram Paulding and Captain Charles Henry Davis were named to the Ironclad Board.

7 August

Advertisements were placed in Northern newspapers soliciting seagoing ironclad designs.

16 August

President Abraham Lincoln proclaimed that the inhabitants of the Confederate States were in a state of insurrection and outlawed all trade with them.

29 August

John Ericsson, in a letter to President Abraham Lincoln, offered to construct a "vessel for the destruction of the rebel fleet at Norfolk."

8 September

Cornelius S. Bushnell accidentally met Cornelius Delameter of Delameter Ironworks on the steps of the Willard Hotel in Washington, D.C. Delameter advised Bushnell to meet with Swedish American inventor John Ericsson to find answers to questions about the stability of the USS *Galena*, an ironclad designed by Naval Constructor Samuel Pook. This ironclad concept featured 3½-inch plating in a tumble dome design to help deflect shot. The form of rows of interlocking iron plates did not have wood backing, and the deck was not plated. This caused the *Galena* to be vulnerable to plunging fire. The *Galena*'s armament consisted of two one-hundred-pounder Parrott rifles and four IX-inch Dahlgrens.

9 September

C.S. Bushnell gave plans of the *Galena* to John Ericsson.

10 September

C.S. Bushnell met with John Ericsson and reviewed plans for his ironclad battery.

USS *Monitor* Chronology

11 September
C.S. Bushnell presented a pasteboard model of Ericsson's "battery" to the navy's Ironclad Board and President Lincoln. After reviewing the model, the Ironclad Board invited Ericsson to come to Washington to answer additional questions about his design.

C.S. Bushnell met with John A. Griswold and John F. Winslow to solicit support of Secretary of State William Seward and Secretary of the Navy Gideon Welles for the battery's design.

13 September
President Abraham Lincoln and Cornelius Bushnell met with the Ironclad Board. The three officers (Smith, Paulding and Davis) were "all…surprised at the novelty" of Ericsson's design. When the meeting concluded, Lincoln looked at the model and remarked, "All I have to say is what the girl said when she stuck her foot into the stocking: 'It strikes me there's something in it.'"

C.S. Bushnell met with John Ericsson in New York and persuaded him to go to Washington, D.C., to discuss his ironclad plans.

14 September
C.S. Bushnell and John Ericsson met with the Ironclad Board. Despite an initial rejection, Ericsson persuaded the board otherwise. At 3:00 p.m., the *Ericsson's Battery* concept was accepted. After meeting with the board, Ericsson was instructed by Secretary of the Navy Gideon Welles to "go ahead at once!" Ericsson had one hundred days to complete the vessel.

16 September
The Ironclad Board submitted a report that noted, "For river and harbor service we consider ironclad vessels of light draught, or floating batteries, thus shielded, as very important." The Ironclad Board reviewed the plans and proposals for seventeen ships, of which three were recommended for construction: *Galena*, *Ericsson's Battery* (*Monitor*) and *New Ironsides*.

22 September
John Ericsson received letter of confirmation that his "ironclad gunboat" design had been accepted by the Ironclad Board.

Appendix I

27 September
Cornelius Bushnell, John A. Griswold, John Winslow and John Ericsson formed a partnership to build *Ericsson's Battery*. John Ericsson completed all blueprints for the ironclad's construction. Although the government contract had not yet been drawn up, Continental Iron Works in Brooklyn's Greenpoint section began rolling the keel plates.

October
Ericsson began seeking qualified foundries and ironworks to manufacture materials needed to construct his battery.

4 October
Government contract issued to John Ericsson and partners for the construction of *Ericsson's Battery* was signed. The total sum for the battery was $275,000, to be paid in five installments of $50,000 and one of $25,000. The Navy Department stipulated that 25 percent of each payment would be held back to ensure the timely completion of the vessel.

Chief Engineer Alban Crocker Stimers, a twelve-year naval veteran and the last chief engineer of the USS *Merrimack*, was named the U.S. Navy's superintendent for the *Ericsson's Battery* project. He was responsible for overseeing construction, keeping an eye on the navy's interests and approving the release of payments as sufficient work was completed.

7 October
Contract for the construction of the ship house at Greenpoint Ship Yard, Long Island, New York, was issued to Allan Deckerman.

8 October
H. Abbot and Son of Baltimore, Maryland, was contracted to produce all of the one-inch iron plate for the *Ericsson's Battery* project.

19 October
Worthington Pumps purchased for *Ericsson's Battery*.

23 October
Construction of *Ericsson's Battery*'s rudder was begun.

25 October
Thomas Fitch Rowland of Continental Iron Works signed a contract with John Ericsson to construct the single-turreted ironclad at .075 cents per pound.

Ericsson's Battery's keel was laid at Continental Iron Works.

November
The primary work for iron plate, castings, fittings, etc., was contracted out to three New York mills. Holdane & Co. contracted for 125 tons of plate, and Albany Ironworks and Rensselaer Iron Works manufactured hundreds of additional tons of plate and castings. H. Abbott & Sons of Baltimore rolled the one-inch-thick iron plates for the turret to be shipped to Novelty Iron Works in New York for assembly into Ericsson's "shot proof" tower. Delameter Ironworks and the Clute Brothers Foundry cast and assembled most of the components of the ship's machinery. Throughout the month, work on the iron battery continued to proceed at a "feverish pace." Materials from the foundries were assembled as soon as they arrived at Continental Iron Works. A ship house 180 feet long was constructed over the ways to allow work to continue in bad weather and throughout the night.

11 November
Thomas Rowland of Continental Iron Works advised John Ericsson that he had 175 men at work on the ironclad project.

15 November
Chief Engineer Alban C. Stimers authorized the first payment of $50,000 for work on the *Ericsson's Battery* project.

16 November
Iron deck beams were installed on *Ericsson's Battery*.

29 November
Lieutenant John L. Worden reported to Washington, D.C., following his seven-month imprisonment in a Montgomery, Alabama jail as a POW.

4 December
Ericsson's Battery investors received second payment of $37,500.

Appendix I

Acting Master's Mate George Frederickson was assigned to the *Ericsson's Battery*.

5 December
Wooden bulwark installed on the *Ericsson's Battery*.

Ericsson received a letter from Commodore Joseph Smith informing him that he had been made aware of serious delays in material production. "I beg of you to push up the work. I shall demand heavy forfeiture for delay over the stipulated time of completion. You have only thirty-nine days left."

17 December
Boilers and auxiliary machinery arrived at Continental Iron Works.

31 December
Ericsson's Battery had its boilers installed and hull painted. Chief Engineer Alban Stimers tested the ironclad's engines and propellers.

1862

4 January
Acting Assistant Paymaster William F. Keeler was assigned to the *Ericsson's Battery*.

12 January
The one-hundred-day construction timetable for *Ericsson's Battery* expired.

13 January
Lieutenant John L. Worden detailed as commander of *Ericsson's Battery*. He was placed in charge of mounting the ship's ordnance and assembling a crew.

16 January
Lieutenant John L. Worden, still weak and unwell from his time in a Southern prison, arrived in New York and assumed command of the U.S. steamer built by Captain Ericsson. Even though Worden recognized the ironclad as an experiment, he was "induced to believe that she may prove a success. At all events, I am quite willing to be an agent in testing her capabilities."

USS *Monitor* Chronology

20 January
Ericsson's Battery was officially named the USS *Monitor*. Assistant Secretary of the Navy Gustavus Vasa Fox had asked Ericsson for a suggestion, to which the inventor had replied:

> *In accordance with your request, I now submit for your approbation a name for the floating battery at Greenpoint. The impregnable and aggressive name of this structure will admonish the leaders of the Southern rebellion that the batteries on the banks of their rivers will no longer present barriers to the entrance of Union forces. The ironclad intruder will thus prove a severe monitor to those leaders. But there are other leaders who will also be startled and admonished by the booming of the guns from the impregnable iron turret. "Downing Street" will hardly view with indifference this last Yankee notion, this monitor. To the Lords of the Admiralty the new craft will be a monitor...On these and many similar grounds I propose to name the new battery* Monitor.

24 January
Lieutenant Samuel Dana Greene, less than three years out of the Naval Academy and who as an acting midshipman had cruised the seas of the Far East until the Civil War broke out, was assigned to the *Monitor* as executive officer. He was responsible for assigning crew to their watches and quarters and, as gunnery officer, training the crew on the two Dahlgren guns in the turret.

28 January
Lieutenant Samuel Dana Greene detailed to the *Monitor* as executive officer.

30 January
John Ericsson informed that the *Monitor*'s two XI-inch Dahlgren guns would be transferred from the USS *Dacotah*.

10:00 a.m.
USS *Monitor* launched at Greenpoint, Continental Iron Works. Assistant Secretary of the Navy G.V. Fox telegraphed Ericsson, "I congratulate you and trust that she will be a success. Ready her for sea, as the *Merrimack* is nearly ready at Norfolk and we wish to send her there."

4:30 p.m.
All of the *Monitor*'s turret armor was placed aboard the ironclad.

Appendix I

31 January
Acting Master Louis N. Stodder was assigned to the *Monitor*.

7 February
Second Assistant Engineer Albert Campbell, Third Assistant Engineer Robinson Hands, Acting Assistant Surgeon Daniel C. Logue, First Assistant Engineer Isaac Newton, Third Assistant Engineer William Park and Third Assistant Engineer Mark Sunstrom were assigned to the *Monitor*.

14 February
USS *Galena* launched at the Maxson Ship Yard in Mystic, Connecticut.

17 February
Third Assistant Engineer William Park was dismissed.

19 February
Monitor is turned over to the U.S. Navy for testing. She experienced steering problems during her trial run in New York Harbor. Chief Engineer Alban Stimers reported that the steering required significant repairs and that the ship only made three and a half knots.

20 February
Lieutenant John L. Worden received orders from Secretary of the Navy Gideon Welles to: "Proceed with the USS *Monitor*, under your command, to Hampton Roads, Virginia."

By this date, First-Class Fireman George Geer had been assigned to the *Monitor*.

Coal Heaver Thomas Feeney deserted.

24 February
Captain of the Hold Thomas Carroll #1 was assigned to the *Monitor*.

25 February
The USS *Monitor* was commissioned into the U.S. Navy as a third-rate steamer and transferred to Brooklyn Navy Yard.

Coal Heaver William Durst, First-Class Fireman Patrick Hannan and Seaman Thomas Viall were assigned to the *Monitor*.

USS *Monitor* Chronology

26 February
First-Class Fireman John Garety and Quartermaster Moses Stearns were assigned to the *Monitor*.

27 February
Monitor developed a steering malfunction due to an unbalanced rudder. Ericsson and Alban Stimers corrected the problem by installing a series of pulleys between the tiller and the steering wheel drum.

3 March
Flag Officer F.H. Gregory, Chief Engineer B.F. Garvin and Naval Constructor E. Hartt were aboard the *Monitor*'s trial to test her seaworthiness. Acting Master Louis Stodder reported in the *Monitor*'s log: "2:15 with 30 lbs. steam making 50 Revolutions turned with helm hard a starboard turned in 4 min 15 sec within a compass of 3 times her length & proceeded towards the yard against a strong ebb tide vessel going at the maximum speed of 6 & ¼ knots an hour."

4 March
Monitor successfully completed her sea trials and ordnance tests. Master's Mate George Frederickson reported in the *Monitor*'s log about the trials:

> *First of firing blank cartridges, second a stand of grape, third with canister with a full charge of powder 2:15 with 30 pounds steam making 50 revolutions turned with helm hard a starboard turned in four minutes and fifteen seconds within a compass of three times her length and proceeded towards the yard against strong ebb tide vessel going at the maximum speed of G one quarter knots an hour. Greatest number of revolutions attained was 64.*

Acting Master John Joshua Nathaniel Webber had been assigned to the ship by this date. However, the *Monitor* lost two crew members, as George Frederickson also noted in the *Monitor*'s log: "Norman McPherson and John Atkins deserted taking the ship's cutter and left for ports unknown so ends this day."

5 March
As the *Monitor* prepared to leave New York Harbor another steering malfunction delayed departure.

Appendix I

By this date, Ship's Steward Robert Hubbell and Ship's Cook Henry Sinclair had been assigned to the *Monitor*.

First-Class Fireman Hugh Fisher and Ship's Cook Henry Sinclair deserted.

6 March
USS *Monitor* was ready to leave New York for Hampton Roads.

By this date, Gunner's Mate Joseph Crown; Captain's Steward David Cuddeback; First-Class Firemen John Driscoll, Thomas Joice, Mathew Leonard, John Mason and William Richardson; Second-Class Fireman Robert Quinn; Seamen Thomas Loughran and Charles Sylvester; Quartermasters William Marion and Peter Williams; Wardroom Cook Edward Moore; Wardroom Steward Lawrence Murray; Landsman William Nichols; Coal Heavers Christy Price and James Seery; Master at Arms John Rooney; Boatswain's Mate John Stocking; and Coxswain Daniel Walsh had been assigned to the *Monitor*.

Quartermasters Hans A. Anderson, Richard Anjier and Peter Truscott; Seamen Anton Basting, Anthony Connoly and James Fenwick; Yeoman William Bryan; First-Class Boy Thomas Carroll #2; Carpenter's Mate Derick Bringman; Quarter Gunner John P. Conklin; Coal Heaver David Ellis; and First-Class Fireman Abraham Tester were assigned to the *Monitor*.

Quartermaster Anderson was the enlisted member who would be in charge of the watch-to-watch navigation during the forthcoming trip.

11:00 a.m.
Monitor taken under tow by the USS *Seth Low*.

4:00 p.m.
Monitor and *Seth Low* joined gunboats USS *Currituck* and *Sachem*. This flotilla rounded Sandy Hook en route to Hampton Roads just as Secretary of the Navy Gideon Welles changed the *Monitor*'s orders and directed the ironclad to steam to Washington, D.C. Since the *Monitor* was out of communication range when the message arrived, the orders were transmitted from New York to Hampton Roads, where they awaited her arrival.

USS *Monitor* Chronology

7 March
Monitor was struck by a heavy gale along the New Jersey coast. Commander Worden and several crewmen were seasick enough to have to take to the top of the turret for relief. The blowers providing air to the vessel failed, and the engine room crew was overcome by gas fumes. Executive Officer S.D. Greene led a rescue team to evacuate the engine room. He stated later in the day "that there was imminent danger that the ship would founder." When Greene hailed and ordered the *Seth Low* to tow the *Monitor* toward calmer waters, the ironclad was saved. However, George Geer remembered the trip, recalling, "I was not a bit sea sick and stood the trip well. Our only difficulty was that the water washed into us and kept us all soaking wet and did not give us any chance to sleep."

8 March
3:00 a.m.
Monitor struck by the final force of the heavy gale that had previously threatened the ironclad. Waves were once again crashing over the *Monitor*'s deck. Executive Officer Greene "began to think that the *Monitor* would never see daylight" as the wheel rope jumped off the steering wheel and the ironclad began to sheer, stressing its towline with the *Seth Low*.

6:00 a.m.
Monitor was able to ride out the heavy seas and continued her course toward Hampton Roads.

3:00 p.m.
Monitor, towed by the *Seth Low*, passed Cape Henry and entered the Chesapeake Bay.

4:00 p.m.
Monitor neared Cape Henry Lighthouse. Acting Master Louis Stodder recorded in the *Monitor*'s logbook that he "heard heavy firing in the distance."

9:00 p.m.
The *Monitor* entered Hampton Roads. Worden met Captain John Marston of the USS *Roanoke*, acting commander of Union naval forces in Hampton Roads, who rescinded the orders he had received from Secretary of the Navy Gideon Welles to immediately send the *Monitor* to Washington, D.C. He recognized that the best way to stop any Confederate ironclad assault

Appendix I

against Washington was to defend the wooden frigates in Hampton Roads. Marston ordered Worden to station the *Monitor* near the *Minnesota* and to protect that warship from the *Virginia*.

10:00 p.m.
Lieutenant John L. Worden wrote his wife, "The *Merrimac* has caused sad work amongst our vessels. She can't hurt us."

11:00 p.m.
The *Monitor* anchored next to the *Minnesota*. Worden accepted the volunteer services of Acting Master Samuel Howard to pilot the *Monitor* through the shallows of Hampton Roads. First-Class Fireman Edmund Brown also arrived on the *Monitor*.

As the *Congress* continued to send an eerie glow across Hampton Roads, Catesby ap Roger Jones, executive officer and de facto commander of the CSS *Virginia* since the wounding the previous day of Flag Officer Franklin Buchanan, was informed by one of the pilots that he "chanced to be looking in the direction of the *Congress* when there passed a strange-looking craft, brought out in bold relief by the brilliant light of the burning ship, which at once he proclaimed to be the *Ericsson*."

11:05 p.m.
Worden and Greene went on board the *Minnesota* and met with Captain G.J.H. Van Brunt, who doubted that the ironclad could aid the *Minnesota*. Paymaster Keeler recalled, "The idea of assistance or protection being offered to the huge thing by the little pygmy at her side seemed absolutely ridiculous."

9 March
12:00 a.m.
Executive Officer Greene boarded the *Minnesota* and returned to the *Monitor* about 12:30 a.m.

1:00 a.m.
The USS *Congress* exploded. "Certainly a grander sight was never seen, but it went right to the marrow of our bones."

USS *Monitor* Chronology

2:00–3:00 a.m.
Captain Van Brunt attempted to float the USS *Minnesota* at high tide, but the frigate remained stuck in the mud. During the maneuver, Worden and Greene kept the crew mustered in case the *Monitor* needed to move out of the *Minnesota*'s way.

5:30 a.m.
According to a crewman, the CSS *Virginia* "began the day with two jiggers of whiskey and a hearty breakfast."

The *Monitor*'s officers and some of the crew remained on alert, unsure when the *Virginia* would reappear. Other crew members tried to catch some sleep, the first they had had in forty-eight hours.

6:00 a.m.
Virginia slipped her mooring at Sewell's Point but could not steam into Hampton Roads due to heavy fog.

7:00 a.m.
Worden ordered breakfast for the crew. In Washington, President Lincoln met with Captain John Dahlgren and lamented the news of the Confederate ironclad's stunning victory. A cabinet meeting was held later in the morning. Gideon Welles observed the "frantic" Secretary of War Edwin Stanton. Stanton believed that "the *Merrimac*…would destroy every city on the coast under contribution, could take Fortress Monroe; McClellan's mistaken purpose to advance must be abandoned."

7:20 a.m.
The crew of the *Monitor*'s breakfast of canned roast beef, a navy biscuit (hard tack) and coffee was interrupted by a call to quarters—the *Virginia* had entered Hampton Roads and moved toward the *Minnesota*.

8:00 a.m.
At a range of one thousand yards, the *Virginia* fired her forward seven-inch Brooke gun at the *Minnesota*, her shells "exploding on the inside of the ship, causing considerable destruction and setting the ship on fire." It appeared that the *Virginia* would make short work of the *Minnesota*. The *Monitor* then moved from alongside of the *Minnesota* and blocked the *Virginia*'s approach to the stranded frigate. A crewman of the CSS *Patrick Henry* noted, "Such a

Appendix I

craft as the eyes of seamen never looked upon before—an immense shingle floating in the water, with a gigantic cheesebox rising from its center."

8:20 a.m.
The *Monitor* opened fire on the *Virginia*.

8:30 a.m.
Onboard the *Monitor*, the speaking tube meant to allow communication between the ship's captain in the pilothouse and the gunnery officer in the turret was not working, so Paymaster William Keeler and Captain's Clerk Daniel Toffey relayed messages between Lieutenants Worden and Greene.

8:40 a.m.
The *Monitor*'s gunners discovered that the turret itself turned freely but the reversing wheel used to rotate it had rusted from its drenching in seawater on the stormy trip south. Acting Master Louis Stodder was originally stationed at the reversing wheel, but he couldn't turn it, so Chief Engineer Alban Stimers, onboard as an inspector, took over. The turret proved difficult to start revolving and, once moving, difficult to stop quickly. Executive Officer Greene, firing each shot himself, decided to aim and fire on the fly, not stopping the turning turret but slowly and constantly revolving until the *Virginia* entered his sights. Greene and his gun crews managed to fire a gun every six to eight minutes.

8:45 a.m.
When the first shots hit the *Monitor*'s turret, the gunners expressed concern at the dents made in the armor. Chief Engineer Stimers answered, "Of course it made a big dent—that is just what we expected, but what do you care about that as long as it keeps out the shot?"

Monitor and *Virginia* began circling each other in concentric circles, testing their opponent's armor. The battle was primarily fought at a range of fewer than one hundred yards. Often the ships almost touched each other as each ironclad endeavored to gain an advantage.

10:00 a.m.
When a shot from the *Virginia* hit the *Monitor*'s turret while Chief Engineer Stimers, Acting Master Stodder and Quartermaster Peter Truscott were leaning against the wall, all three were thrown to the floor. Stimers recovered

quickly, but Stodder and Truscott were knocked unconscious and taken to the ship's doctor for treatment when the turret was stilled during reloading.

10:05 a.m.
Monitor broke off action and steamed onto a shoal to reload ammunition in the turret. Worden hoped that by firing his heavy shot, 187-pound spherical projectiles using 15 pounds of powder from his XI-inch Dahlgrens, such pounding would loosen or break the *Virginia*'s iron plates.

Worden also ordered the spirit locker opened and liquid fortification distributed to the crewmen.

10:10 a.m.
Lieutenant Catesby Jones, commander of the *Virginia*, realized that he had the wrong ammunition with which to fight another ironclad. This was a tremendous disadvantage. The *Virginia* had only explosive shells, hot shot and cannister to use against wooden ships. Her chief engineer, Ashton Ramsay, wrote, "If we had known we were to meet her [the *Monitor*], we would have at least been supplied with solid shot for our rifled guns." Therefore, while the *Monitor* replenished its ammunition, the *Virginia* steamed toward the *Minnesota*.

10:15 a.m.
Virginia began shelling the *Minnesota*, but due to leaking from the loss of her ram, she ran aground.

10:30 a.m.
Monitor began bombarding the *Virginia* with shot testing "every chink in [her] armor."

11:15 a.m.
Chief Engineer Ashton Ramsay of the *Virginia* coaxed enough power out of the old engines to free the ironclad from the shoal. Lieutenant Jones decided to ram the *Monitor* and maneuvered the *Virginia* into position.

11:45 a.m.
Monitor eluded ramming and was only hit with a glancing blow. This action caused no damage to the *Monitor*; however, the *Virginia* developed a new leak at her bow. The *Monitor* also fired both of her XI-inch Dahlgrens at the

Appendix I

Virginia when she was rammed. The shot struck just above the stern pivot gun port, which forced the *Virginia*'s iron shield in three inches.

11:50 a.m.
When the *Monitor* avoided the *Virginia*'s attempt to ram her, the Union ironclad moved away from the action. The *Virginia* steamed toward the *Minnesota*. The *Minnesota* and the tug *Dragon* were shelled. Several shells struck the *Minnesota* and ignited a fire. One shell struck the *Dragon*. The *Dragon*'s boiler burst, and the tug, which had been alongside the *Minnesota* to tow that vessel to safety, sunk.

12:10 p.m.
Monitor attempted to ram the *Virginia*. Lieutenant Worden decided to strike the larger ironclad's vulnerable propeller and rudder. A steering malfunction caused the *Monitor* to miss the fantail of the *Virginia*; as the *Monitor* passed the stern of the *Virginia*, a shell from the seven-inch Brooke rifle commanded by Lieutenant John Taylor Wood struck the *Monitor*'s pilothouse. Worden, peering out of the observation slit at that moment, was wounded in the face and eyes, and the *Monitor* broke off action.

12:25 p.m.
Catesby Jones believed that the "*Monitor* has given up the fight and ran into shoal water." As Jones considered how to strike at the *Minnesota*, Worden was taken to his cabin. Lieutenant Greene conferred with the seriously wounded Worden. Worden advised, "Gentlemen, I leave it with you, do what you think best. I cannot see, but do not mind me. Save the *Minnesota* if you can."

12:30 p.m.
The *Virginia* could not get closer than one mile to the *Minnesota*. The pilots warned Lieutenant Jones that the tide was falling fast. Jones concurred with the ironclad's officers and decided to return to the Elizabeth River.

12:40 p.m.
Lieutenant Greene assumed command of the *Monitor* and brought her back into action. Greene mistook the *Virginia*'s course toward Sewell's Point as a sign of defeat and proclaimed, "We had evidently finished the *Merrimac*." The *Monitor* did not pursue the *Virginia* and steamed to a defensive position near the *Minnesota*.

USS *Monitor* Chronology

The crew, who had been confined in the ship for over half a day and some for nearly twenty-four hours, was allowed up on deck for fresh air. Visiting inspector Stimers assessed the damage to the ship: twenty-one dents with the only damage to the pilothouse.

1:30 p.m.
Assistant Secretary of the Navy G.V. Fox, who observed the entire engagement from the *Minnesota*, was received on board the USS *Monitor*. Fox told the *Monitor*'s officers, "Well, gentlemen, you don't look as though you just went through one of the greatest naval conflicts on record."

Major General George Brinton McClellan, planning an attack on the Confederate capital of Richmond via the Virginia Peninsula, telegraphed Fox and asked, "Can I rely on the *Monitor* to keep the *Merrimack* in check, so that I can make Fort Monroe a base of operations?" Fox replied, "The *Monitor* is more than a match for the *Merrimack*, but she might be disabled in the next encounter…The *Monitor* may, and I think will, destroy the *Merrimack* in the next fight; but this is hope not certainty."

2:00 p.m.
Lieutenant John L. Worden was taken to Fort Monroe for medical attention.

By this date, Hospital Steward Jesse Jones had been assigned to the *Monitor*.

3:00 p.m.
Chief Engineer Alban Stimers telegraphed John Ericsson: "You have saved this place to the nation by furnishing us with the means to whip an ironclad frigate that was, until our arrival, having it all her way with our most powerful vessels."

Captain John A.B. Dahlgren noted when he learned about the *Monitor–Virginia* engagement, "Now comes the reign of iron—and cased sloops are to take the place of wooden ships."

10:00 p.m.
Acting Volunteer Lieutenant William Flye reported on board the *Monitor* to take over as executive officer, relieving Lieutenant Greene of those duties.

Appendix I

10 March

Lieutenant John Mercer Brooke began producing wrought-iron, steel-tipped armored piercing "bolts" for 6.4- and 7-inch Brooke rifles. John Brooke was "one of the few men in the Confederate Navy who showed genius during the Civil War." He was the son of Brigadier General George Mercer Brooke and graduated from the U.S. Naval Academy at Annapolis in 1847. Brooke was assigned to oceanographic survey work under the tutelage of Mathew Fontaine Maury, during which time he invented a deep-sea sounding apparatus for topographical mapping of the ocean floor. His fathometer won the Gold Medal of Science from the Academy of Berlin. He was promoted to lieutenant in 1855 but resigned his commission when Virginia left the Union. Brooke was commissioned lieutenant in the CS Navy on 2 May 1861. His most notable work involved developing Confederate ironclad casemates sloped to effectively deflect shot and the concept of submerged ends to enhance seaworthiness. Brooke's greatest achievement during the war was the invention of a rifled gun for use by the Confederate Navy. The Brooke gun, primarily produced in 6.4- and 7-inch models, became the standard rifled gun for sea coast fortifications as well as aboard warships. While the Brooke gun looked like a Parrott rifle, the method of reinforcement of the breech was different. Whereas an iron band was welded to the tube of a Parrott in one piece, Brooke heated several heavy wrought-iron rings. When expanded, they were placed tightly around the breech. Once the iron cooled, it contracted over the breech and forged a very strong band. The reinforced breech, created by the shrinkage hoops around the barrel, provided the metal with resilience to the internal force of exploding powder. Brooke's rifle was one of the most successful weapons produced during the Civil War.

8:00 a.m.

USS *Monitor* steamed through the Union fleet in a "victory procession." Lieutenant Greene described the scene: "Cheer after cheer went up from the frigates and small craft for the glorious little *Monitor* and happy, indeed, we did all feel. I was the captain of the vessel that had saved Newport News, Hampton Roads, Fortress Monroe, (as General Wool himself said) and perhaps your Northern ports."

10:00 a.m.

Major General John Wool visited the *Monitor*, advising the ship's officers of the *Virginia*'s capabilities.

USS *Monitor* Chronology

11:00 a.m.
President Abraham Lincoln visited the wounded Lieutenant Worden at the Washington, D.C. home of Lieutenant Henry A. Wise. Lincoln told Worden, "You have done me more honor, sir, than I can ever do to you."

12:00 p.m.
Gideon Welles sent a telegram stating, "It is directed by the President that the *Monitor* be not too much exposed, and that in no event shall any attempt be made to proceed with her unattended to Norfolk."

6:00 p.m.
Lieutenant Thomas Oliver Selfridge Jr., with a naval career of over eight years and fresh from the sinking of the USS *Cumberland* two days previously during the battle against the CSS *Virginia*, was named commander of the USS *Monitor*. In his short command, he ordered that a new pilothouse be built to replace the one destroyed in battle.

Acting Master Edwin V. Gager also reported for duty.

12 March
Lieutenant William Nicholson Jeffers, a twenty-one-year naval officer and gunnery expert who had published a manual on the subject, was named commander of the USS *Monitor*. After a little over two months in command, he wrote an analysis of the ship, especially noting faults in the area of gunnery and proposing many well-argued, factual improvements to the ship, which were nonetheless ignored by the Navy Department.

14 March
The *Monitor* received one hundred hand grenades to help the crew repel boarders.

The final payment for the *Monitor*—$68,750, the 25 percent held in reserve—was made by the U.S. Navy.

16 March
The *Monitor* received additional ammunition: ten solid and seven hollow shot.

Appendix I

17 March
Major General George B. McClellan initiated the transfer of the Army of the Potomac to Fort Monroe. Eventually, 389 vessels delivered 121,500 men, 14,592 animals, 1,244 vehicles, 44 artillery batteries "and the enormous quantity of equipage…required for an army of such magnitude."

Chief Engineer Alban Stimers completed repairs to the *Monitor*'s pilothouse. Stimers placed a shell of solid oak covered with three inches of wrought iron, laid in three layers around the structure. The pilothouse's sides were reconfigured from perpendicular to a slope of thirty degrees to deflect shot.

18 March
George Geer wrote his wife: "We are laying on top of an Oyster Bed and our boys are very much provoked that they have no Oyster Rake but they will contrive some way to get one."

21 March
The tugboat the *Young America* placed an anchor buoy for the *Monitor* off the Hampton Bar in five fathoms of water.

26 March
During a heavy snow squall, the *Monitor* received fifteen tons of coal.

28 March
Major General John Ellis Wool visited the *Monitor*.

31 March
The *Monitor* Gunners Department received seven wrought-iron shot, as well as Ross sponges and rammers.

Vice President Hannibal Hamlin and a large party of gentlemen and ladies toured the *Monitor*.

3 April
Major General George McClellan came aboard the *Monitor* to confer about the *Virginia*.

6 April
The *Monitor* received nineteen tons of coal.

USS *Monitor* Chronology

8 April

Assistant Secretary of the Navy Gustavus Vasa Fox toured the *Monitor* with Senator Grime and Congressman Sedgwick.

11 April

7:10 a.m.
Virginia entered Hampton Roads. The Federal transports fled the harbor to the protection of Fort Monroe. The *Monitor*, reinforced by the iron-hulled USRMS *Naugatuck* (*Stevens Battery*) armed with a 100-pounder Parrott rifle, stayed in the channel between Fort Monroe and the Rip Raps. The Union ironclad had strict orders not to engage the *Virginia* unless the Confederate ironclad moved into the open waters of the Chesapeake Bay. Flag Officer Josiah Tattnall, commander of Confederate naval forces in Virginia waters, refused to take his ironclad out of Hampton Roads, and the *Monitor* would not accept the *Virginia*'s challenge.

15 April

William Keeler reflected on the *Monitor*'s inactivity: "Lunch at twelve, of whiskey and crackers of which I don't partake, but am sorry to say all the rest do. Dinner at two which we draw out as long as possible, after which we loaf around deck (those who have cigars smoke them) and wish for the *Merrimac*."

17 April

Chief Engineer Alban Stimers and Captain's Clerk Daniel Toffey left the *Monitor*.

27 April

The *Galena* arrived in Hampton Roads.

4 May

Acting Master Louis Stodder observed signal lights from Craney Island.

6–7 May

President Lincoln reviewed the naval forces in Hampton Roads: the USS *Minnesota*, *Vanderbilt*, *Monitor* and USRMS *Naugatuck* (*Stevens Battery*).

1:30 p.m.
William Flye observed the *Merrimac* steaming off Craney Island.

Appendix I

8 May

Monitor and the iron-hulled *Naugatuck* (*Stevens Battery*), supported by the USS *Susquehanna*, *San Jacinto*, *Dacotah* and *Seminole*, began shelling the batteries on Sewell's Point. The *Virginia* steamed down the Elizabeth River from Gosport Navy Yard to contest the Union advance. While it appeared a second conflict between the two ironclads might occur, Flag Officer Louis M. Goldsborough, commanding the Atlantic Blockading Squadron, ordered the Federal squadron to withdraw to its anchorage beyond Fort Monroe. The *Virginia* stayed out in Hampton Roads for several hours hoping to engage the *Monitor*. When no action ensued, Flag Officer Tattnall ordered the *Virginia* into the Elizabeth River.

9 May

Monitor received onboard President Lincoln, who met with Lieutenant Jeffers and inspected the ship.

11 May

4:58 a.m.

The *Virginia*, scuttled and set afire by her crew after the Confederate evacuation of Norfolk trapped her within Federal reach, exploded. Louis Stodder, aboard the *Monitor*, described it: "Nearly everybody that could came on deck to watch her. Immediately after we had caught sight of her it seemed as if the whole vessel was suddenly lifted from the water and, bursting apart, scattered in every direction."

11 a.m.

Monitor, followed by the steamer USS *Baltimore* carrying Squadron Commander Goldsborough and President Lincoln, visited abandoned batteries along the Elizabeth River, the city of Norfolk and the Gosport Naval Shipyard.

Landsman James Slover was assigned to the *Monitor*.

12–14 May

Commander John Rodgers's James River Flotilla, which consisted of the *Galena*, *Monitor*, *Aroostook*, *Port Royal* and *Naugatuck* (*Stevens Battery*), left on a mission to shell Richmond, Virginia.

USS *Monitor* Chronology

15 May
6:30 a.m.
Rodgers's flotilla left its anchorage near the mouth of Kingsland Creek and steamed toward Drewry's Bluff, also called Fort Darling.

7:45 a.m.
With the *Galena* in the lead, Rodgers placed his flotilla within six hundred yards of the bluff. The river was very narrow at this point and also blocked by obstructions placed by the Confederates only a few days before. Rodgers swung the *Galena*'s broadside toward the Confederate batteries. Confederate Charles H. Hasker was amazed by how Rodgers placed the *Galena* into action with such "neatness and precision." Hasker called the maneuver "one of the most masterly pieces of seamanship of the whole war." The *Galena* received two hits while completing the maneuver and quickly became the primary target of the Confederate batteries. The Confederates encountered significant problems. The ten-inch Columbiad, loaded with a double charge of powder, recoiled off its platform when the first shot was fired. The mud and log casemate protecting the seven-inch Brooke gun collapsed from the first shot's vibrations.

9:00 a.m.
Confederate plunging shot had begun to take effect upon the *Galena*. Lieutenant William Jeffers moved the *Monitor* virtually abreast of the *Galena* in an effort to draw some of the Confederate shot away from the larger ironclad. The *Monitor*'s turret, however, did not permit the ironclad to elevate her two XI-inch Dahlgrens sufficiently to hit the Confederate batteries. The *Monitor* was struck three times by Confederate shot before she backed downstream.

10:00 a.m.
The *Naugatuck*'s Parrott rifle burst and forced the vessel out of action.

11:15 a.m.
An eight-inch shell crashed through the *Galena*'s bow gun port and exploded. The shell ignited a cartridge then being handled by a powder monkey, killing three men and wounding several others. The explosion sent smoke billowing out of the ironclad's gun ports.

Appendix I

11:30 a.m.
The *Galena* slipped her cables and retreated downriver. Confederate gunners on the bluff gave her three hearty cheers when the *Galena* broke off action. As the *Monitor* passed Confederate sharpshooters along the riverbank, Lieutenant John Taylor Wood, former gunnery officer from the CSS *Virginia*, hailed the *Monitor*'s pilothouse and shouted, "Tell Captain Jeffers that is not the way to Richmond."

11:45 a.m.
The Battle of Drewry's Bluff (Fort Darling) was over. The *Galena* had suffered significant damage. She was hit forty-three times, and thirteen shots had penetrated the iron plate. Her railings were shot away, the smokestack riddled and she suffered twenty-four casualties. The fight demonstrated that the *Galena* was not shot-proof. The *Monitor*'s limited gun elevation and firepower made the ironclad ineffective during the battle. The U.S. Navy had reached to within eight miles of the Confederate capital. Commander John Rodgers noted that if he had had infantry support, then Richmond would have fallen.

16 May
Acting Master E.V. Gager reported that several crew members reported sick due to drinking river water and inhaling foul air in the ship.

17 May
Federal naval forces, including the ironclads *Galena* and *Monitor*, silenced Confederate lower James River fortifications.

19 May
Monitor, anchored at City Point, Virginia, sent a boat ashore in conjunction with a humanitarian mission set up by Commander William Smith, head of the City Point Squadron. Smith's boat crew was captured by Confederate infantry, but the *Monitor*'s crewmen escaped capture and returned to their ship. The *Monitor* shelled a warehouse and a private home. Siah Carter, a runaway slave, enlisted on the *Monitor* as a first-class boy.

24 May
George Geer of the *Monitor* reflected about some of his "old salt" shipmates to his wife: "I wish you could see the body's [*sic*] of some of these old sailors; they are regular Picture Books. They have India Ink pricked all over their

body. One has a snake coiled around his leg, some have splendid done pieces of Coats of Arms of states, American Flags, and most of all have the Crusifiction [*sic*] of Christ on some part of their Body."

11 JUNE
The schooner *Mary Haley* arrived and commenced coaling the *Monitor*.

12 JUNE
William Flye reported that at 1:00 p.m., the thermometer stood at 140 degrees inside the galley.

14 JUNE
Thermometer in the galley with both blowers running stood at 132 degrees and 128 degrees, but as soon as the blower stopped, it rose to 150 degrees.

18 JUNE
While George Geer often complained about his food, he advised his wife, "Our boat went onshore with three officers and five men on what they call Foraging Expedition, but what I call Thieving Expedition, and killed five very hadson [*sic*] Pigs, and we had Rost [*sic*] Pig for Dinner and have some cooking for Dinner today."

23 JUNE
1:30 a.m.
Crew members discovered a fire around the stovepipe of the galley. The fire was extinguished, but the galley was unusable for several weeks thereafter.

26 JUNE
Monitor and the City Point flotilla participated in a covert mission to blow up a railroad bridge over Swift Creek, which ran into the Appomattox River below Petersburg. The mission was not successful.

28 JUNE
While Acting Master's Mate George Frederickson noted that the weather was pleasant, he recorded the following temperatures at noon: Berth Deck—89°, Turret—91°, Deck (shade)—86°, Galley—121°, Engine Room—124° and Front of Blower—111°.

Appendix I

1 July
In the last engagement of the Seven Days' Battles, General McClellan's army fought at Malvern Hill, Virginia, supported by artillery support provided by gunboats including the *Galena*, *Mahaska* and *Monitor*. The *Monitor* fired off a few shots but was unable to elevate her muzzles enough to be of help to the retreating Union army and returned to City Point.

4 July
The *Monitor* and *Maratanza*, captained by Commander Thomas H. Stevens, captured the CSS *Teaser*, along with a reconnaissance balloon, ultra-secret papers concerning torpedoes in the James River and detailed information about the ironclad CSS *Richmond*.

8 July
Acting Master Edwin Gager resigned.

9 July
While stationed off Harrison's Landing, Virginia, *Monitor* was host to President Lincoln, Assistant Secretary of War Franklin Blair and Squadron Commander Louis Goldsborough.

13 July
Commodore Charles Wilkes, noted Antarctic explorer and the near instigator of war with Great Britain in the Trent Affair, assumed command of the flotilla from Goldsborough. He inspected the fleet and found the *Monitor* "defective."

14 July
Congress passed an act that ended grog rations in the U.S. Navy. According to George Geer of the *Monitor*, "The Grog is whiskey and they give a Gil cup twice each day, and it is equal to a good stiff horn each time."

18 July
Louis Stodder reported that the *Monitor* received the following stores from the USS *Satellite*: "Engineers' Dept.: 12 tin lanterns, 8 belt punches, 1 box Potash, 1 bag of waste, 2 half barrels oatmeal and 6 sides ___ leather. Ship stores: 1 keg butter, 1 barrel beef, 1 barrel pork, 5 half barrels flour, 1 keg pickles, 1 keg dried apples, 4 boxes p'd beef, 1 box coffee, 1 barrel rice, 9 barrels bread, and 5 boxes soap. The ironclad also raised a cooking stove and utensils."

USS *Monitor* Chronology

19 July
Seaman Thomas Viall was discharged, as his enlistment had expired.

30 July
George W. Burrows came on board *Monitor* for temporary service as landsman from *Port Royal*.

31 July
Seamen John Hardy and Henry Harrison joined the crew of the *Monitor*.

7 August
William Keeler complained about the uncomfortable and unhealthy conditions aboard monitors. On July 14 he had written, "If this sweltering weather continues it will curtail letter writing as you will find, for our state rooms are nearly as bad as the black hole of Calcutta and I find a letter is written at the cost of a large amount of perspiration." On this date, Keeler reinforced his complaints and noted, "Hot, hotter, hottest—could stand it no longer, so last night I wrapped my blanket round me and took to our iron deck—if the bed was not soft it was not so insufferable hot as my pen…what with heat, mosquitoes and a gouty Captain I have nearly gone distracted."

9 August
Commander Thomas Holdup Stevens Jr., a naval officer for twenty-five years who spent much of his service assigned to the Pacific Squadron and had distinguished himself during the Mexican-American War, was named captain of the USS *Monitor*. He was in command for less than two months, and during that time, the *Monitor* saw almost no action.

18 August
Acting Third Assistant Engineer George White was assigned to the *Monitor*.

19 August
General McClellan was the last Union soldier to leave Harrison's Landing. The majority of the supporting flotilla steamed downriver, leaving only ten ships, the *Monitor* among them. The U.S. Army's Peninsula Campaign was over.

20 August
Monitor's acting chief engineer, Isaac Newton, was transferred to New York to oversee construction of the new ironclads.

Appendix I

24 August
Monitor's black hull was repainted "battleship gray."

28 August
Monitor fired the final shots of the Navy's Peninsula Campaign, bombarding a plantation on the tip of City Point when pickets fired on the Federal flotilla as it withdrew.

29 August
Monitor was left to guard the mouth of the James River while Commodore Wilkes's squadron set sail for the Potomac River. The crew of the *Monitor* received shore leave.

30 August
Monitor returned to Hampton Roads, taking anchorage off Newport News Point between the wrecks of the *Cumberland* and the *Congress*.

31 August
Seamen John Hardy and Henry Harrison were assigned to the *Monitor*.

2 September
Wardroom Steward Lawrence Murray drowned off Newport News, Virginia.

10 September
Commander John Pyne Bankhead, with twenty-four years of service in the U.S. Navy, serving in the Mexican-American War and most recently on blockade duty at Charleston, South Carolina, was assigned command of the USS *Monitor*.

15 September
Commander John P. Bankhead reported aboard the *Monitor*, replacing Commander Stevens.

Throughout September, the *Monitor* remained on blockade duty in Hampton Roads.

30 September
The *Monitor* was ordered to Washington Navy Yard for repairs.

USS *Monitor* Chronology

1 October
By this date, Officer's Steward William Jeffrey had been assigned to the *Monitor*.

3 October
9:00 a.m.
The *Monitor* arrived at the Washington Navy Yard.

4 October
The *Monitor* became a tourist attraction while in the yard. Paymaster William Keeler wrote, "Our decks were covered and our ward room filled with ladies…There appeared to be a general turn out of the sex in the city, there were women with children and women without children, and women—hem—expecting, an extensive display of lower extremities was made going up and down our steep ladders."

Acting Master Louis Stodder said that after these visits, the visitors had taken souvenirs. He wrote, "When we came up to clean that night there was not a key, doorknob, escutcheon—there wasn't a thing that hadn't been carried away."

5 October
The *Monitor* was turned over to the yard superintendent, and repairs were immediately begun. Commander John P. Bankhead authorized leave of two weeks to thirty days for the crew. While on furlough, Seaman James Fenwick was married to Mary Ann Duffy on October 10. When he drowned with the *Monitor* in December, she was carrying his child.

While at Washington Navy Yard, the *Monitor* had her bottom cleared of marine growth, iron patches were placed over dents caused by Confederate shot, new davits and cranes were installed to hold cutters and telescoping smokestacks were added.

7 October
Acting Assistant Surgeon Daniel C. Logue resigned.

30 October
Third Assistant Engineer Samuel A. Lewis and Acting Assistant Surgeon Thomas W. Meckly were assigned to the *Monitor*.

Appendix I

Acting Volunteer Lieutenant William Flye transferred to command the USS *Underwriter*.

2 November
Landsman Francis Butts was assigned to the *Monitor*.

Monitor crew member George Geer wrote his wife about the repairs to the Union ironclad: "They are fixing the *Monitor* up much bettor [*sic*] than she was before. They will make a perfect little palace of her. The workmen work nights and Sundays. I can hear them hammering away as I am writing. They have named her guns, 'Worden and Ericsson,' and have had the names engraved on them in very large letters."

3 November
Seaman Jacob Nicklis wrote his father about the *Monitor* when he arrived at the Washington Navy Yard: "The *Monitor* lies in the Yard at present for repairs and she will probably take some of us, as her crew ran away when they landed in the yard on account of her not being sea worthy. But since then they have altered her so I think there will be no danger."

6 November
Hospital Steward William Halman was assigned to the *Monitor*. Seaman Anton Basting, Quarter Gunner John P. Conklin, Coal Heaver William Durst and Seaman Thomas Loughran deserted.

7 November
Landsmen William Allen, James Coleman, James Malone and John Monaghan; Coal Heavers George Littlefield, William Morrison, William Remington, Charles Smith and James Smith; Ordinary Seamen John Brown, Jacob Nicklis and Isaac Scott; First-Class Boy Siah Carter; and Second-Class Fireman Michael Mooney were assigned to the *Monitor*.

By this date, First-Class Boy Robert Cook and Landsman William Egan had been assigned to the *Monitor*.

Carpenter's Mate Derick Bringman deserted.

8 November
Assistant Surgeon Grenville M. Weeks was assigned to the *Monitor*.

12:00 p.m.
The *Monitor* left Washington Navy Yard.

9 NOVEMBER
Acting Ensign Norman K. Atwater was assigned to the *Monitor*.

10 NOVEMBER
12:00 p.m.
The *Monitor* entered Hampton Roads and anchored at Newport News.

13 NOVEMBER
First-Class Boy Edward Cann and Landsman Daniel Moore joined the *Monitor*.

15 NOVEMBER
Officer's Cook Robert Howard joined the *Monitor*.

16 NOVEMBER
Monitor crew member Jacob Nicklis wrote his father, "I do not like the boat because of her accommodations."

17 NOVEMBER
Acting Assistant Paymaster William Keeler recorded the *Monitor*'s repairs:

> *Our vessel has undergone a variety of changes. A large telescopic smoke pipe capable of being run up some thirty feet, takes the place of the two low square box like things you see in the photograph. The fresh air funnels have been replaced by two much higher. Our old boats were all left behind and we were furnished with others better adapted to our wants and large iron cranes and davits to raise them out of the water and carry them on, instead of dragging them up to our decks to be in the way, or dragging them in the water after us. The ragged shot marks in our sides have been covered with iron patches and the places marked, "Merrimac," "Merrimac's Prow," "Minnesota," and "Fort Darling," to indicate the source from whence the blow was received. New awnings have been furnished, ventilators for our deck lights and many other little conveniences which would have added greatly to our comfort last summer could we have had them then.*

Appendix I

28 November
Coal Heaver James Seery was discharged.

30 November
The *Passaic* arrived, joining the *Monitor*, *New Ironsides* and *Galena* and making Hampton Roads temporarily home to the most powerful ironclads afloat in the world.

8 December
President Abraham Lincoln sent a recommendation of thanks to the Congress on behalf of Commander J.L. Worden in honor of his service as commanding officer of the USS *Monitor* during the 9 March 1862 Battle of Hampton Roads.

23 December
All hands were mustered onboard the *Monitor*, and an order from the War Department was read forbidding any army or navy personnel from communicating any news about military operations.

25 December
Monitor's officers and crew celebrated Christmas Day in Hampton Roads. The officers enjoyed a three-hour dinner. William Keeler noted, "In fact we arrived at the conclusion that the Star Spangled Banner next to us 'ironclads' was about the 'biggest thing' to be found just now outside Barnum's Museum."

Monitor was ordered to proceed to Beaufort, North Carolina, to help blockade Wilmington, North Carolina, in the Cape Fear River. This news was not well received by the crew members who had experienced the ironclad's voyage from New York to Hampton Roads in March. Lieutenant S.D. Greene warned, "I do not consider this steamer a seagoing vessel. She has not the steam power to go against a head wind or sea, and…would not steer even in smooth weather, and going slow she does not mind her helm readily."

28 December
Monitor readied for her sea voyage. "The turret and sight holes were caulked," remembered Dr. Grenville Weeks, "and every possible entrance for water made secure, only the smallest opening being left in the turret top."

USS *Monitor* Chronology

Acting Chief Engineer Albert Campbell injured his leg in the piston linkage of a pump he was inspecting. Although his leg was not broken, he was put ashore in a hospital, possibly saving his life since he would not be aboard the *Monitor* when she foundered. Second Assistant Engineer Joseph Watters came aboard to replace him.

Seaman Jacob Nicklis wrote his father about the *Monitor*'s upcoming voyage to Beaufort, North Carolina: "They say we will have a pretty rough time going around Hatteras, but I hope that it will not be the case."

29 December
2:30 p.m.
The *Monitor*, with the weather "clear and pleasant, and every prospect of its continuation," left Hampton Roads en route to Beaufort, North Carolina. The powerful, 236-foot-long side-wheeler, the USS *Rhode Island*, was detailed to tow the ironclad. The *Monitor* was secured to the *Rhode Island* by two towlines. The *Rhode Island* and *Monitor* were to be accompanied on their voyage south by the USS *State of Georgia* towing the improved monitor USS *Passaic*.

6:00 p.m.
The *Monitor* passed Cape Henry, Virginia.

30 December
5:00 a.m.
Commander John Bankhead reported:

> We began to experience a swell from the southward with a slight increase of wind from the southwest, the sea breaking over the pilothouse forward and striking the base of the tower, but not with sufficient force to break over it. Found that the packing of oakum under and around the base of the tower had loosened somewhat from the working of the tower as the vessel pitched and rolled. Speed at this time was about five knots; ascertaining from the engineer of the watch that the bilge pumps kept her perfectly free, occasionally sucking. Felt no apprehension at the time.

6:40 a.m.
The *Monitor* made signal to the *Rhode Island* to stop. This action enabled adjustments to be made to the hawser connecting the two ships.

Appendix I

7:30 a.m.
The *Rhode Island* and the *Monitor* proceeded on their course.

12:00 p.m.
Winds increased during the *Monitor*'s voyage south, and waves began crashing over the ironclad's pilothouse. George Geer remembered that "soon the sea commenced to break over us and wash up against the tower with a fearful rush, and the sea was white with foam, but I was satesfide [*sic*] she would stand it out unless the storm should increase."

1:00 p.m.
The *Rhode Island* reported that she arrived at Cape Hatteras Lighthouse.

3:00 p.m.
Winds increased to gale force, and the sea had grown rough. Bilge pumps started to remove small amounts of water coming into the ironclad.

4:00 p.m.
George Geer noted that the *Monitor* was "in sight of the Hatras [*sic*] lighthouse and I thought as soon as we got past the cape it would clear up."

5:00 p.m.
Dinner was served to officers and crew. The *Monitor* had weathered well the rising swells. Commander Bankhead sent the *Rhode Island* a message indicating that if the *Monitor* needed help during the storm, he would signal with a red light.

6:00 p.m.
Commander Bankhead recorded, "Toward evening the swell somewhat decreased, the bilge pumps being found amply sufficient to keep her clear of the water that penetrated through the sight houses of the pilot house, hawse hole, and base of the tower (all of which had been well caulked previous to leaving)."

7:30 p.m.
The storm increased in its ferocity. Waves dashed across the deck and broke against the turret and pilothouse with violent force.

Commander Bankhead recalled that he "found the vessel towed badly, yawing very much, and with the increased motion making somewhat more water around the base of the tower. Ordered engineer to put on the Worthington pump and bilge injection and get the centrifugal pump ready and report to me immediately if he perceived any increase of water."

8:00 p.m.
The *Monitor* was suddenly struck by a series of fierce squalls. The ironclad was now "in heavy weather, riding one huge wave, plunging through the next, as if shooting straight for the bottom of the ocean." The *Monitor*'s helmsman, Francis Butts, recalled the effects of the heavy gale on the *Monitor*, stating that the ironclad would drop into a wave "with such force that her hull would tremble, and with a shock that would sometimes take us off our feet."

Commander Bankhead noted that "the sea about this time commenced to rise very rapidly causing the vessel to plunge heavily, completely submerging the pilot house and washing over and into the turret and at times into the blower pipes. Observed that when she rose to the swell, the flat under surface of the projecting armor would come down with great force, causing considerable shock to the vessel and turret, thereby loosening still more packing around its base."

9:00 p.m.
The Worthington steam pump and Adams centrifugal steam pump had failed to stop the flow of seawater. The water had risen over a foot deep in the engine room when Commander Bankhead put the crew to work on the hand pumps and organized a bucket brigade. The ironclad struggled "in a sea of hissing, seething form." William Keeler recalled that "her bow would rise on a huge billow and before she could sink into the intervening hollow, the succeeding wave would strike her under her heavy armor with report like a thunder and violence that threatened to tear apart her thin sheet iron bottom and heavy armor which it supported."

10:00 p.m.
The pumps were having little effect combating the incoming water. George Geer reported that "the pump threw a stream as large as your body, and for about one hour the water did not gain. Nor did we gain on it much."

Commander Bankhead ordered the red lantern displayed and tried to signal the *Rhode Island* to stop in order to ascertain if the *Monitor* would ride easier or the inflow of water would decrease.

Appendix I

10:30 p.m.
The *Rhode Island* and the *Monitor* drew nearer to each other, and Commander Bankhead requested boats to evacuate the crew. The ironclad's commander also ordered the lines cut and dropped anchor to stop the ironclad's pitching.

11:00 p.m.
Lifeboats were launched to begin rescue operations. It was a difficult task, and several of the *Monitor*'s crewmen were carried overboard as they tried to enter the lifeboats.

11:30 p.m.
Commander Bankhead recognized that the *Monitor* was in serious trouble and the ironclad was in danger of sinking: "My engines working slowly, and all the pumps in full play, but water gaining rapidly, sea very heavy and breaking entirely over the vessel, rendering it extremely hazardous to leave the turret in fact, several men were supposed to have been washed overboard at this time."

31 December
12:00 a.m.
Rescuers returned to the *Rhode Island*, and survivors were disembarked. Second Assistant Engineer Joseph Watters, who had come aboard to replace Albert Campbell as chief engineer, reported to Bankhead, who included in his official report: "While waiting for the boats to return, the engineer reported that the engines had ceased to work, and shortly after all the pumps stopped; also, the water putting out the fires and leaving no pressure of steam."

12:30 a.m.
When the lifeboats were readied to save those who remained aboard the *Monitor*, one was shattered by the *Rhode Island*'s hull. The single remaining boat returned and took the last crew members, including Captain Bankhead, away from the sinking ship. Dr. Grenville Weeks was severely injured when his arm was crushed between the lifeboat and the *Rhode Island* during rescue operations.

1:00 a.m.
The *Monitor* sank with the loss of four officers and twelve sailors in 220 feet of water.

1:30 a.m.
Commander Stephen Trenchard, captain of the *Rhode Island*, noted that the *Monitor*'s "light had unfortunately disappeared."

1863

11 January
Surgeon Grenville Weeks of the *Monitor* wrote to the sister of crew member Jacob Nicklis:

> *I am too unwell to dictate more than a short sad answer to your note. Your brother went down with the other brave souls, and only a good providence prevented my accompanying him. You have my warm sympathies, and the assurance that your brother did his duty well, and has I believe gone to a brighter world, where storms do not come.*

31 January
Second-Class Fireman Christy Price deserted.

5 May
Second Assistant Engineer Albert Campbell resigned, possibly due to sickness.

10 July
Third Assistant Engineer William D. Park died in service at Port Hudson, Mississippi River.

1864

15 September
First-Class Fireman Mathew Leonard died at POW Camp Groce, Waller County, Texas.

1865

15 January
First-Class Fireman John Mason died in service, lost at sea off Charleston, South Carolina.

Appendix I

1866

13 September
First Assistant Engineer Joseph Watters died in service at New Orleans, Louisiana.

1867

27 April
Captain John P. Bankhead died at sea near Arabia on the ship that was bringing him home.

1869

17 April
Ship's Cook David Cuddeback died at sea.

1875

24 October
Second Assistant Engineer Mark T. Sunstrom died at Baltimore, Maryland.

1876

3 June
Chief Engineer Alban C. Stimers died at Staten Island, New York.

1878

22 March
Acting Third Assistant Engineer Abraham Tester died at Brooklyn, New York.

1879

25 May
Master at Arms John Rooney died at Brooklyn, New York.

1880

27 June
Quartermaster Charles Peterson died at Russia, New York.

1882

2 June
First-Class Fireman John Garety died at New York City, New York.

1883

23 July
Commodore William N. Jeffers III died at Washington, D.C.

1884

22 February
Coal Heaver Thomas Carroll #2 died at New York City, New York.

25 September
Chief Engineer Isaac Newton Jr. died by his own hand at New York City, New York.

11 December
Commander Samuel Dana Greene died by his own hand at the Portsmouth (New Hampshire) Navy Yard.

APPENDIX I

1886

27 FEBRUARY
Acting Assistant Paymaster William F. Keeler died at Mayport, Florida.

1890

2 FEBRUARY
Dr. Thomas W. Meckly died at Jersey Shore, Pennsylvania.

1892

15 JANUARY
First-Class Fireman Edmund Brown died at Bath, New York.

2 MARCH
First-Class Fireman Patrick Hannan died at New York City, New York.

12 APRIL
First-Class Boy Siah Hulett Carter died at Philadelphia, Pennsylvania.

9 OCTOBER
Acting Second Assistant Engineer George S. Geer died at Charleston, South Carolina.

1893

9 FEBRUARY
Alderman Daniel Toffey died at Jersey City, New Jersey.

1896

15 MAY
Rear Admiral Thomas H. Stevens Jr. died at Rockville, Maryland.

1897

5 May
Quartermaster Moses M. Stearns died at Chelsea, Massachusetts.

1898

12 June
Acting Volunteer Lieutenant Commander William P. Flye died at Ashland, Kentucky.

7 December
Seaman Francis A. Riddey died at Philadelphia, Pennsylvania.

1899

21 May
Landsman James T. Slover died at St. Michaels, Maryland.

1900

14 January
USRM Lieutenant Samuel Howard died at Washington, D.C.

1905

15 June
Acting Third Assistant Engineer Darius F. Gallagher died at New York City, New York.

8 September
Paymaster's Clerk Francis B. Butts died at Cleveland, Ohio.

11 October
Seaman Thomas Carroll #1 died at Iola, Texas.

Appendix I

11 November
Mate Thomas Viall died at Warwick, Rhode Island.

1906

9 April
Coxswain Daniel Walsh died at Lawrence, Massachusetts.

1908

28 August
Reverend David R. Ellis died at Annville, Pennsylvania.

1909

20 April
Chief Boatswain Hans A. Anderson died at Brooklyn, New York.

October
Acting Master John J.N. Webber died at New York City, New York.

1910–1914

Ordinary Seaman John Brown died.

1911

13 July
Second-Class Fireman Michael Mooney died at Los Angeles, California.

8 October
Captain Louis N. Stodder died at Brooklyn, New York.

1914

3 February
Dr. Daniel C. Logue died on Long Island, New York.

12 July
Acting Master Edwin V. Gager died at Newark, New Jersey.

1915

10 February
Seaman Anton Basting died at Jersey City, New Jersey.

1916

4 January
Coal Heaver William Durst died at Philadelphia, Pennsylvania.

28 September
Second-Class Fireman Christy Price died at Stockbridge, Wisconsin.

1919

26 April
Dr. Grenville M. Weeks died at Kearny, New Jersey.

5 December
Coal Heaver William H. Remington died at Osceola, Michigan.

1924

4 February
Rear Admiral Thomas O. Selfridge Jr. died at Washington, D.C.

Appendix I

1927

17 May
U.S. Customs Inspector Isaac H. Scott, the last documented *Monitor* survivor, died at Brooklyn, New York.

1932

7 March
Former slave Thomas L. Taylor, purportedly a *Monitor* crewman, died a free man at Putnam, Connecticut.

1945

19 April
Andrew Fenton died at 101 years of age at Vineland, New Jersey, claiming to be the last *Monitor* survivor.

Appendix II
USS *Monitor* Statistics and Dimensions

Built: Greenpoint Ship Yard, Long Island, New York
Designer: John Ericsson
Laid down: 25 October 1861
Launched: 30 January 1862
Commissioned: 25 February 1862
Cost: $275,000

DIMENSIONS
   Length: 173 ft.
   Beam: 41 ft., 6 in.
   Draft: 10 ft., 6 in.
Tonnage: 987
Speed: 7 knots

Engine: Double trunk, cylinders (2-in-1 casting), 36-in. diameter, 27-in. stroke
Boilers: 2; return tube "box" boilers

Armor: 9 in. on pilothouse, 8 in. on turret, 4.5 in. on hull, 2 in. on deck
Armament: 2 XI-in. Dahlgren smoothbores
   Weight: 15,700 lb.
   Range: 3,400 yd.

Projectiles:
1. Case shot

Appendix II

    Weight of shot: 168 lb.
    Weight of powder charge: 15 lb.
2. Solid shot
    Weight of shot: 187 lb.

Crew: 65
    Officers: 16
    Enlisted: 49

Appendix III
USS *Monitor* Casualties

March 9—Battle of Hampton Roads
WIA:
Quartermaster Moses M. Stearns: hernia caused by passing shot up into turret.
Chief Engineer Alban C. Stimers: knocked to the floor when shot hit turret while he was resting his hand against it; recovered quickly; remained in command of the turret after Greene left to attend to Worden.
Acting Master Louis N. Stodder: knocked unconscious when shot hit turret while he was resting against it; removed from turret to surgeon's care; regained consciousness one hour later.
Ammunition Bearer Thomas L. Taylor: possibly received shrapnel wounds caused by iron flakes broken loose by shots hitting turret; claimed to have been treated at hospital at Fortress Monroe.
Quartermaster Peter Truscott: concussion caused by shot hitting turret as he leaned against it; removed from turret to surgeon's care; returned to duty several hours later.
Commander John L. Worden: minor concussion, facial burns with imbedded iron filings and wooden splinters and left eye blinded, caused by shot hitting the pilothouse as he was looking out the observation slit; removed from pilothouse to his cabin for treatment and later evacuated from the ship.

Appendix III

December 31 – Foundering of the *Monitor*
WIA:
Dr. Grenville Weeks: right hand was severely injured when caught between rescue boats; shoulder dislocated, three of his fingers amputated, never regained use of right arm.

KIA: drowned with ship
Landsman William Allen: The *Monitor* was his first and last service aboard a ship.
Acting Ensign Norman K. Atwater: One eyewitness claims he lost his grip on the rope hanging from the rescue ship and fell into the sea. The *Monitor* was the only ship on which he ever served.
Seaman William Bryan: An experienced sailor, he served on several ships before the *Monitor*.
First-Class Boy Robert Cook, African American: The *Monitor* was his first and last posting.
Landsman William H. Egan: The *Monitor* was his first and last posting aboard a ship.
Seaman James R. Fenwick: One eyewitness claims he saw him swept overboard attempting to cut the hawser between the *Monitor* and *Seth Low*. He served on several ships before the *Monitor*.
Acting Master's Mate George Frederickson: The *Monitor* was the first ship he served aboard.
Third Assistant Engineer Robinson Woollen Hands: The *Monitor* was the only ship on which he served.
Landsman Robert H. Howard, African American: The *Monitor* was his first and last posting.
First-Class Fireman Thomas Joice: The *Monitor* was his first and last service at sea.
Third Assistant Engineer Samuel Augee Lewis: One eyewitness claims he was last seen seasick in his bunk. The *Monitor* was the only ship upon which Lewis served.
Coal Heaver George Littlefield: The *Monitor* was his first and last ship's posting.
Landsman Daniel Moore, African American: The *Monitor* was the only ship he served aboard.
Ordinary Seaman Jacob Nicklis: The *Monitor* was the first and last ship on which he served.
Boatswain's Mate John Stocking: One eyewitness claims he saw him swept overboard attempting to cut the hawser between the *Monitor* and the *Seth Low*. He had served aboard another ship before the *Monitor*.
First-Class Fireman Robert Williams: An experienced sailor, the *Monitor* was his first and last sea posting in the war.

Appendix IV
USS *Monitor* Officers' Assignment Dates

George Frederickson, 4 December 1861
William Frederick Keeler, 4 January 1862
John Lorimer Worden, 16 January 1862
Samuel Dana Greene, 24 January 1862
Louis Napoleon Stodder, 31 January 1862
Albert Bogart Campbell, 7 February 1862
Daniel C. Logue, 7 February 1862
Isaac Newton Jr., 7 February 1862
Robinson Woollen Hands, 7 February 1862
William Dunlap Park, 7 February 1862
Mark Trueman Sunstrom, 7 February 1862
John Joshua Nathaniel Webber, by 4 March 1862
Robert Knox Hubbell, 5 March 1862
Samuel Howard, 9 March 1862
William P. Flye, 10 March 1862
Edwin Velie Gager, 10 March 1862
Thomas Oliver Selfridge Jr., 10 March 1862
William Nicholson Jeffers III, 12 March 1862
Thomas Holdup Stevens Jr., 9 August 1862
John Pyne Bankhead, 10 September 1862
George H. White, 14 October 1862
Samuel Augee Lewis, 30 October 1862

Appendix IV

Thomas W. Meckly, 30 October 1862
Grenville Mellen Weeks, 8 November 1862
Norman Knox Atwater, 9 November 1862
Joseph Watters, 28 December 1862

Appendix V
USS *Monitor* Officers' and Crew Biographies

Officers are marked in bold

Abbreviations:

Adm. = Admiral
ASq = Asiatic Squadron
assgn. = assigned
asst. = assistant
b. = born
Brig. = Brigadier
BSq = Brazil Squadron
btn. = battalion
btry., btries. = battery, batteries
bur. = buried
c. = circa
capt. = captain
cav. = cavalry
cem. = cemetery
cmdr. = commander
co. = company or county
CO = commanding officer
comp. = complexion
Conf. = Confederate
corp. = corporal

Appendix V

CSS = Confederate States Ship
d. = died
Dept. = Department
desc. = described as, description
det. = detachment
disch. = discharged
dsrtd. = deserted
EGBSq = East Gulf Blockading Squadron
EISq = East India Squadron
engr. = engineer
enl. = enlisted
ESq = European Squadron
Ft., fts. = Fort, forts
GAR = Grand Army of the Republic
Gen. = General
Hosp. = Hospital
KIA = killed in action
LOA = leave of absence
Lt. = Lieutenant
mdshp. = midshipman
MSq = Mediterranean Squadron
NABSq = North Atlantic Blockading Squadron
NFR = no further record
occ. = occupation
ORN = Official Records of the Union and Confederate Navies
POW = prisoner of war
prev. = previous
prmtd. = promoted
PSq = Pacific Squadron
QM = Quartermaster
R-Adm. = Rear-Admiral
reenl. = reenlisted
regt. = regiment
RTD = returned to duty
SABSq = South Atlantic Blockading Squadron
sgt. = sergeant
St. = Street
thru = through
trnsf. = transferred

USS *Monitor* Officers' and Crew Biographies

USF = United States Frigate
USN = United States Navy
USNA = United States Naval Academy
USRMS = United States Revenue Marine Ship
USRS = United States Receiving Ship
USS = United States Ship
vol., vols. = volunteer, volunteers
w/ = with
w/out = without
WGBSq = Western Gulf Blockading Squadron
WIA = wounded in action
WISq = West India Squadron
XO = Executive Officer

Allen, William: b. c. 1838, England; enl. 10/6/62 in New York for a 1-year term; desc. 5'10", gray eyes, black hair, light comp.; trnsf. to Washington Navy Yard, 10/31/62; received on the USS *Monitor* as a landsman, ship's no. 70, 11/7/62; KIA—d. by drowning w/ the sinking of the ship off Cape Hatteras, North Carolina, 12/31/62.[356]

Anderson, Hans A.: b. 7/13/24, Gothenberg (Gottenberg, Guttenberg), Sweden; father, Andreas Anderson; married to Priscilla Gladden (or Gladding), 3 children; occ. mariner, served in Swedish Merchant Marine, '45 (or '46)–'47; immigrated to the U.S., served in the American Merchant Marine, '52–'56; enl. 12/29/56 on the USRS *North Carolina* for a 3-year term as a seaman; served on the USS *Falmouth*, *Princeton* and *Congress* until disch. 1/13/62 w/ the rank of coxswain; reenl. 2/17/62 for a 3-year term as a seaman; trnsf., 3/6/62, from the *North Carolina* to the USS *Monitor*, ship's no. 9; acting QM on the perilous trip from New York to Hampton Roads, 3/6–8/62; participated in the battle w/ CSS *Virginia*, 3/9/62, as "shot man" for one of the *Monitor*'s two guns; was "shanghaied" and apparently dsrtd. 11/6/62, but escaped and surrendered 2/2/63 to the civil authorities at NY w/ no information in the records that he was tried by court-martial or that any penalty was imposed, save that of short confinement and forfeiture of pay due during his absence from duty; RTD, sailing on the USRS *North Carolina*, USS *Catskill*, *Home*, *Wabash* and *Princeton* '63–'65, serving as chief boatswain on the *Catskill*; participated in assault of Ft. Fisher, North Carolina, as member of landing party from the USS *Gettysburg*, WIA (hip and pelvis), 1/16/65; disch. in Philadelphia, 1/25/65; postwar: occ. watchman after

Appendix V

the war, the owner of a "milk, bread and wood small way" by '98; resided 93 Hall St., Brooklyn, New York; pension #6449, #5028 Navy, $30/month; desc. then as 5'7", blue eyes, brown hair, florid comp.; member, GAR and the Monitor Association of Naval Veterans; d. 4/20/1909 of gastric duodenal hemorrhage; bur. 4/23/1909 Evergreen Cem., Brooklyn.[357]

ANJIER, RICHARD (ALSO ANGIER, ANGUE, ANGUR, ANJIOR, ANTIER): b. c. 1825, England; enl. in New York 1/28/62 for a 3-year term; desc. 5'6½", hazel eyes, brown hair, dark comp.; trnsf. from the USRS *North Carolina* to the USS *Monitor* about 3/6/62, served as QM, ship's no. 2; as the ship sank on 12/31/62, he heroically remained aboard ship until his CO left; received a commendation for his actions during the loss of the vessel from CO J.P. Bankhead, 1/1/63; prmtd., acting master's mate, 1/12/63, possibly due to his actions during the sinking; assgn. to NABSq; NFR.[358]

ATKINS, JOHN: b. c. 1826, Baltimore, Maryland; occ. sailor; enl. in Boston 2/1/62 for a 3-year term in the 26th Maryland Vols.; desc. 5'9¾", hazel eyes, brown hair, fair comp.; assgn. to the USRS *North Carolina*; trnsf. to the USS *Monitor* w/ rank of seaman, ship's no. 6; dsrtd., 3/4/62; NFR.[359]

ATWATER, NORMAN KNOX (ALSO ATTWATER): b. 3/15/30, New Haven, Connecticut, a descendent of one of the original settlers; father, John Knox Atwater; mother, Charlotte Atwater; occ. sailor; resided 14 College, New Haven, Connecticut; acting ensign officer, 9/12/62; joined the USS *Monitor* 11/9/62; KIA—d. by drowning during attempted rescue by the USS *Rhode Island* during the sinking of the *Monitor* 12/31/62 off Cape Hatteras, North Carolina.[360]

BANKHEAD, JOHN PYNE (ALSO JOHN PINE, JOHN PAYNE AND JOHN F.): b. 8/3/21, Ft. Johnston, Charleston Harbor, South Carolina; father, Brig. Gen. James Bankhead, hero of the Mexican-American War, who was son of James Bankhead, soldier of the Revolutionary War; mother, Anne Pyne; single; occ. sailor; entered USN as mdshp. 8/10/38 on the USS *Macedonian*, WISq; served on the USS *Concord*, BSq, until it was run aground and the capt. drowned; trnsf. to the USS *Independence*, flagship of the Home Squadron; graduated 2nd in his class, Naval School, Philadelphia, 5/20/44, prmtd., passed mdshp.; ordered to coast survey, '45–'46; assgn. to naval btry. cooperating w/ Gen. Winfield Scott's army at Vera Cruz; served as chief of staff for his father's brigade at Vera Cruz during the Mexican-American War; onboard the USS

USS *Monitor* Officers' and Crew Biographies

Truxtun when run aground and captured by Mexicans, 8/14/46; extended LOA, possibly due to illness, '46–'50; RTD, the USS *Vandalia*, PSq, '50–'52; master, 5/8/51; lt., 4/7/52; LOA due to illness, '53; RTD, the USS *Columbia*, flagship of Home Squadron, '54, then the USS *Constellation*, '55–'58, both MSq; cmdr., the USS *Cranford*, Coast Survey; first station in the war: aboard the USS *Susquehanna*, SABSq; participated in naval bombardment of the fts. guarding Hatteras Inlet, North Carolina, as army troops landed 8/28/61, until surrender on 8/29/61; cmdr., the USS *Pembina*, SABSq, 10/8/61; participated in attack on Port Royal Sound, South Carolina, 11/5/61, and subsequent operations—including some w/ the USS *Ottawa*, commanded by Lt. T.H. Stevens—along the southern Atlantic coast, until 8/22/62, when he was ordered to New York City; mentioned as "a superior officer" by Flag Officer DuPont; prmtd. to cmdr., 7/16/62; cmdr., the USS *Monitor*, 9/10/62; towed to Washington Navy Yard, 9/30–10/3/62 for dry dock refurbishment; LOA, 3 weeks; the *Monitor*, a popular public attraction in Washington, returned to Hampton Roads by the end of 11/62; departed Hampton Roads for Beaufort, North Carolina, shortly after 2:00 p.m., 12/29/62, towed by the USS *Rhode Island*; survived ship sinking in heavy weather off Cape Hatteras, North Carolina, at 1:00 a.m., 12/31/62 w/ loss of 4 officers and 12 men; LOA, suffering from exposure; RTD, 3/9/63, cmdr., the USS *Florida* and blockade duty off Wilmington, North Carolina, capturing blockade runners *Calypso* and *Hattie*; LOA, sick, 8/63–1/64; RTD, cmdr., the USS *Iosco*; cmdr., the USS *Otsego*, NABSq, 2/6/64; cmdr., Union naval forces in the sounds, 6/21/64; cmdr., the USS *Wyoming*, EISq, 2/10/65, and sent in search of the CSS *Shenandoah*; began a 3-month cruise on the China coast, 2/21/66; capt., 7/25/66; resigned, '67, due to poor health; d. at sea on the steamer *Simla* that was bringing him home, 4/27/67; bur. Aden, Arabia.[361]

BASTING, ANTON (ALSO BASTON): b. 3/2/35, Germany; married to Ellen Spriggs, 2 children; resided New York City, New York; enl. as a seaman for a 3-year term, 2/1/62; trnsf. from the USRS *North Carolina* to the USS *Monitor* 3/6/62, ship's no. 7; allegedly wounded, confined to a hosp.; dsrtd., 11/6/62; postwar: occ. grocer; resided 153 Van Horne St., Jersey City, New Jersey; applied for pension, denied 2/6/1907 because of the desertion; desc. then as 5'10", gray eyes, brown hair, fair comp.; d. 2/10/1915; bur. Holy Name Cem., Jersey City.[362]

BRINGMAN, DERICK (ALSO BORRIGMAN, DERICK OR BRINGMAN, GIRICK): b. c. 1834, Bremen, Germany; enl. as a seaman, 1/28/62, in New York for a

Appendix V

3-year term; desc. 5'4½", hazel eyes, brown hair, florid comp.; trnsf., on or before 3/6/62, from the USRS *North Carolina* to the USS *Monitor*, served as carpenter's mate, ship's no. 3; dsrtd., 11/7/62; NFR.[363]

BROWN, EDMUND: b. c. 1818, New York City, New York; occ. engineer; resided Brooklyn, New York; enl. 5/27/61 for a 3-year term as a first-class fireman; desc. 5'6", dark eyes, black hair, dark comp.; served on the USRS *North Carolina* to 6/20/61; USS *Roanoke* until 1/7/62; USS *Whitehead* until 3/7/62; on the USS *Monitor* from 3/8–11/7/62; claimed "rheumatism from exposure" contracted onboard ship; LOA, sick, Navy Hosp., Washington, D.C., 11/8–11/24/62; disch., 11/24/62, as "unfit for service on account of physical disability"; postwar: occ. machinist, later watchman due to disability; resided 298 Wyckoff St., Brooklyn; naval invalid pension #8309, filed 9/19/82, rejected 2/14/88, "failure to establish claim"; d. 1/15/92, New York State Soldiers' and Sailors' Home, Bath, New York.[364]

BROWN, JOHN: b. c. 1841, Germany; occ. laborer; desc. 5'6¼", blue eyes, light hair, florid comp.; enl. in New York for a 3-year term as a landsman on the USRS *North Carolina*, 10/8/62; trnsf. to Washington Navy Yard, 10/28/62; received on the USS *Monitor*, 11/7/62, w/ rating of ordinary seaman, ship's no. 72; survived her sinking, 12/3/62; trnsf. to the USS *Jacob Bell*; dsrtd., 10/2/64; NFR; postwar: may have resided Rolfe, Iowa; d. c. 1910–14; likely bur. Clinton-Garfield Cem., Rolfe.[365]

BRYAN, WILLIAM: b. c. 1830, New York City, New York; occ. sailor; enl. at Boston 7/19/61, for a 2-year term as a seaman; desc. 5'7¾", blue eyes, auburn hair, fair comp.; served on the USS *Ohio* and *Sabine*; trnsf., 3/6/62, to the USS *Monitor*, served as yeoman, ship's no. 39; KIA—d. by drowning, 12/31/62, w/ the sunken ship off Cape Hatteras, North Carolina.[366]

BURROWS, GEORGE W.: 7/30/62 came on board *Monitor* as landsman for temporary service from *Port Royal*; transferred back to *Port Royal*; NFR.[367]

BUTTS, FRANCIS (FRANK) BANISTER: b. 1/27/44, Providence, Rhode Island; father, John Wood Butts; mother, Mehitable Wentworth; occ. farmer; resided Cranston, Rhode Island; enl. Providence, 9/16/61, as a corp. in Btry. E, 1st Regt., Rhode Island Light Artillery; desc. 5'9½", blue eyes, auburn hair, florid comp.; enl. in USN in New York, 10/3/62, for a 1-year term as a landsman, serving on the USRS *North Carolina*; trnsf. to Washington Navy Yard, 10/31/62;

assgn. to the USS *Monitor*, ship's no. 63; survived the ship's sinking, 12/31/62; trnsf., 1/3/63, to the USS *Brandywine*; disch. from the USS *Stepping Stones* 10/2/63; paymaster's steward on the USS *Flag*, SABSq, 1/25/64; paymaster's clerk, 3/14/65; disch., 4/23/65; postwar: occ. clerk; resided Cleveland, Ohio; married to Helen Francis Battey; widow's pension #21561; d. 9/8/1905 of nephritis and arteriosclerosis; bur. Swan Point Cem., Providence.[368]

CAMPBELL, ALBERT BOGART: b. New York; enl. 8/26/59 in New York, as third asst. engr.; second asst. engr., 10/21/61; assgn. to special duty, New York, 11/20/61; trnsf. to the USS *Monitor*, 2/7/62; trnsf. to the USS *Saranac*, 6/4/62; LOA, possibly due to sickness, 1/29/63; resigned, 5/5/63.[369]

CANN, EDWARD: arrived on the USS *Monitor* 11/13/62; desc. brown eyes, black hair, dark comp.; served as first-class boy, ship's no. 75 or 76, until the ship's foundering, 12/31/62, off Cape Hatteras, North Carolina; NFR.[370]

CARROLL #1, THOMAS: b. c. 1833, Boston, Massachusetts; married to Eliza Stanley, at least 4 children, possibly 7; resided Charlestown, Massachusetts; enl. as a seaman for a 2-year term, 7/22/61; desc. 5'6½", blue eyes, brown hair, fair comp.; served on the USS *Ohio* until 8/25/61, the USS *Sabine* until 2/24/62; trnsf. to the USS *Monitor*, served as capt. of the hold, ship's no. 19, until 12/31/62, when he survived the sinking of the vessel; served as seaman until 1/31/63; trnsf. from the *Monitor* to the *Ohio*, 4/27/63; trnsf. to the USS *Niphon* 4/28/63, served as capt. of forecastle 4/29–7/24/63; seaman 7/25/63; trnsf. to the USRS *North Carolina*, sailed w/ her 7/26–8/1/63; disch., Brooklyn Navy Yard; postwar: occ. farmer; resided Iola, Texas; pension #23432, 9/2/92; d. 10/11/1905; bur. Salem Church Cem., near Antioch, Texas; widow's pension #24688.[371]

CARROLL #2, THOMAS: b. c. 1843, Ireland; married to Margaret Walsh, 3 children; resided New York City, New York (at enlistment, claimed birthplace as Ritchfield Spring, New York); enl. 2/11/62 in New York for a 3-year term; desc. 5'6½", gray eyes, brown hair, fair comp.; served on the USRS *North Carolina* from enlistment date until 2/20/62; trnsf. to the USS *Monitor*, 2/26/62, served as first-class boy, ship's no. 44; prmtd., coal heaver, 11/7/62; survived the sinking of the vessel, 12/31/62; on the *Monitor*'s rolls until 1/31/63; served on the *North Carolina* from 2/1–6/11/63; trnsf. to the USS *Dai Cheng*, 6/12/63; served as coal heaver, 6/15/63, until disch., 7/18/63; postwar: occ. coachman; suffered from rheumatism said to be from exposure while awaiting rescue from the sinking *Monitor*; d. 2/22/84, St. Vincent's Hosp., New York City.[372]

CARTER, SIAH HULETT: b. 10/4/39; father, John Hulett; mother, Molly; both slaves of Hill Carter, Shirley Plantation, Charles City Co., Virginia; occ. carpenter; resided Chesterfield Co., Virginia; enl. on the USS *Monitor* on the James River, 5/19/62, for a 3-year term as first-class boy, ship's no. 53; first slave to desert and enlist from Shirley plantation, he enl. using his master's last name—Carter—but his family's name was Hulett; desc. 5'6½"–5'8½", brown eyes, black hair, dark comp.; survived the *Monitor*'s sinking, 12/31/62, and was officially disch. from the ship 1/31/63; also served on the USS *Brandywine, Florida, Belmont, Wabash* and *Commodore Barney*; suffered frostbite when on the *Commodore Barney*; disch., 5/19/65, St. Mary, Maryland; postwar: married to Eliza Tarrow Hulett, 13 children; occ. laborer; resided 1349 Rose (now Kenilworth) St., Philadelphia, Pennsylvania; pension #4542, granted 4/9/91; d. of peritonitis, 4/12/92, Philadelphia; bur. 4/16/92, Olive Cem., Philadelphia; cem. condemned in 1923, bodies reinterred, Eden Cem., Philadelphia; widow's pension #11878.[373]

COLEMAN, JAMES (ALSO WILLIAM): b. c. 1841, Ireland; occ. cotton spinner; enl. 9/20/62 in New York for a 1-year term as landsman; desc. 5'5½", blue eyes, brown hair, florid comp.; served on the USRS *North Carolina*; trnsf. to Washington Navy Yard, 10/28/62; assgn. to the USS *Monitor* 11/7/62, ship's no. 65; survived the ship's sinking, 12/31/62, off Cape Hatteras, North Carolina; later served on the USS *Brandywine* and the USS *Stepping Stones*; disch., 9/23/63, from the Washington Navy Yard.[374]

CONKLIN, JOHN P. (ALSO CONKLING): b. c. 1832, Watertown, New York; occ. farmer; enl. as a seaman, 7/18/61, in Boston for a 2-year term; desc. 5'9", hazel eyes, black hair, dark comp.; served on the USS *Ohio* and USS *Sabine*; trnsf. to the USS *Monitor* 3/6/62, serving as quarter gunner, ship's no. 40; dsrtd., 11/6/62; NFR.[375]

CONNOLY, ANTHONY: b. c. 1835, Ireland; enl. in Boston as a seaman, 7/18/61, for a 2-year term; desc. 5'6", blue eyes, brown hair, fair comp.; served on the USS *Ohio*; trnsf. from the USS *Sabine* to the USS *Monitor*, ship's no. 47, by 3/6/62; survived the ship's sinking off Cape Hatteras, North Carolina, 12/31/62; later served on the USS *Arizona*, from which he dsrtd. 7/31/63; NFR.[376]

USS *Monitor* Officers' and Crew Biographies

COOK, ROBERT: b. c. 1844, Gloucester Co., Virginia, probably as a slave; enl. Hampton Roads, Virginia, 9/8/62 for a 3-year term as a first-class boy; desc. 5'6½", brown eyes, black hair, dark comp.; assgn. to the USS *Monitor* before 11/7/62, ship's no. 59; KIA—d. by drowning, 12/31/62, off Cape Hatteras, North Carolina, when the ship sunk.[377]

CROWN, JOSEPH (ALSO JAMES): b. c. 1823, New York City, New York; enl. as a seaman for a 3-year term, 1/20/62; desc. 5'7¼", blue eyes, black hair, swarthy comp.; trnsf., by 3/6/62, from the USRS *North Carolina* to the USS *Monitor*, serving as gunner's mate, ship's no. 10; participated in the 3/9/62 battle w/ the CSS *Virginia*, manning one of the *Monitor*'s two guns; survived the sinking of the vessel, 12/31/62; trnsf. to the USS *Catskill*, 2/24/63; disch. 1/25/65, Philadelphia Navy Yard.[378]

CUDDEBACK, DAVID (ALSO CUTTEBACK): b. 3/28/40, Port Jervis, New York; enl. as a landsman for a 3-year term, 6/29/61; desc. 5'10", hazel eyes, dark hair, fair comp.; trnsf., by 3/6/62, from the USRS *North Carolina* to the USS *Monitor*, served as capt.'s steward, ship's no. 18; lost rating, 7/15/62; prmtd., 11/7/62, to ship's cook; survived the ship's sinking, 12/31/62, off Cape Hatteras, North Carolina; served on the USS *Ohio*; disch., 5/20/63; postwar: d. onboard a ship 4/17/69.[379]

DRISCOLL, JOHN AMBROSE: b. c. 1836, Co. Cork, Ireland; abandoned his wife, Abigail Sweeney, and children to immigrate to America; enl. 2/15/62 in New York for a 3-year term as first-class fireman; desc. 5'4", blue eyes, dark hair, swarthy comp.; trnsf., by 3/6/62, from the USRS *North Carolina* to the USS *Monitor*, ship's no. 20; participated in the battle w/ the CSS *Virginia* 3/9/62, stationed at the foot of the turret ladder, passing up shot; dsrtd. or trnsf. off the *Monitor* by 11/7/62; enl. on the USS *Connecticut*, 12/15/62; disch., 6/11/63; postwar: occ. landscaper; resided Buffalo, New York.[380]

DURST, WILLIAM (ALSO WILHELM, ALIAS WALTER DAVID/DAVIS): b. 5/6/36, Tarnow, Austria; immigrated to America, 11/61; occ. porter; enl. in New York, 2/14/62, for a 3-year term; served as a coal heaver on the USRS *North Carolina*; trnsf., 2/25/62, to the USS *Monitor*, ship's no. 29; dsrtd., 11/6/62; reenl. 2/16/63 as a second-class fireman under the alias "Walter David," claiming he was 32 years old and formerly a machinist; desc. 5'6", hazel eyes, brown hair, dark comp., "WD" tattoo on right wrist; served on the *North Carolina* to 2/24/63, the USS *Catskill* until 5/18/64, the USS *Princeton*

Appendix V

to 5/31/64; postwar: married to Ester Wallensteine, 2 children; second marriage to Anna Goronozy Neuman, 4 children; occ. jewelry peddler, oyster stand keeper, waterworks employee; resided 1315 George St., Philadelphia, Pennsylvania; net worth, 1870, $2,000; USN dropped desertion charge 8/1906; member, GAR; d. 1/4/1916 of pneumonia.[381]

EGAN, WILLIAM H.: b. c. 1841, Ireland; enl. in New York as landsman, 10/6/62, for a 3-year term; desc. 5'6", blue eyes, sandy hair, fair comp.; assgn. to Washington Navy Yard; served on the USS *Monitor*, ship's no. 73, from 11/7/62; KIA—d. by drowning when the ship was lost off Cape Hatteras, North Carolina, 12/31/62.[382]

ELLIS, DAVID ROBERTS (ALIAS ELLIS ROBERTS): b. 11/12/40, Carmarthen, Wales; father, Roberts Ellis; enl. as a coal heaver in New York, 2/14/62, for a 3-year term under the name of Ellis Roberts because he was wrongly identified by another crew member; desc. 5'5½", blue eyes, auburn hair, fair comp.; assgn. to the USRS *North Carolina*; trnsf. to the USS *Monitor*, 3/6/62, ship's no. 28; prmtd. to paymaster's steward, 10/7/62; survived the *Monitor*'s sinking, 12/31/62, continuing on payroll until 1/31/63; disch., 3/2/63; served on the USS *Florida* as paymaster's steward, disch., 8/31/63; appointed capt.'s clerk on the *Florida*, disch., 3/16/64; paymaster's steward, 6/20/64, serving on board the USS *Tunxis* until 11/64; on the USS *Sassacus* as paymaster's clerk until 5/13/65; postwar: married to Martha Jane Keck, 6 children; occ. minister, United Brethren Church; resided Rockwood and Annville, Pennsylvania; pension #21634, 11/7/91; member, GAR and Independent Order of Odd Fellows; d. 8/28/1908 of chronic Bright's disease w/ heart complications; bur. 8/31/1908, Mt. Annville Cem., Annville.[383]

FEENEY, THOMAS (ALSO FEENY): b. c. 1841, Brooklyn, New York; enl. in New York on 2/13/62 for a 3-year term as a coal heaver; desc. 5'7½", blue eyes, dark hair, swarthy comp.; trnsf. from the USRS *North Carolina* to the USS *Monitor*, ship's no. 26; dsrtd., 2/20/62; NFR.[384]

FENWICK, JAMES R.: b. c. 1838, Scotland; father, Charles Fenwick; mother, Elizabeth; enl. w/ rank of ordinary seaman for a 2-year term in Boston, 7/22/61; desc. 5'5", blue eyes, auburn hair, light comp., a "J.R.F. Dundee" tattoo on his right forearm; served on the USS *Ohio*; trnsf., 3/6/62, from the USS *Sabine* to the USS *Monitor*, where he served as seaman, ship's no. 48; prmtd. to quarter gunner, 11/7/62; arrested for fighting w/ a fellow crewman; married to Mary Ann Duffy while on furlough, 10/10/62; KIA—d. by drowning when swept off the deck of the sinking *Monitor*, 12/31/62, off Cape Hatteras, North Carolina; his wife was pregnant as of 4/14/63; widow's pension applied for 7/22/63.[385]

FISHER, HUGH: b. c. 1823, Ireland; occ. fireman; enl. for a 3-year term as first-class fireman in New York, 1/2/62; desc. 5'7½", blue eyes, black hair, dark comp.; assgn. to the USRS *North Carolina*; trnsf. to the USS *Monitor*, ship's no. 34; dsrtd., 3/5/62; NFR.[386]

FLYE, WILLIAM P.: b. 10/25/14, Newcastle, Lincoln Co., Maine; married to Mary Elizabeth Perkins, 2 children; occ. sailor; desc. 5'7", brown hair, blue eyes, light comp.; appointed Professor of Mathematics, USNA, 12/7/41, ordered to the USS *John Adams*; trnsf. to the USS *Jamestown*, 12/12/44; assgn. to the Naval Observatory, '48–'54; resigned, 3/7/57; net worth, 1860, $7,000; prmtd. to acting vol. lt., 12/6/61; first station in the war: the USRS *North Carolina*, on which he served from 12/14/61–1/10/62; trnsf. to the USS *R.B. Forbes* as vol. lt.-commanding from 1/11/62 until the ship was wrecked and burned 2/25/62; serving in Hampton Roads on the USS *Roanoke* 3/8/62 when the CSS *Virginia* destroyed the USS *Cumberland* and the USS *Congress*, and apparently suffered the disability for which he received a pension; trnsf. to the USS *Monitor* as acting vol. lt., 3/9–10/29/62; participated in the attack on Confed. btries., Sewell's Point, Virginia, 5/8/62 and in the engagement at Ft. Darling (Drewry's Bluff) 5/15/62; served on the USS *Underwriter* as acting vol. lt.-commanding, 10/30/62–6/30/63; on the *North Carolina* as acting vol. lt., 7/1–8/31/63; on the USS *Kensington*, 9/1/63–3/29/64; assgn. to Naval Yard, Memphis, Tennessee, 3/30–5/24/64; served on the USS *Osage* 5/25–7/11/64; back to the Naval Yard, Memphis, 7/12–8/16/64; served on the USS *Benton* 8/17–11/14/64; the USS *Lexington* 11/15/64–9/3/65; acting vol. lt. cmdr., 7/18/65; disch., 12/24/65; net worth, 1870, $3,000; postwar: occ. various; partially disabled due to deafness, invalid pension claim #3703, dated 10/2/83; member, GAR; d. of uremia 6/12/98, Ashland, Kentucky; bur. Riverside Cem.[387]

FREDERICKSON, GEORGE: b. c. 1834, island of Møn, Denmark; married to Magdalena E. Heobst (or Holst), 2 children; resided Philadelphia, Pennsylvania; enl. 12/4/61 in New York as acting master's mate on the USS *Monitor*; desc. 5'5", light blue eyes, light brown hair, light comp.; KIA—d.

Appendix V

by drowning 12/3/62, when the ship was lost off Cape Hatteras, North Carolina; postwar: widow's pension, $10 per month, dated 8/13/63; NFR.[388]

GAGER, EDWIN VELIE: b. 10/14/33, Dutchess Co., New York; father, Joseph Gager; mother, Hanna M. Velie; married to Julia P. Werner, no children; second marriage to Rose A. Morley, 5 children; occ. sailor; enl. Brooklyn, New York, 4/61; desc. 5'7", gray eyes, dark hair, fair comp.; acting master, 5/17/61; served on the USS *Monticello* and participated in the rescue of an Indiana regiment at Hatteras, North Carolina, 8/28/61; came aboard the USS *Monitor* 3/10/62; resigned, 7/8/62; postwar: resided Newark, New Jersey; net worth, 1870, $20,000; member, Military Order of the Loyal Legion of the United States; pension #18294, dated 9/16/1904; d. 7/12/1914.[389]

GALLAGHER, DARIUS FARRINGTON: b. Manhattan, New York; father, Darius F. Gallagher; mother, Marshal W. Higgins; married to Margaret A. Lynch, 5 children; served onboard the USS *Monitor* during her battle w/ CSS *Virginia* 3/9/62; acting third asst. engr., 11/5/62; apparently trnsf. off the *Monitor* before her sinking; dismissed, 7/15/63; postwar: occ. leading man, Brooklyn Naval Yard; resided New York City; d. 6/15/1905.[390]

GARETY, JOHN (ALSO GARETTY, GARITY, GARRETY): b. c. 1837, Ireland; father, John Garety; mother, Mary; enl. for a 3-year term as a first-class fireman in New York, 1/20/62; desc. 5'7", hazel eyes, dark hair, light comp.; served on the USRS *North Carolina*; trnsf., 2/26/62, to the USS *Monitor*, ship's no. 36; survived the ship's sinking, 12/31/62; returned to the *North Carolina* until 3/2/63; trnsf. to the USS *Keokuk*, participating in R-Adm. DuPont's attack on Charleston Harbor, South Carolina, surviving the ship's sinking 4/8/63; trnsf. to the USS *Vermont*, serving on her until 4/20/63; trnsf. to the USS *Ohio* and served until 5/20/63, when disch.; postwar: married to Bridget Davis, no children; occ. laborer; resided 431 E. 16th St., New York City, New York; d. of phthisis (consumption) pulmonalis 6/2/82; bur. Calvary Cem., 6/4/82; widow's pension #9793, filed 12/20/90.[391]

GEER, GEORGE SPENCER: b. 5/17/36, Troy, New York; father, Gilbert Geer; mother, Adna Spencer; married to Martha Clark Hamilton, 6 children; occ. machinist—worked in machine shop where the USS *Monitor*'s boiler was built; enl. at New York, 2/15/62, as a first-class fireman; desc. 5'7½", blue eyes, brown hair, fair comp.; trnsf., by 2/20/62, from the USRS *North*

Carolina to the *Monitor*, ship's no. 24; engr. yeoman, 5/62; survived the ship's sinking, 12/31/62; acting third asst. engr., 1/19/63, began service aboard the USS *Galena*; granted LOA, 5/25/63; RTD, ordered to Port Royal, South Carolina, 6/3/63; assgn. to the USS *Vermont* 10/1/63; acting second asst. engr., 6/1/64; trnsf. to the USS *Philadelphia* 8/18/64; LOA, 8/28/65; honorably disch., 12/1/65; postwar: occ. engineer (of steamer *Merrimac*), paving, charity commissioner; net worth, 1870, $1,000; member, GAR; pension #6219, dated 10/3/67; d. of "swamp fever" 10/9/92, St. Francis Xavier Infirmary, Charleston, South Carolina; bur. Oakwood Cem., Troy, New York; widow's pension #6338, dated 11/28/92.[392]

GREENE, SAMUEL DANA: b. 2/11/40, Cumberland, Maryland; father, Brig. Gen. George Sears Greene, graduated 2nd in his class at West Point Military Academy in 1823, hero of Culp's Hill at Gettysburg and founder of the American Society of Civil Engineers and Architects; mother, Martha Barrett Dana; married to Mary Willis Dearth, 3 children; second marriage to Mary Abby Babbitt; occ. sailor; appointed acting mdshp., USNA, 9/21/55; graduated 7th in a class of 20, 6/9/59; warranted as mdshp. 6/9/60 on the USS *Hartford*, while serving '59–'61 at the China Station; at the outbreak of Civil War, 4/61, the *Hartford* was ordered home; lt., 8/31/61; *Hartford* arrived Philadelphia, 12/2/61; LOA, 12/3/61–1/23/62; assgn. to the USS *Monitor*, as XO, 1/24/62–12/31/62; led team to rescue engine room crewmen overcome by carbon dioxide when blowers failed during near swamping of ship on trip from New York to Hampton Roads, 3/6–7/62; during engagement w/ the CSS *Virginia* 3/9/62, commanded the turret and its guns, personally firing every shot until assuming command when the ship's CO was WIA; participated in the engagement w/ rebel btries. Ft. Darling (Drewry's Bluff), James River, Virginia, 5/62; served under Adm. Wilkes in the James River Flotilla, 6–9/62; participated in the attempt to blow up a railroad bridge near the Appomattox River, 6/26/62; survived the sinking of the *Monitor* 12/31/62, after taking charge of evacuating the crew into the boats sent by the rescue ship, *Rhode Island*; received a commendation for his actions during the loss of the vessel from CO J.P. Bankhead, 1/1/63; served as XO to Bankhead on the USS *Florida*, blockading coast of North Carolina, 1–9/63, serving as temporary CO when Bankhead fell ill, 8/63; LOA, 9/20/63, to get married 10/9/63; assgn. special duty as assistant inspector, New York Navy Yard, 11/63–2/64; trnsf. as XO to the USS *Iroquois*, serving in NABSq and European waters, '64–'65; lt. cmdr., 8/11/65; postwar: asst. professor, mathematics, USNA, 10/65–10/68; served on practice vessel

Appendix V

USS *Marblehead*, '66; *Macedonian*, '66; *Savannah*, '68; assgn. to PSq, onboard the USS *Ossipee*, *Saranac*, *Saginaw*, *Nyack* and the *Pensacola*, '68–'71; prmtd. to cmdr., 12/12/72; head, Dept. of Navigation and Astronomy, USNA, 6/71–6/73; superintendent of grounds, USNA, 12/74; cmdr., the USS *Juniata*, European station '75, and along the coast of U.S., '76; cmdr., USS *Monongahela*, home coast, '77; trnsf., USNA, 9/15/77; assgn. to special duty, Washington, D.C., 4/10/82; cmdr., the USS *Despatch* 7/14/82–84; XO, Portsmouth Navy Yard; d. 12/11/84 of suicide (a pistol shot to the head) at Portsmouth (New Hampshire) Navy Yard, apparently due to depression ("temporary insanity") over questions concerning his service on the *Monitor*; desc. then as 5'11", black hair, hazel eyes, dark comp.; widow's pension #3921, granted 12/20/84; USS *Greene* (DD-266), a *Clemson*-class destroyer commissioned 5/9/1919, was named for him.[393]

HALMAN, WILLIAM S.: became the USS *Monitor*'s hosp. steward 11/6/62; survived the sinking of the ship 12/31/62; NFR.[394]

HANDS, ROBINSON WOOLLEN: b. c. 1835, Baltimore, Maryland; father, Washington Hands; mother, Jane Woollen; occ. student of mechanical engineering; third asst. engr., 2/1/62; assgn. to the USS *Monitor* 2/7/62; KIA—d. by drowning off Cape Hatteras, North Carolina, 12/31/62, when the *Monitor* foundered.[395]

HANNAN, PATRICK (ALSO HANNON): b. c. 1826, Ireland; father, Patrick Hannan; mother, Nora; married to Mary, no children; second marriage to Bridget Kenery, 2 children; occ. machinist; enl. in New York as a first-class fireman for a 3-year term, 2/18/62; desc. 5'6½", hazel eyes, brown hair, fair comp.; served on the USRS *North Carolina*; trnsf. to the USS *Monitor* 2/25/62, ship's no. 25; survived her sinking 12/31/62, remaining on payroll until 1/31/63; assgn. to the USS *Brandywine*; trnsf. to the *North Carolina*, served there until 3/4/63; served on the USS *Keokuk*, participating in DuPont's attack on Charleston, until 4/8/63, when he survived the vessel foundering off Morris Island, Charleston, South Carolina; trnsf. to the USS *Vermont*; served onboard a steamer captured at St. John's; served on the USS *Ohio*; disch., Boston, 5/20/63; postwar: occ. engineer; d. of pneumonia 3/2/92; bur. Calvary Cem., New York City, New York.[396]

HARDY, JOHN: b. c. 1838, Portland, Maine; enl. 7/11/61 in New York for a 3-year term as a seaman; desc. 5'5½", blue eyes, brown hair, light comp.;

trnsf. from the USS *Wachusetts* to the USS *Monitor*, ship's no. 54, 7/31/62; served on the ship until her sinking 12/31/62 off Cape Hatteras, North Carolina; NFR.[397]

HARRISON, HENRY: b. c. 1824, Sweden; occ. sailor; enl. 7/10/62 in New York for a 3-year term as a seaman; desc. 5'8", blue eyes, brown hair, florid comp.; served on the USRS *North Carolina*; trnsf. from the USS *Wachusetts* to the USS *Monitor* 7/31/62, ship's no. 55; survived the *Monitor*'s sinking, 12/31/62, off Cape Hatteras, North Carolina; served on the USS *Brandywine*; the USS *Mount Vernon* by 6/30/63; trnsf. to the USS *Niphon*, dsrtd., 12/31/63 at either New York or Boston; NFR; postwar: resided 7 Washington St., New York City; became American citizen 4/24/94.[398]

HOWARD, ROBERT H. (ALSO W.H.): b. c. 1824, a probable slave, Howard Co., Virginia; enl. in Washington for a 1-year term as a landsman on 11/7 or 8/62; desc. 5'6½", brown eyes, black hair, "negro" comp.; served on the USS *Monitor* as officer's cook from 11/15/62 until 12/31/62, when he was KIA by drowning off Cape Hatteras, North Carolina, when the ship sunk.[399]

HOWARD, SAMUEL: b. 3/12/21, near Dublin, Ireland; resided Newport, Rhode Island; married to Mary Dugan, 4 children; net worth, 1860, $500; acting master, 10/9/61; gunnery instruction, Brooklyn Navy Yard; served on the USS *Amanda* 10/19/61–7/9/63; brought the *Amanda* to Norfolk, Virginia, 3/8/62, in time to witness the CSS *Virginia*'s destruction of the USS *Congress* and the USS *Cumberland*; volunteered to serve as acting master and pilot for the USS *Monitor* during her battle w/ the CSS *Virginia* 3/9/62; returned to the *Amanda*, acting vol. lt., 6/5/62; served on the USS *Neosho*, Mississippi Squadron, 7/16/63–8/22/65; granted LOA; RTD, ordered to the USS *Vermont* 10/14/65; trnsf. to the USS *Pensacola*, serving 8/16/66–9/4/68; honorable disch., 11/4 or 14/68; postwar: commissioned lt., U.S. Revenue Marine, stationed at New Orleans, Savannah and Mobile; resided 2307 M St. NW, Washington, D.C.; invalid pension #26988, dated 6/27/98; member GAR, A.F. & A.M.; d. 1/14/1900, of concussion caused by fall; bur. Arlington National Cem., Washington, D.C.[400]

HUBBELL, ROBERT KNOX: b. St. Louis, Missouri; father, Charles Benjamin Hubbell Jr.; mother, Mary Adeline Knox; enl. in New York the week ending 9/14/61; served on board the USS *Whitehall*; trnsf. to the USS *Monitor* c. 3/5/62 as ship steward, ship's no. 46, probably serving as Paymaster

Appendix V

Keeler's steward; prmtd. to acting ensign, 10/1/62; possibly trnsf. from the *Monitor* before 11/7/62; served on the USS *Osage*, commanded by Thomas Selfridge, as acting ensign, '63–'64; dismissed, 4/22/64; however, participated in the engagement at Deloach's Bluff, Red River, Louisiana, 4/26/64, volunteering to retrieve the body of a fellow crewman who had been shot while on picket duty; NFR.[401]

JEFFERS, WILLIAM NICHOLSON, III: b. 10/16/24, Swedesboro, New Jersey; father, John Ellis Jeffers; mother, Ruth Westcott; married to Lucie LeGrand Smith, 2 children; occ. sailor; enl. as acting mdshp., 9/25/40, ordered to the USRS *North Carolina*; trnsf. 9/41 to the USF *United States*, flagship for Commodore Thomas ap Catesby Jones; served four years on the *United States* and the USS *Congress*, PSq and BSq; participated in the seizure of Monterey, California, when Commodore Jones erroneously believed the U.S. was at war w/ Mexico; entered USNA, 10/10/45; graduated no. 4 out of 47, prmtd., passed mdshp. 7/11/46, specialist in ordnance and gunnery; ordered to the USS *Vixen*, served thru the War w/ Mexico; present at attacks on fts. of Alvarado under Commodore Connor, at two attacks on and capture of Tabasco, of Tuspan, Coatzacoalcos and Laguna de Terminos; covered the landing of the U.S. Army at siege of Vera Cruz, took part w/ others of the Mosquito Fleet in bombardment and capture of Vera Cruz and the castle of San Juan d'Ulloa; on duty at Naval Academy '48–'49 as acting master and asst. professor of mathematics; assgn. to the USS *Morris*, Coast Survey, 12/49–10/50; sailed mail steamer service between New York and Aspinwall, Havana, Kingston and New Orleans, 10/50–3/52, a part of the time in command; trnsf. to the USS *Princeton* as acting master 3/52; trnsf. to the USS *Macedonian* 11/52; participated in exploration of Isthmus of Honduras 12/52–53; trnsf. to the USS *Alleghany* 9/53; to the USS *Germantown* 10/53; master, 6/12/54; lt., 1/30/55; trnsf. as XO to the USS *Water Witch* for survey of La Plata and Parana Rivers, South America, '55–'56; engaged w/ the ft. at Paso de la Patria, causing the expedition to Paraguay by Commodore Shubrick; presented w/ a sword w/ gold hilt and scabbard by Her Majesty the Queen of Spain for saving the *Cartagenera* (or *Carthagena*), 10/55; preliminary survey of the Isthmus of Honduras for Inter-Oceanic Railway, '57; served on the USS *Plymouth*, '58–'59, under Lt. Cmdr. John A. Dahlgren—designer of the smoothbore gun—and XO Catesby ap Roger Jones—later XO of the CSS *Virginia*; served on the USS *Brooklyn* and USS *Saratoga*, West Indies, 1/59–'60; while attached to the *Brooklyn*, was brought up on charges by Captain David G.

USS *Monitor* Officers' and Crew Biographies

Farragut for failure to salute a commanding officer, for which Jeffers wrote what amounted to an apology and requested a transfer; also while aboard the *Brooklyn*, appointed as hydrographer and surveyed Chiriqui Isthmus for a possible canal route; immediately applied for service when the rebellion broke out, even though on sick leave; first service in the war: ordered to take charge of the ordnance facility at the Norfolk Navy Yard, but the yard was evacuated and partially destroyed before he could; cmdr., USS *Philadelphia* on the Potomac River 4–5/61; assgn., 5/12/61, USS *Roanoke* for blockade duty, Atlantic seaboard and Charleston, South Carolina; engaged w/ btries., Sewell's Bluff, Virginia; cmdr., the USS *Underwriter*, Pamlico Sound, 11–12/61; participated in battles of Roanoke Island and Elizabeth City while under Commodore L.M. Goldsborough and Cmdr. S.C. Rowan 1–2/62, receiving commendation for "zeal and intelligence"; expedition to Currituck Sound; cmdr., the USS *Monitor* 3/12/62; participated in bombardments and battles of Drewry's Bluff, James River, Virginia, 5/15/62, under Cmdr. John Rogers; lt. cmdr., 7/16/62; detached from *Monitor*, possibly due to ill health, 8/15/62; trnsf. to ordnance duty, Philadelphia, 9/62; assgn. as inspector of ordnance, Washington Navy Yard, 9/63; assisted in preparing powder ship *Louisiana* for explosion, 12/24/64, off Ft. Fisher, Florida; prmtd. to cmdr., 3/3/65; postwar: cmdr., USS *Swatara* 7/65, sailed to Bermuda, West Indies, Mediterranean and Africa; returned escaped Lincoln assassination conspirator John Harrison Surrett to the U.S.; assgn. to the Naval Observatory, '68; assgn. to the Navy Dept., '68; Board of Examiners, '69–'70; capt., 7/13/70; assgn. as asst. to the chief, Bureau of Ordnance, 9/30/70; cmdr., the USS *Constellation*, gunnery practice ship, West Indies and coast, 10/71; chief, Bureau of Ordnance, 4/10/73 (served until '83) w/ relative rank of commodore; in '76, doubled the power of the Dahlgren M.L. 11-inch smoothbore by converting it into an 8-inch rifle; renominated as chief, Bureau of Ordnance, 4/10/77; author of *The Armament of Our Ships of War* ('46), *Short Rules in Navigation* ('49), *Theory and Practice of Naval Gunnery* ('50), *Inspection and Proof of Cannon* ('64), *Nautical Surveying* ('71), *Care and Preservation of Ammunition* ('74), *Ordnance Instruction for U.S. Navy* (eds. '66 and '80); prmtd., commodore, 2/26/78; d. 7/23/83 of kidney disease, Washington, D.C.; bur. Naval Cem., Annapolis, Maryland; USS *Jeffers* (DD-621), a *Gleaves*-class destroyer commissioned 11/5/1942, was named in his honor.[402]

JEFFREY, WILLIAM H.: b. c. 1837, Philadelphia, Pennsylvania; enl. in New York for a 3-year term as a landsman, 10/15/61; desc. 5'4", "mulatto" comp.; served on the USS *Pembina*; assgn. to Washington Navy Yard; trnsf.

APPENDIX V

to the USS *Monitor* by 10/1/62, serving as officer's steward, ship's no. 58; survived the *Monitor* sinking off Cape Hatteras, North Carolina, 12/31/62; NFR; postwar: invalid, resided with his mother in Philadelphia.[403]

JOICE, THOMAS: b. c. 1839, Ireland; enl. in New York for a 3-year term 1/15/62 as a first-class fireman; desc. 5'9½", blue eyes, brown hair, fair comp.; trnsf., by 3/6/62, from the USRS *North Carolina* to the USS *Monitor*, ship's no. 35; KIA—d. by drowning when the ship sank off Cape Hatteras, North Carolina, 12/31/62.[404]

JONES, JESSE M.: served onboard the USS *Monitor* as hosp. steward during the fight w/ the CSS *Virginia* 3/9/62; may have left the ship at Washington Navy Yard, 10 or 11/62, since William S. Halman became the *Monitor*'s hosp. steward on 11/6/62; NFR; postwar: d. sometime before 1877.[405]

KEELER, WILLIAM FREDERICK: b. 6/9/21, Utica, New York; res. Bridgeport, Connecticut; father, Roswell Keeler; mother, Mary Eliza Plaut; married to Anna Eliza Dutton, daughter of Henry Dutton, head of Yale Law School and later governor of Connecticut, 3 children; occ. iron foundry owner and partner; resided La Salle, Illinois; acting asst. paymaster, 12/17/61; ordered to Navy Yard, New York, 1/4/62, assgn. the USS *Monitor*; participated in the battle w/ CSS *Virginia* 3/9/62, transmitting messages between the CO in the pilothouse and the XO in the turret; survived the ship's sinking, 12/31/62; detached from the *Monitor* 1/6/63; trnsf. to the USS *Florida* 2/7/63–10/26/65; WIA 2/10/64 (shell fragment from enemy shore btries.), blockade of Wilmington, North Carolina; LOA, sick, 12/25/65; honorably disch., 4/26/66; postwar: pension #2177, $10 per month, dated 6/2/75; desc. then as 5'8", light comp.; d. of heart disease 2/27/86, Mayport, Florida.[406]

LEONARD, MATHEW (ALSO MATTHEW): b. c. 1827, Ireland; married to Catherine Dailey, 1 child; occ. sailor; enl. 6/26/63 for a 1-year term at New York as a first-class fireman; desc. 5'6½", blue eyes, sandy hair, fair comp.; trnsf. from the USRS *North Carolina* to the USS *Monitor* by 3/6/62, ship's no. 33; disch. sometime after surviving the sinking of the vessel off Cape Hatteras, North Carolina, 12/31/62; served on the USS *Granite City*; d. 9/15/64, POW Camp Groce (also Camp Gross), Waller Co., Texas.[407]

LEWIS, SAMUEL AUGEE: b. Baltimore, Maryland; enl. 10/28/62; warranted third asst. engr., 10/29/62; ordered to the USS *Monitor* 10/30/62;

USS *Monitor* Officers' and Crew Biographies

KIA—d. by drowning 12/31/62 when the ship foundered off Cape Hatteras, North Carolina.[408]

LITTLEFIELD, GEORGE: b. c. 1837, Saco, Maine; occ. stonecutter; enl. in New York 9/27/62 for a 1=year term; desc. 5'7½", hazel eyes, light hair, light comp.; assgn. to Washington Navy Yard, 10/28/62; served on the USS *Monitor* as coal heaver, ship's no. 67, from 11/7/62; KIA 12/31/62—d. by drowning w/ the loss of the vessel off Cape Hatteras, North Carolina.[409]

LOGUE, DANIEL C.: b. 9/27/32, Otisville, New York, which was named after his maternal grandfather; father, Dr. John Logue; mother, Ruth Otis; married to Elizabeth A. Cassidy, 2 children; occ. physician; resided 303 Hudson St., New York City, New York; acting asst. surgeon, 1/25/62; ordered to New York Navy Yard, 2/7/62, assgn. to the USS *Monitor*; resigned, 10/7/62; postwar: occ. surgeon; resided Bellmore, Long Island, New York; pension #20941, dated 12/5/1905; invalid #52771, dated 11/16/1905; desc. then as 5'7", blue eyes, brown hair, blonde comp.; d. 2/3/1914.[410]

LOUGHRAN, THOMAS: b. c. 1828, New York City, New York; enl. in Boston as a seaman 8/1/61 for a 3-year term; desc. 5'7", hazel eyes, brown hair, fair comp.; served on the USS *Ohio*; by 3/6/62, trnsf. from the USS *Sabine* to the USS *Monitor* as ship's no. 41; served as 1 of 2 gun captains in the turret during the battle w/ CSS *Virginia*, 3/9/62; dsrtd., 11/6/62; NFR.[411]

MALONE, JAMES: b. c. 1834, Ireland; occ. chandler; enl. in New York for a 1-year term as a landsman, 9/6/62; desc. 4'4½", hazel eyes, black hair, dark comp.; served on the USRS *North Carolina*; assgn. to Washington Navy Yard, 10/28/62; served on the USS *Monitor*, ship's no. 68, 11/7–12/31/62, surviving the vessel's loss off Cape Hatteras, North Carolina; trnsf. to the USS *Brandywine* 8/31/63, disch. from the USS *Ceres*.[412]

MARION, WILLIAM (ALSO MARIAM, MARRION): b. c. 1826, Ireland; occ. sailmaker; enl. in Boston as a seaman for a 3-year term, 8/5/61; desc. 5'1½", blue eyes, dark hair, light comp.; served on the USS *Ohio*; trnsf. from the USS *Sabine* to the USS *Monitor* by 3/6/62, serving as QM, ship's no. 42; survived when the vessel sank off Cape Hatteras, North Carolina, 12/31/62; dsrtd. from the *Ohio* w/ the rank of seaman 2/18/63; NFR.[413]

Appendix V

MASON, JOHN: b. c. 1841, Providence, Rhode Island; enl. 2/27/62 in New York for a 3-year term as a first-class fireman; desc. 5'3½", hazel eyes, brown hair, fair comp.; trnsf. from the USRS *North Carolina* to the USS *Monitor* by 3/6/62, ship's no. 32; survived the sinking off Cape Hatteras, North Carolina, 12/31/62; served on the USS *Patapsco* (on roll, 12/31/64) until that vessel hit a torpedo and sank off Charleston, South Carolina, on 1/16/65; MIA—listed as lost at sea; NFR.[414]

MCPHERSON, NORMAN: enl. as a seaman 1/18/62; trnsf. from the USRS *North Carolina* to the USS *Monitor*, ship's no. 13; dsrtd. 3/4/62; NFR.[415]

MECKLY, THOMAS W.: b. 8/27/40, Milton, Pennsylvania; father, Dr. John Meckly; mother, Rebecca Martin; occ. physician; resided Williamsport, Pennsylvania; graduated w/ honor from Pennsylvania College Medical Dept., '61; appointed surgeon on the USS *Tuscarora* running to Europe; appointed to Medical Corps, U.S. Army as asst. surgeon, 7/62; served in the VI Army Corps, Army of the Potomac, from the Campaign of the Peninsula to Antietam 9/17/62; resigned to accept appointment to USN as acting asst. surgeon, 10/22/62; trnsf. to the USS *Monitor* 10/30/62, served until 11/8/62; trnsf. to the USS *Lodona*, SABSq 12/12/62, assisted in engagements w/ Fts. Moultrie, Sumter, Wagner, Gregg and Btry. Bee 4/7–9/17/63; ship sent to Philadelphia for repairs, subsequently ordered to cruise until her recall after peace was declared, 4/65; resigned 5/13/65, in Philadelphia; postwar: married to Elizabeth E. Frederick, 3 children; resided Jersey Shore, Pennsylvania; occ. ophthalmic surgeon; net worth, 1870, $4,000; member, GAR; invalid pension: #5302 for acute bronchitis; desc. then as 5'6", blue eyes, dark hair, light comp.; d. 2/2/90, Jersey Shore, Pennsylvania.[416]

MONAGHAN, JOHN: b. c. 1841, New York City, New York; enl. in New York as a landsman, 9/13/62, for a 1-year term; desc. 5'9", blue eyes, brown hair, light comp.; assgn., Washington Navy Yard; trnsf. to the USS *Monitor*, ship's no. 69, 11/7/62; served until 12/31/62, when the ship sunk off Cape Hatteras, North Carolina; NFR.[417]

MOONEY, MICHAEL: b. 9/25/37, Ireland; father, James Mooney; mother, Mary Delany; occ. grocery store clerk; resided New York City, New York; enl. 4/61 in Co. H, 12th Regt., New York Infantry Vols., served until 9/61; enl. 2/14/62 in New York for a 3-year term as a coal heaver; desc. 5'7½", blue eyes, brown hair, fair comp.; trnsf., 2/25/62, from the USRS *North Carolina* to the USS *Monitor*, ship's no. 27; second-class fireman, 11/7/62; survived the *Monitor* sinking off Cape Hatteras, North Carolina, 12/31/62; served as a coal heaver until 1/31/63, perhaps on the *North Carolina*; served as coal heaver on the *North Carolina* until 2/24/63; assgn. as second-class fireman to the USS *Catskill* until 8/2/63; casualty, 7/10/63, "exhaustion from heat of fire room"; served on the USS *Wabash* to 11/30/63; the *North Carolina*; disch., Brooklyn Navy Yard, from the USS *August Dinsmore*, 12/28/63; postwar: resided in New York until relocating to 527 2nd St., San Francisco, California; occ. gardener; pension #2800; d. 7/13/1911, Soldier's Home, Los Angeles, California; bur. Los Angeles National Cem.[418]

MOORE, DANIEL: b. Prince William, Virginia; father, Henry, a slave; mother, Sarah; single; arrived on the USS *Monitor* 11/13/62; served as a landsman, probably ship's no. 75 or 76; KIA—d. by drowning 12/31/62, off Cape Hatteras, North Carolina, when the ship was lost; deceased pensioner #1249 for Sarah Moore, dated 3/13/79.[419]

MOORE, EDWARD: b. c. 1837, Scotland; enl. in New York for a 3-year term as a landsman, 2/5/62; desc. 5'4½", gray eyes, dark hair, florid comp.; trnsf., by 3/6/62, from the USRS *North Carolina* to the USS *Monitor*, served as wardroom cook, ship's no. 15; may have been WIA before 11/7/62; disch., 11/28/62.[420]

MORRISON, WILLIAM: b. c. 1842, England; enl. as a landsman in New York, 10/4/62, for a 1-year term; desc. 5'5", hazel eyes, auburn hair, light comp.; assgn., Washington Navy Yard, 10/28/62; served on the USS *Monitor* as coal heaver, ship's no. 64, from 11/7/62 until 12/31/62 when he survived the sinking of the vessel off Cape Hatteras, North Carolina; trnsf. to the USS *Brandywine*; disch., 10/3/63, from the USS *Stepping Stones*.[421]

MURRAY, LAWRENCE: b. c. 1828, New York City, New York; occ. cook; possibly had a wife and child in California; enl. in New York for a 3-year term as a landsman, 2/15/62; desc. 5'6", blue eyes, bald head, fair comp.; trnsf. from the USRS *North Carolina* by 3/6/62 to the USS *Monitor*, serving as wardroom steward, ship's no. 16; when detained in chains for being drunk on duty, he leapt overboard and d. by drowning off Newport News, Virginia, 9/2/62; body recovered 9/5/62 and interred.[422]

Appendix V

Newton, Isaac, Jr.: b. 8/4/37, New York City, New York; father, Isaac Newton Sr., steamboat builder/owner; mother, Anna H. (or Hannah); single; occ. pilot/ engineer/ boat builder/ machinist/mechanic; degree, civil engineering, University of the City of New York, '56; asst. engr. on ocean liner sailing between New York and Liverpool, '57–'58; Engineers' Certificate, State of New York, '59; first asst. engr., 6/15/61; assgn. to the USS *Roanoke* 6/24/61, served on blockade duty, Charleston Harbor; assgn. to New York for special duty, 11/29/61; trnsf. to the USS *Monitor* 2/7/62; superintendent of construction, Office of the General Inspector of Ironclads, New York, 8/14/62; assgn. to special duty, Port Royal, 1/2/63; trnsf. to New York, 3/16/63, to superintend the building of ironclad vessels; detached to work w/ the *Monitor*'s designer, John Ericsson, on the USS *Madawaska* 1/23/65; resigned, 2/8/65; postwar: resided New York City; occ. chief engineer, Croton Aqueduct, Public Works, City of New York; member, Society of Civil Engineers and Society of Mechanical Engineers; d. of suicide (cut throat) due to depression over ill health, New York City, 9/25/84; bur. Greenwood Cem., 9/27/84.[423]

Nichols, William H.: b. c. 1843, Brooklyn, New York; occ. soldier; enl. for a 3-year term as a landsman in New York 2/13/62; desc. 5'4", dark eyes, dark hair, dark comp.; trnsf. from the USRS *North Carolina* to the USS *Monitor* by 3/6/62, ship's no. 17; prmtd. to officer's steward, 11/7/62; survived the ship's sinking off Cape Hatteras, North Carolina, 12/31/62; did not report for duty 2/1/63 on the *North Carolina*; claimed honorable disch., '63, at Fortress Monroe, Virginia; applied, invalid pension #444468; NFR.[424]

Nicklis, Jacob (also Nickles): b. c. 1841, Buffalo, New York; father, William Nicklis Sr.; mother, Catherine; occ. sailor; reenl. in Buffalo 10/13/62 for a 1-year term as an ordinary seaman; desc. 5'7½", gray eyes, light hair, ruddy comp.; assgn. to New York Navy Yard, 10/22/62; assgn. to Washington Navy Yard, 10/31/62; trnsf. to the USS *Monitor*, ship's no. 61, 11/7/62; KIA—d. by drowning when the vessel was lost off Cape Hatteras, North Carolina, 12/31/62; memorial plaque, Nicklis-Leonard family obelisk, Forest Lawn Cem., New York City, New York.[425]

Park, William Dunlap: appointed third asst. engr., 2/4/62; assgn. to the USS *Monitor* 2/7/62; dismissed, 2/17/62; third asst. engr., 12/17 or 23/62 ; assgn. to the USS *Pensacola* 1/15/63; the USS *Richmond* 5/16/63; d. 7/10/63, off Port Hudson, Louisiana; bur. Profit Island, Mississippi River.[426]

USS *Monitor* Officers' and Crew Biographies

PETERSON, CHARLES/PHILIP (OR PHILLIP): it is probable that a "Charles P[illegible], ship's no. 5," who appears on the USS *Monitor* muster roll 3/6/62, and the husband of Mrs. Emma J. Peterson named Philip Peterson—whom she claimed served on the *Monitor* during the engagement w/ the CSS *Virginia*—are the same person: both quartermasters, both supposed to be on the *Monitor* during the battle w/ the *Virginia*, both Scandinavian; Philip Peterson was b. 11/2/34, Christinia, Norway; married to Emma Jean Newman, 5 children; enl. in New York, 6/24/61; volunteered & trnsf. to the *Monitor*; present at the battle w/ the *Virginia* 3/9/62; served on the USS *Currituck*; disch., 6/23/64; reenl. on the *Currituck* 7/8/64; disch., 1/19/65; NFR; postwar: occ. laborer; suffered from consumption; d. 6/27/80, Poland district, village of Russia, New York.[427]

PRICE, CHRISTY (ALSO CHRISTOPHER): b. c. 1830, Ireland; at least 2 children; enl. in New York as a coal heaver, 2/27/62, for a 3-year term; desc. 5'6½", blue eyes, brown hair, fair comp.; trnsf. from the USRS *North Carolina* to the USS *Monitor* by 3/6/62, ship's no. 38; second-class fireman, 11/7/62; survived the sinking of the *Monitor* off Cape Hatteras, North Carolina, 12/31/62; trnsf. to the *North Carolina*; dsrtd, 1/31/63; NFR; postwar: resided Stockbridge, Wisconsin; d. 9/28/1916.[428]

QUINN, ROBERT (ALSO QUIN): b. c. 1840, Ireland; enl. in New York as a coal heaver, 2/18/62, for a 3-year term; desc. 5'7¾", blue eyes, brown hair, fair comp.; trnsf. from the USRS *North Carolina* to the USS *Monitor* by 3/6/62, ship's no. 31; second-class fireman, 11/7/62; survived the *Monitor*'s sinking off Cape Hatteras, North Carolina, 12/31/62; trnsf. to the *North Carolina* but dsrtd. 1/31/63; NFR; postwar: as of '89, resided 661 Washington St., New York City, New York; occ. patrolman, North Police Precinct, New York.[429]

REMINGTON, WILLIAM HENRY (ALSO REMMINGTON): b. 6/7/44, Syracuse, New York; father, William T. Remmington; mother, Caroline; occ. farmer; married to Emoline Cook, died of cancer; second marriage to Mary Bates McCartney, 4 children; enl. 10/30/62 at Washington, D.C., for a 1-year term, rating landsman; trnsf. from Washington Navy Yard to the USS *Monitor*, 11/7/62, served as coal heaver, ship's no. 62; survived the vessel foundering 12/31/62 off Cape Hatteras, North Carolina; probably trnsf. to the USS *Brandywine*; assgn. to Norfolk Navy Yard; dsrtd., 7/8/63; reenl., 12/16/63; disch. near Brandy, Virginia, 12/28/63; postwar: net worth, 1870, $1,000; pension #834554; d. of cancer of the duodenum, 12/5/1919, Osceola, Michigan; bur. 12/7/1919, Davenport Cem., Reed City, Michigan.[430]

Appendix V

RICHARDSON, WILLIAM: b. c. 1836, Philadelphia, Pennsylvania; enl. in New York as a first-class fireman 2/14/62 for a 3-year term; desc. 5'7½", gray eyes, brown hair, florid comp.; trnsf. from the USRS *North Carolina* to the USS *Monitor* by 3/6/62, ship's no. 22; survived the *Monitor*'s sinking off Cape Hatteras, North Carolina, 12/31/62; trnsf. to the USS *Keokuk* 3/7/63, participating in R-Adm. DuPont's attack on Charleston Harbor, South Carolina, surviving the ship's sinking 4/8/63; disch. from the USS *Boston* 5/20/63.[431]

RIDDEY, FRANCIS A. (ALIAS FRANK RYEDAY): b. c. 1830, Philadelphia, Pennsylvania; occ. sailor; married to Margery Buchanan, 2 children; enl. as a seaman at New York, 2/2/62, for a 3-year term; desc. 5'4", blue eyes, brown hair, fair comp.; trnsf. from the USRS *North Carolina* to the USS *Monitor*, ship's no. 8, 30, 37 or 49; dsrtd., 2/21/62; reenl. under the name "Ryeday" at Philadelphia as a seaman for "another" 3-year term 3/29/64; served as gunner's mate on the USS *Princeton*; trnsf., 4/8/64, to the USS *Pocahontas*; dsrtd., 7/29/65; desertion charge removed and a certificate of discharge issued at Brooklyn, New York, 7/29/65; postwar: occ. painter; pension #12988; d. 12/7/98, 1719 Snyder Ave., Philadelphia; bur. Fernwood Cem.[432]

ROONEY, JOHN: b. c. 1832, Brooklyn, New York; occ. sailor, 9 years prev. service in navy; enl. for a 3-year term as an ordinary seaman, New York, 6/6/61; desc. 5'7¼", blue eyes, brown hair, fair comp.; served on the USS *Savannah* until that vessel was disabled; assgn. to the USS *Ottawa*; trnsf. to the USS *Monitor* by 3/6/62, serving as master at arms, ship's no. 14; survived her sinking off Cape Hatteras, North Carolina, 12/31/62; sent to Brooklyn Navy Hosp.; served on the USRS *North Carolina*; disch., 5/29/63; postwar: resided Brooklyn; married to Catherine Mitchell, 2 children; pension #7239; d. 5/25/79, College Hosp., Brooklyn; bur. Holy Cross Cem.[433]

SCOTT, ISAAC H.: b. c. 10/16/36–39, Quebec, Canada; father, John Scott; mother, Mary Rye; occ. cabinetmaker; resided Buffalo, New York; enl. in Buffalo, 10/21/62, for a 1-year term as an ordinary seaman on the USS *Michigan*; desc. 5'11", gray eyes, brown hair, ruddy comp.; assgn. to New York Navy Yard, 10/22/62; trnsf. to the USRS *North Carolina*; assgn. to Washington Navy Yard, 10/31/62; trnsf. to the USS *Monitor*, 11/7/62, ship's no. 60; survived the ship's sinking off Cape Hatteras, North Carolina, 12/31/62; served on the USS *Brandywine* until 2/12/63; trnsf. to the USS *Mount Vernon* off Wilmington, North Carolina, served as a yeoman; disch., 10/23/63, Philadelphia, Pennsylvania; reenl., 12/16/63, Philadelphia, appointed paymaster's steward; served on the USS *Kansas* until resigning at Hampton Roads, Virginia, 4/10/65; postwar: resided Brooklyn, New York; occ. deputy sheriff, coroner, U.S. Customs inspector; married to Catherine Sausse, 9 children; net worth, 1870, $250; pension #26933; member, GAR; d. of arteriosclerosis, 5/17/1927.[434]

SCOTT, WILLIAM: b. c. 1840, a probable slave, Petersburg, Virginia; enl. as first-class boy on the James River, Virginia, 8/10/61, for a 3-year term; desc. 5'2", "light mulatto" comp.; served on the USS *Monitor*, ship's no. 56; survived the sinking of the vessel off Cape Hatteras, North Carolina, 12/31/62; trnsf. to the USS *Brandywine*; trnsf. to the USS *Ben Morgan*; dsrtd., 5/31/63; NFR.[435]

SEERY, JAMES: b. c. 1838, Ireland; enl. in New York 2/18/62 for a 3-year term as a coal heaver; desc. 5'5", hazel eyes, brown hair, fair comp.; trnsf., by 3/6/62, from the USRS *North Carolina* to the USS *Monitor*, ship's no. 8, 30, 37 or 49; disch., 11/28/62.[436]

SELFRIDGE, THOMAS OLIVER, JR.: b. 2/6/36, Charlestown, Massachusetts; father, R-Adm. Thomas Oliver Selfridge Sr.; mother, Louisa Cary Soley; entered USN as cadet mdshp., 10/3/51; graduated 6/53 at the head of a class of six, first officer to receive a diploma under the permanent Naval Academy system; mdshp., 6/10/54; served on the USS *Independence*, PSq, under Josiah Tattnall, later commander of the CSS *Virginia*, doing coast survey work until 9/56; passed mdshp., 11/22/56; acting master, Coast Survey, to 10/56; trnsf., 1/57, USS *Nautilus*, surveying Rappahannock River, Virginia, and Hudson River, New York; master, 1/22/58; served on the USS *Vincennes*, coast of Africa, to suppress slave trade until 4/60; LOA, sick, 1860; lt., 2/15/60; ordered to the USS *Cumberland*, flagship of Home Squadron, 9/60; first action in the war: led a detail to destroy equipment and facilities of Gosport Navy Yard, Virginia, when abandoned by the Union 4/20/61; cmdr., the USS *Yankee*, reconnaissance duty; participated in engagement w/ Conf. btries. on Gloucester Point, Virginia, the first shot fired in the war in defense of the Commonwealth of Virginia 5/3/61; trnsf. back to the *Cumberland*, participated in bombardment and capture of Hatteras Inlet fts. 9/61; volunteered for command of cutting out expedition of boats at Newport News, Virginia, 2/62; second lt. on the *Cumberland*, in command of the forward gun btry., during the engagement w/ the CSS *Virginia* 3/8/62, saving himself when the ship sank by jumping from a gun port and swimming

Appendix V

to a rescue launch; witnessed the battle between the USS *Monitor* and the *Virginia* 3/9/62, from the shore of Hampton Roads; cmdr., the USS *Monitor* 3/10/62, after the wounding of CO Worden; replaced 3/12/62; LOA, 3–4/62; cmdr., USS *Illinois*, 4/62; assgn., flag lt. under Commodore L.M. Goldsborough, cmdr. NABSq; present at recapture of Norfolk and engaged in destroying rebel defenses in the waters of Virginia until 6/62; volunteered to command the submarine torpedo boat USS *Alligator*, destined for service against CSS *Virginia II*; lt. cmdr., 7/16/62; cmdr., ironclad USS *Cairo*, 12/12/62, which was blown up by a torpedo while in command of flotilla of gunboats forcing the passage of the Yazoo River—the first Union vessel sunk by a torpedo in the war—all hands were saved; cmdr., gunboats USS *Conestoga* and USS *Manitou*; cmdr., btry., siege of Vicksburg 5/63, manned by guns and men from the *Manitou*; cmdr., 7/63, flotilla of gunboats, captured the steamers CSS *Louisville* and CSS *Elmira* and engaged in skirmishes w/ guerillas and btries. obstructing navigation of the Mississippi River; the USS *Conestoga* was sunk 3/8/64 by collision w/ the ram CSS *General Price*; cmdr., ironclad USS *Osage*, Red River Expedition, during which ship ran aground and was attacked by a btry. and a brigade of dismounted cav.—he defeated them in an unusual fight between infantry and gunboats, 4/12/64; cmdr., the USS *Vindicator* and 5th division, Mississippi Fleet, between Vicksburg and Natchez; selected by Adm. Porter to accompany him to the east, assgn. to command gunboat USS *Huron* and join the blockade of Cape Fear River, North Carolina; took part in the bombardments of Ft. Fisher, commanding the 3rd division of the assaulting columns of sailors and Marines; took part in the bombardment of Ft. Anderson, Cape Fear River and subsequent capture of Wilmington, North Carolina; on duty, James River, Virginia, when war ends; ordered to Key West, Florida, to intercept Conf. President Davis if he tried to flee the country; three times recommended for promotion by Adm. Porter, selected for a promotion of thirty numbers at the close of the war; postwar: ordered as XO to *Hartford*, preparing to leave for the ASq; requested shore duty to get married to Ellen F. Shepley, 4 children; USNA, instructor, Dept. of Seamanship, '65–'68; cmdr., the USS *Macedonian*, practice cruises, '67–'68; cmdr., the USS *Nipsic*, WISq, '68–'69; prmtd., cmdr., 12/31/69; selected to take charge of the expedition for the surveys of the Isthmus of Darien (now Panama) for inter-oceanic canal,' 69–'74; LOA, managed a mining co., '75–'77; RTD, Navy Yard, Boston, '77–'78; cmdr., the USS *Enterprise*, selected to make a survey of the Amazon and Madeira Rivers, South America, '78; joined ESq, '79–'80; invited as special delegate by Ferdinand de Lesseps to International Canal Congress, 5/79,

USS *Monitor* Officers' and Crew Biographies

Paris, France; presented w/ the Legion of Honor by the French government, made an honorary member of the Royal Geographical Society of Belgium, for his survey work at Isthmus of Darien; capt., 11/26/80; cmdr., Torpedo Station, Newport, Rhode Island, '80–'84, where experiments w/ torpedoes and torpedo nets were conducted; cmdr., the USS *Omaha*, ASq, '85–'87; tried and acquitted by court-martial for carelessness and neglect of duty by conducting target practice in Japanese territorial waters, 6/88; member, Board of Inspection and Survey, '89–'90; commandant, Boston Navy Yard, '90–'94; member, Dry Dock Board, 10/21/93; commodore, 4/11/94; pres., Board of Inspection and Survey, 3/28/94; cmdr., ESq, 11/12/95, w/ rank of acting r-adm., flagship USS *San Francisco*; prmtd., r-adm., 2/28/96; attended coronation, Tsar Nicholas II, 5/96, received Gold Coronation Medal; received in audience by Pope Leo XIII, 3/97; retired, 2/6/98; resided Washington, D.C.; second marriage to Gertrude Wildes; d. 2/4/1924; USS *Selfridge* (DD-357), a *Porter*-class destroyer commissioned 11/25/1936, was named for both Thomas O. Selfridge and his father, R-Adm. Thomas O. Selfridge Sr. [437]

SINCLAIR, HENRY: enl. in New York; served as the USS *Monitor*'s ship's cook, ship's no. 8, 30, 37 or 49; trnsf. from the USRS *North Carolina* to the *Monitor* before 3/5/62, the date on which he dsrtd.; NFR. [438]

SLOVER, JAMES T. (ALSO SLOFER): b. 10/30/17, St. Michaels, Maryland; occ. sailor; married to Sarah Ann Hopkins, 6 children; enl. as a landsman on the USS *Pennsylvania* 7/23/43, serving until 8/19/43; trnsf. to the USS *Union*, serving until 10/31/43; assgn. to Philadelphia Navy Yard until 11/15/43; disch. from the USS *Raritan*, 8/12/46; desc. 5'7½", blue eyes, dark hair, florid comp.; enl. at Baltimore, 4/19/62, onboard the tug USS *Dragon*, serving there until she was sunk by the CSS *Virginia* on 3/9/62; trnsf. to the USS *Monitor*, 5/11/62; served on the USS *Commodore Morris* 12/26/62–8/64, on the USS *Commodore Barney* until 10/64; honorably disch., Appomattox River, 4/2/65; postwar: net worth, 1870, $650; pension # 26943; d. 5/21/99; bur. St. Luke's United Methodist Cem., St. Michaels. [439]

SMITH, CHARLES: b. c. 1834, Rome, New York; enl. as a landsman in New York, 9/27/62, for a 3-year term; desc. 5'3½", gray eyes, brown hair, dark comp.; served on the USRS *North Carolina*; trnsf. to the USS *Monitor* as a coal heaver, ship's no. 67, from 11/7–12/31/62, surviving the sinking of the *Monitor* on that date; trnsf. to the USS *Stepping Stones*; disch., 9/30/63. [440]

Appendix V

SMITH, JAMES: b. c. 1829, Haverstraw, New York; enl. as a landsman in New York, 10/1/62, for a 1-year term; desc. 5'7", blue eyes, brown hair, fair comp.; assgn. to Washington Navy Yard, 10/28/62; served on the USS *Monitor* as a coal heaver from 11/7–12/31/62, when he survived the vessel's sinking; trnsf. to the USS *Brandywine* and then to the USS *Stepping Stones*, finally to the USS *Allegheny*; disch., 9/30/63.[441]

STEARNS, MOSES M.: b. c. 1825, New Hampshire; occ. machinist; married to Ruth Gardiner, 4 children; enl. 7/24/61 for a 3-year term as a seaman at Boston; served on the USS *Ohio* and the USS *Sabine*; prmtd., QM, 9/1/61; trnsf., 2/26/62, to the USS *Monitor*, ship's no. 45; WIA 3/9/62 (hernia, caused by passing shot during battle w/ CSS *Virginia*); assgn. to Washington Navy Yard, 11/8/62; disch. "unfit for service" 11/28/62; postwar: resided 31 Mulberry St., Chelsea, Massachusetts; pension #2279, dated 12/17/75; desc. then as 5'7¼", hazel eyes, dark hair, dark comp.; d. 5/5/97.[442]

STEVENS, THOMAS HOLDUP, JR.: b. 5/27/19, Middletown, Connecticut; father, Commodore Thomas Holdup Stevens Sr., one of the heroes of the Battle of Lake Erie; mother, Elizabeth Read Sage; married to Anna Maria Christie, 9 children; entered USN as mdshp., 12/14/36; cruised on the USS *Independence*, BSq and Coast Survey, '38–'41; studied at Philadelphia Naval School, graduated 3rd in his class; passed mdshp., 7/1/42; naval aide to Pres. John Tyler, '42; survey duty, Gulf of Mexico, '42–'43; assgn. acting master to the USS *Michigan*—U.S. Navy's first iron-hulled warship—on Lake Erie, '43–'44; naval storekeeper, Honolulu, Hawaii, '45–'48; shipwrecked on Christmas Island 1/4/48, while returning home w/wife and daughter aboard the Chilean ship *Maria Helena*, rescued three months later; master, 7/25/48; RTD, assgn. to the Naval Station, Sacketts Harbor, New York, '49; lt., 5/10/49; served on the *Michigan* and *Germantown*, Great Lakes, '49–'51; cmdr., USS *Ewing*, west coast survey work, '52–'55; on USN inactive list, '55–'58, possibly unable to secure an appointment to a naval vessel; USS *Colorado*, Home Squadron, 1/58; trnsfr., USS *Roanoke*, 8/58–'60; net worth, 1860, $11,000; trnsfr., *Michigan*, '61, his station when Civil War broke out; cmdr., the USS *Ottawa*, SABSq under Adm. DuPont, 9/61–'62; took part in the battle of Port Royal, South Carolina, and the capture of Fts. Walker and Beauregard, 11/4–7/62; participated in the Battle of Port Royal Ferry, 1/1/62, and the engagement w/ Flag Officer Josiah Tatnall's Sunboat Squadron, 2/62; commanded the leading vessel in the attack of navy and land forces on Ft. Clinch, Florida, 3/3/62; part of a small force that captured

Fernandina and St. Mary's, Florida, 3/4/62, eliciting a commendation from Cmdr. Percival Drayton; cmdr., first expedition up St. John's River, 3/7–25/62, resulting in occupation of Jacksonville, Florida, and capture of the CSS *America*; LOA, 3/26–4/14/62; cmdr., USS *Maratanza*, NABSq, 4/15/62; participated in the Battle of Eltham's Landing, Virginia, 5/8/62; opened up the Pamunkey River, 5/12/62, in support of the advance of Gen. George McClellan in his Peninsula Campaign; participated in the first expedition to Cumberland and White House to open the James River; engaged a small force of Conf. infantry and artillery near City Point, Virginia, driving them from their positions 5/18/62; accompanied the USS *Aroostook* in a sortie up the river to within sight of Drewry's Bluff to determine Conf. strength, 5/30/62; took part in the Battle of Malvern Hill, 7/1/62; captured gunboat CSS *Teaser*, 7/4/62; prmtd., cmdr., 7/16/62; named cmdr., the USS *Monitor*, 8/9/62, reported onboard 8/16/62; supported McClellan's withdrawal from the Virginia Peninsula, firing the final shots of the navy's Peninsula Campaign, bombarding a plantation on the tip of City Point when pickets fired on the Federal flotilla as it withdrew 8/28/62; replaced as cmdr. of the *Monitor* 9/10/62, mentioned as "a valuable officer" by R-Adm. John Rodgers; assgn. to USS *Sonoma*, "Flying Squadron" under Commodore Charles Wilkes, Caribbean, '62–'63; seized blockade runners *Virginia* and British bark *Springbok*; senior officer, 7/63; cmdr., monitor USS *Patapsco* in actions around Charleston, South Carolina, 8–9/63, including leading an amphibious assault on Ft. Sumter, 9/8/63, which was repulsed w/ casualties; participated in the siege of Ft. Sumter, 10/26–12/6/63, w/ two ship's casualties from a burst gun; joined naval force off Savannah, Georgia, 1/64; captured blockade runner, 2/9/64; 3/64, trnsf. to special duty, New York, under R-Adm. Gregory; cmdr., the USS *Oneida*, WGBSq, 5/27/64; temporary trnsf. to the monitor USS *Winnebago*, Battle of Mobile Bay, 8/2–5/64; resumed command of *Oneida*, 8/18/64, sent to New Orleans for six mos. for repairs; ordered to join blockading force off Texas, 3/65; witnessed Conf. Generals Kirby Smith and J. Bankhead Magruder surrender the last Southern armies, and the surrender of Galveston, Texas, 6/5/65; cmdr., Texas division, Gulf Squadron, 7/65; mentioned as "skillful and brave in all the duties pertaining to your command" by Adm. S.F. DuPont; capt., 7/25/66; Lighthouse Inspector, '66–'70; cmdr., the USS *Guerriere*, flagship of the ESq, '70–'71; Court of Inquiry conducted to investigate the grounding of the *Guerriere*, 7/71, found Stevens at fault, suspending him from rank and duty for three years, reinstated 11/26/72; commodore, 2/73, retroactive to 11/20/72; commandant, Norfolk Navy Yard, '73–'76; U.S. Advisory Board

Appendix V

to Harbor Commissioners of Norfolk and Portsmouth, Virginia; special duty, Norfolk Harbor, '78–'80; prmtd., r-adm., 10/27/79; cmdr., PSq, '80–'81; traveled aboard flagship USS *Pensacola* to Guatemala, met w/ Pres. Justo Rufino Barrios 2/81; pres., Board of Visitors, USNA; retired, 5/27/ 81; post-service: resided Washington, D.C.; occ. author; d. 5/15/96, Rockville, Maryland; bur. Arlington National Cem., Washington, D.C.[443]

STIMERS, ALBAN CROCKER: b. 6/5/27, Smithfield, New York; father, James Stimers; mother, Imogene; married to Julia Ann Appleby, 5 children; appointed in New York as third asst. engr., 1/11/49; assgn. to the USS *Water Witch* 4/3/49; LOA, 5/11/50; RTD to the USS *Michigan* 11/5/50; second asst. engr., 2/26/51; trnsf. to the USS *Powhatan* 5/15/52; the USS *Walker* 7/21/52; assgn. to Office of Engineer in Chief, 6/9/53; first asst. engr., 5/21/53; trnsf. to the USS *San Jacinto* 7/22/54; the USS *Merrimack* 12/8/55; assgn. to special duty, New York, 12/7/57; trnsf. to the USS *Arctic* 5/22/58; assgn. to special duty, New York, 7/24/58; chief engr., 7/26/58; first station in war: the USS *Roanoke*; superintendent for the Ericsson's Btry. Project, 10/4/61, in which he oversaw the construction of the USS *Monitor*; served as "technical passenger" during the ship's trial trip from New York to Hampton Roads, Virginia 3/6–8/62; operated the turret during the battle w/ the CSS *Virginia* 3/9/62, during which he was thrown to the floor when a shell hit the turret while he was leaning his hand against it; left *Monitor*, 4/17/62; collaborated w/ Ericsson building the next classes of ironclad, *Passaic* and *Canonicus*; on board monitor *Passaic* during attack on Ft. McAllister, South Carolina, 3/5/63; observed and reported on the ironclad monitors in action at Ft. Sumter, South Carolina, 4/7/63; tried and acquitted by a Court of Inquiry for spreading falsehoods and conduct unbecoming an officer, 5–9/63; in charge of failed project to construct light-draft monitors of the *Casco* class '63–'64—an investigating committee ruled that he had committed a professional error; assgn. to the monitor USS *Tunxis*; trnsf., the USS *Wabash* 10/15/64; LOA w/ permission to visit Europe 3/9/65; RTD to the USS *Powhattan* 7/28/65; resigned, 8/3/65; postwar: occ. civil engineer; net worth, 1870, $15,000; d. of smallpox 6/3/76, Staten Island, New York; widow's pension #3048, dated 5/29/83.[444]

STOCKING, JOHN (ALIAS OF WELLS WENTZ): b. 8/18/30, Binghamton, New York; enl. in New York as a boatswain's mate, 1/25/62, for a 3-year term; desc. 5'8", blue eyes, brown hair, florid comp.; trnsf., by 3/6/62, from the

USS *Monitor* Officers' and Crew Biographies

USS *Sabine* to the USS *Monitor*, ship's no. 43; served as 1 of 2 gun captains in the turret during the battle w/ CSS *Virginia*, 3/9/62; KIA—d. by drowning, one eyewitness claims to have seen him swept from the deck as the vessel sank off Cape Hatteras, North Carolina, 12/31/62; it is unknown why he enl. under an alias.[445]

STODDER, LOUIS NAPOLEON: b. 2/12/37, Boston, Massachusetts; father, John Low Stodder; mother, Almira Fuller; married Watie Howland Alderich, 2 children, separated but never divorced; second marriage Rose B. Champlin, 1 child; occ. sailor; appointed at Boston in USN as acting master, 12/26/61; assgn. to the USS *Monitor* 1/31/62; stationed at the wheel turning the turret during the battle w/ the CSS *Virginia* 3/9/62; WIA on that date (knocked unconscious when a shell struck the turret when he was leaning against it; recovered several hours later); survived the sinking of the *Monitor* 12/31/62; received a commendation for his actions during the loss of the vessel from CO J.P. Bankhead, 1/1/63; trnsf. to the USS *Rhode Island*; acting vol. lt., 1/10/63; cmdr., the USS *Release* 1/20/63; cmdr., the USS *Adela* 5/15/63, EGBSq and Hampton Roads, 6/63–11/64, participating in shelling of Ft. Brooke, Florida, and raid on blockade runners, 10/17/63, w/ casualties among his crew; trnsf., the USS *Niphon* 2/6/65, the USS *Calypso* 2/22/65; postwar: assgn. to the Potomac Flotilla, guarding Washington, D.C., following Pres. Lincoln's assassination, 4/14/65; assgn. to Charleston Harbor, SABSq, 4/24–6/65; LOA, 8/7/65; third lt., U.S. Revenue Cutter Service, 10/27/65; honorably disch., USN, 11/20/65; prmtd. to second lt. 6/6/66; to first lt. 7/20/70; to capt., 11/12/79; cmdr., the USS *Oliver Wolcott*, acted to quell Indian uprising, Ft. Simpson, British Columbia, Canada, 1/11/83; supervisor of anchorages, Port of New York, '92–1901; retired, 4/12/1902; resided 284 Kingston Ave., Brooklyn, New York; donated log book from the *Monitor* to USN, 7/27/1910; d. of cerebral apoplexy and pulmonary edema following a nervous breakdown 10/8/1911, Brooklyn; bur. 10/11/1911, Green-Wood Cem., Brooklyn.[446]

SUNSTROM, MARK TRUEMAN: b. 1844, Baltimore, Maryland; father, Robert C. Sunstrom; mother, Rachel; occ. bookkeeper; married to Rhoda M. Bullock; third asst. engr., 2/1/62; assgn. to the USS *Monitor* 2/7/62; participated in engagement w/ CSS *Virginia* 3/9/62; survived the ship's sinking, 12/31/62; second asst. engr., 10/15/63; served on the USS *Unadillo*, *Pontoosuc* and *Ashuelot*, NABSq; resigned, 11/10/65, on account of disability incurred in service; postwar: occ. clerk; resided 161 Barre St., Baltimore,

Appendix V

Maryland; net worth, 1870, $3,500; member, Masons, Knights of Pythias; d. of consumption of the throat 10/24/75; bur. Mt. Olivet Cem.[447]

SYLVESTER, CHARLES F. (ALSO SYLLIVESTER, SYLLWESTER AND SYLVESTER, CHARLES W.): b. c. 1841, Sweden; enl. in New York as a seaman, 1/22/62, for a 3-year term; desc. 5'7¾", blue eyes, flaxen hair, florid comp.; trnsf., by 3/6/62, from the USRS *North Carolina* to the USS *Monitor*, ship's no. 1; arrested for fighting w/ a fellow crewman; helped raise shot to the cannon in the battle w/ CSS *Virginia* 3/9/62; survived the sinking of the *Monitor* 12/31/62; served on the USS *Mahaska*, *Nipsic* and *Princeton*; disch., 2/4/65, from the USS *Philadelphia*.[448]

TESTER, ABRAHAM: b. c. 1839, England; occ. machinist; enl. as a first-class fireman in New York, 2/14/62, for a 3-year term; desc. 5'1½", blue eyes, brown hair, fair comp.; trnsf., 3/6/62, from the USRS *North Carolina* to the USS *Monitor*, ship's no. 21; survived the ship's foundering off Cape Hatteras, North Carolina, 12/31/62; trnsf. to the USS *Florida*; acting third asst. engr., 2/18/64, ordered to the USS *Montgomery*; LOA, 6/14/65; honorably disch., 8/21/65; postwar: occ. engineer; resided 481 Smith St., Brooklyn, New York; married to Bridget Graham, 1 child; d. of pneumonia 3/22/78; bur. Holy Cross Cem., Brooklyn.[449]

TOFFEY, DANIEL: b. 12/22/37, Pawling, New York; father, George A. Toffey; mother, Mary De Riemer; nephew of *Monitor* CO John Worden; married to Adeline S. Wilson, 3 children; occ. clerk; served as capt.'s clerk onboard the USS *Monitor*; during the 3/9/62 conflict w/ the CSS *Virginia* he acted as messenger between the CO in the pilothouse and the XO in the turret and claims he caught his uncle when he fell, WIA; left *Monitor*, 4/17/62; presented w/ a gold medal by Congress for war service; postwar: occ. cattle broker, later an alderman; resided Bergen Ave., Jersey City, New Jersey; net worth, 1870, $10,000; d. 2/9/93.[450]

TRUSCOTT, PETER (ALIAS OF SAMUEL LEWIS): b. c. 1839, Ireland; occ. sailor, w/ five years prev. service in USN; enl. in New York as a seaman, 1/20/62, for a 3-year term; desc. 5'4¾", gray eyes, brown hair, fair comp.; trnsf. from the USRS *North Carolina* to the USS *Monitor* 3/6/62, ship's no. 11, serving as QM; stationed in the turret, passing shot to one of the two guns during the battle w/ CSS *Virginia* 3/9/62; WIA on that date (concussion caused when a shell hit the turret while he was leaning against it), returned to station after treatment; survived the *Monitor*'s sinking 12/31/62; served on the USS *Princeton* and *Catskill*; prmtd., acting master's mate, 4/13/63; NFR.[451]

VIALL, THOMAS BROWN: b. 1/2/38, Bristol, Rhode Island; father, Ezra B. Viall; mother, Julia (or Juliette); married to Margaret Pearce Phillips; occ. engineer; enl. New Bedford, Massachusetts, for a 1-year term as a seaman, 6/11/61; desc. 5'6½", blue eyes, light hair, dark comp.; assgn. to the USS *Ohio*, Charleston Navy Yard; 6/27/61, the USS *Vincennes*; 11/30/61, the USRS *North Carolina*; trnsf. to the USS *Monitor* 2/25/62, ship's no. 12; claimed to have contracted chills, fever and dysentery, which resulted in kidney disease and rheumatism, 6/14/62; disch., Harrison's Landing, James River, Virginia, 7/19/62; served on the USS *Big Bells* 5/63, after vessel was sold, reenl. as a seaman aboard the USS *Onward* for a 1-year term, 8/6/63; prmtd. to QM, 2/22/64; disch., Montevideo, 2/7/65; acting master's mate, 2/8/65; served as mate on the *Onward*; LOA, 6/16/65; RTD, 7/12/65, the USS *Vermont*; resigned, 11/6/65; acting master's mate revoked, 11/18/65; postwar: occ. steam roller operator; invalid pension #35242, rejected 12/1/04; d. of Bright's disease 11/11/1905; bur. Brayton Cem., Apponaug, Warwick, Rhode Island.[452]

WALSH, DANIEL (ALSO WALCH, WELCH): b. c. 1835, Ireland; father, John; mother, Catherine Moran; occ. carpenter; enl. in Boston, 7/21/61, for a 2-year term as a seaman; possibly served on the USS *Ohio*; trnsf. from the USS *Sabine* to the USS *Monitor* by 3/6/62, likely ship's no. 8, 30 or 37, ranking of coxswain; dsrtd., 11/6/63; NFR; postwar: resided 253 Oak St., Lawrence, Massachusetts; possibly married to Margaret O'Donnell, widowed; possible second marriage to Ellen McAuliff; d. 4/9/1906; bur. Immaculate Conception, Lawrence.[453]

WATTERS, JOSEPH (ALSO WATERS): b. c. 1837, Bordentown, New Jersey; father, John; mother, Mary; occ. machinist; married to Caroline Kelly, 2 children; third asst. engr., 10/22/60; assgn. to the USS *Crusader* and the USS *Winona*; second asst. engr., 12/8/62; served on the USS *Ossipee*, *Monitor*, *Seneca*, *Mingoe*, *Chippewa* and *Chattanooga*; trnsf. to the *Monitor* to replace Acting Chief Engineer Albert Campbell when he was injured 12/28/62; survived the sinking of the ship 12/31/62; first asst. engr., 12/1/64; trnsf. to the USS *Kickapoo*, a twin-turret ironclad, New Orleans, Louisiana, 4/11/66; d. in service, of yellow fever 9/13/66, New Orleans; widow's pension #953, dated 3/20/67.[454]

Appendix V

WEBBER, JOHN JOSHUA NATHANIEL: b. 8/3/29, Brooklyn, New York; father, Edward, from Scandinavia; mother from Wales; married to Nancy, 5 children; began nautical career 1845; serving as acting master on the USS *Monitor* by 3/4/62; participated in the battle w/ CSS *Virginia* 3/9/62, in charge of the powder division on the berth deck; NFR; postwar: res. Kings and Richmond, New York; d. of cardiac asthma, 10/1909, Sailor's Snug Harbor, New York City, New York.[455]

WEEKS, GRENVILLE MELLEN (ALSO GREENVILLE): b. 11/22/37, New York City, New York; father, Dr. Cyrus Weeks; mother, Maria L. Child; occ. physician; College of Physicians and Surgeons (Columbia) '59–'60; MD, University Medical College (NYU) '61; married to Helen Campbell Stuart, 1 child; attached to 72nd Regt. NGNY, volunteered for service in U.S. Army after 1st Battle of Bull Run; appointed asst. surgeon, 7/1/62, ordered to the USS *Valley City*; LOA, sick, 9/3/62; RTD, the USS *Brandywine* 10/31/62; trnsf. to the USS *Monitor* 11/8/62; survived the sinking of the vessel 12/31/62; WIA on that date (right hand caught between rescue boats; shoulder dislocated, three fingers amputated, never regained use of right arm); LOA 1/4/63; prmtd. by Pres. Lincoln, appointed consulting surgeon w/ rank of major; brigade surgeon and acting medical director, Dept. of Florida, at end of war; postwar: occ. secretary, Indian Commission, Pacific Northwest, later returned east and resumed medical practice, '71; second marriage to Maria Oberg; third marriage to Pauline M. Sauer; author, Resolution Recognizing Cuban Independence by U.S. Congress, '98; d. 4/26/1919, Home for Disabled Soldiers, Kearny, New Jersey.[456]

WHITE, GEORGE H.: warranted acting third asst. engr., 5/25/61; assgn. to the USS *Colorado*, 5/31/61; LOA, 6/27/62; RTD, trnsf. to the USS *Monitor* 8/18/62; third asst. engr., 10/29/62, assgn. to the ironclad USS *Nantucket*; participated as senior engr. w/ ironclad fleet in the action around Ft. Sumter, Charleston Harbor, South Carolina, 4/8/63, receiving special mention by D.M. Fairfax, cmdr.; second asst. engr., 1/1/65; served on the USS *Ottawa*; captured by rebel forces 4/7/65 and sent to Baldwin, eventually traded; first asst. engr., 8/28/65; trnsf. to the USS *Swatara* 9/25/65; first asst. engr., 7/18/66; assgn. to the Navy Yard, Mare Island, California, 6/7/69; special duty bringing ironclads to Philadelphia Yard 5/4/71; NFR.[457]

WILLIAMS, PETER: b. c. 1831, Norway; occ. sailor, w/ nine years prev. experience; residence California; enl. as a seaman for a 3-year term in New York, 1/27/62;

desc. 5'4", blue eyes, brown hair, florid comp.; served on a Norfolk packet; trnsf. from the USRS *North Carolina* to the USS *Monitor* by 3/6/62, serving as QM, ship's no. 23; stationed in the pilothouse at the wheel steering the ship during the battle w/ CSS *Virginia* 3/9/62; the only man to receive the Medal of Honor for his service on the *Monitor*; citation stated, "Serving on board the U.S.S. Ironclad Steamer *Monitor*, Hampton Roads, 9 March 1862: During the engagement between the U.S.S. *Monitor* and the C.S.S. *Merrimack*, WILLIAMS gallantly served throughout the engagement as quartermaster, piloting the *Monitor* throughout the battle in which the *Merrimack*, after being damaged, retired from the scene of the battle"; mate, 3/25/62; acting master's mate, 3/28/62, due to that heroic service; survived the sinking of the *Monitor* 12/31/62, commended for his actions on that occasion by CO J.P. Bankhead, 1/1/63; acting ensign, 1/10/63, assgn. to the USS *Florida*; cmdr., the USS *Clematis*, WGBSq, 12/65; honorably disch., 11/9/67.[458]

WILLIAMS, ROBERT: b. 1832, Wales; occ. boilermaker, sailor, w/ seven years prev. experience; enl. in New York as a first-class fireman, 2/15/62, for a 3-year term; desc. 5'8½", hazel eyes, black hair, swarthy comp.; trnsf. from the USS *North Carolina* to the USS *Monitor*, ship's no. 4, by 3/6/62; KIA—d. by drowning when the ship sank off Cape Hatteras, North Carolina, 12/31/62.[459]

WORDEN, JOHN LORIMER: b. 3/12/18, Mount Pleasant Township, Westchester Co., New York; father, Ananias Worden; mother, Harriet Graham; grandson of Surgeon Andrew Graham, member of Committee of Public Safety, Revolutionary War; married to Olivia Aiken Toffey, 4 children; appointed mdshp., 1/10/34, served on the USS *Erie*, BSq, '34–'37 and the USS *Cyane*, MSq, '38–'39; attended Philadelphia Naval School, passed mdshp. 7/23/40; served on the USS *Relief* ('40–'42) and the *Dale* ('42–'43), PSq; assgn. to the Naval Observatory, Washington, D.C., 4/7/44–'46; master, 8/15/46; lt., 11/30/46; served on the West Coast as XO on the USS *Southampton* during the Mexican-American War, afterward served on the USS *Independence*, *Warren* and *Ohio*, PSq; LOA, '49; assgn. to the Naval Observatory, '50–'52; cruises in the MSq (on the USS *Cumberland*) and Caribbean Squadron (on the USS *Levant*) 4/52–2/55; back to the Naval Observatory, 10/55–3/56; trnsf. to Brooklyn Navy Yard, 3/56–7/58; first lt. on the USS *Savannah*, Home Squadron, 7/1/58–11/20/60; first action in the war: reported to Washington, D.C., 4/6/61, accepted mission to take secret orders to Pensacola Squadron for the reinforcement of Ft. Pickens, Florida, accomplished mission 4/12/61; arrested 4/13/61 to become the

Appendix V

Union's first prisoner taken in the war; POW, seven months, Montgomery, Alabama jail, exchanged, 11/22/61; LOA, recovery from incarceration, 11/20–12/2/61; RTD, New York Navy Yard, 12/3/61; offered command, the USS *Monitor*, 1/13/62; reported for duty as cmdr., 1/16/62, supervised her completion at Greenpoint, Long Island; onboard vessel with Ericsson on launch day, 1/30/62; ordered to proceed to Hampton Roads, 2/20/62, but due to needed repairs and bad weather, didn't raise anchor until 3/6/62; arrived Hampton Roads, 3/8/62, after suffering seasickness during the trip; commanded the vessel in the 3/9/62 battle against the ironclad CSS *Virginia*, WIA on that date (minor concussion, shrapnel and severe burns on the face, blinded in the left eye when a shell exploded near the pilothouse observation slit as he was looking out); recovered from wounds in Washington, D.C., visited by Pres. Lincoln and lauded by Congress; presented by the State of New York w/ a ceremonial sword made by Tiffany's; prmtd., cmdr., 7/16/62; asst. to Adm. Gregory in the supervision of building ironclads, Brooklyn, 8/62; cmdr., ironclad USS *Montauk*, SABSq, 10/8/62; returned to Hampton Roads, 12/29/62, the same day the *Monitor* was leaving for her fate off Cape Hatteras, North Carolina; participated in attacks of Ft. McAllister, Georgia, 1/27 and 2/1/63; capt., 2/3/63; destroyed the Conf. privateer CSS *Nashville* 2/28/63; participated in ironclad fleet attack of Charleston, South Carolina, and bombardment of Ft. Sumter, 4/7/63; assgn. to Washington Navy Dept., advising on development of new monitors, 5/63–2/66; postwar: cmdr., the USS *Idaho* 2/1–5/23/66, and the USS *Pensacola* 8/6/66–5/8/67; LOA '67–'69, w/ permission to visit Europe; commodore, 5/27/68; superintendent, USNA 12/1/69–'74, during which the first minority mdshp. was admitted; net worth, 1870, $21,000; prmtd., r-adm., 11/20/72; cmdr. in chief, ESq, '74-'77, concentrating naval forces in Turkish waters after Russia declares war on that nation, 4/24/77; Naval Retiring & Examining Boards '77–'86, pres. '82 & '83; retired, 12/23/86, w/ full sea pay; resided, Washington, D.C.; member and pres., Washington Metropolitan Club; d. of pneumonia 10/18/97, Washington, D.C.; state funeral, St. John's Episcopal Church, Washington, D.C., 10/20/97, w/ Pres. William McKinley, cabinet members and military leaders in attendance; bur. Pawling Cem., Pawling, New York; widow's pension granted by Congress, 1/19/99; USN has named four ships in his honor: USS *Worden* (torpedo boat destroyer #16), 1902–'20; USS *Worden* (destroyer #288, later DD-288), '20–'31; USS *Worden* (DD-352), '35–'44; and USS *Worden* (DLG-18, later CG-18), '63–2000; Ft. Worden in Port Townsend, Washington, was also named in his honor, as was the parade field at the USNA.[460]

Appendix VI
USS *Monitor* Purported Crew Members

The USS *Monitor*, as well as her officers and crew, secured never-ending fame for dueling the Confederate ironclad *Virginia* to a standstill on 9 March 1862. The crew was fêted everywhere they went in 1862, as William Keeler wrote about Daniel Toffey's visit to Washington, D.C.: "The Capt.'s clerk was recently sent to Washington on business—he staid [sic] two or three days at Willard's & returned without having spent a cent. The *Monitor* was enough."[461]

Keeler noted that he and others were treated the same way. During the postwar era, being a *Monitor* Boy placed you as a true war hero. The *Monitor* was the ship that had saved the nation, and those veterans associated with the ironclad were honored for the rest of their lives.

As the years passed, a few naval veterans sought a glorious connection with the USS *Monitor*. Accordingly, three individuals have certain references indicating that they were *Monitor* Boys, but naval records do not support their claims. Peter Brodie and Thomas Taylor were both in Hampton Roads on 9 March 1862, but they were not on the *Monitor*. Taylor was on the USS *Roanoke* on 8 and 9 March 1862. Brodie was a mate on the USS *Cumberland* when the vessel was sunk by the Confederate ironclad *Virginia*. Of course, being there and surviving the eighth of March was a significant occurrence. Perhaps, as these men passed on in years, they expanded their service in Hampton Roads to the *Monitor* since they were there when the little Union ironclad beat back her Confederate foe. As for Andrew Fenton, he has no naval record. Nevertheless, men like Brodie realized that there was true magic in being one of the *Monitor* Boys.

Andrew Fenton's obituary on 19 April 1945 stated that he "served on the ironclad *Monitor* during the Civil War, died in the Soldiers' Home at

Appendix VI

the age of 101. Mr. Fenton was flag bearer when the United States took possession of Alaska."[462] If Fenton had actually served on the *Monitor*, his story would have been an awesome ending to the *Monitor* Boys' story. As the last person to claim to be one of the *Monitor* Boys, he witnessed the evolution of ironclads to steel battleships and then to the advent of naval aviation, which ended the style of warships introduced by the *Monitor*. Unfortunately, there is no proof other than his obituary to document his service on the *Monitor*.[463]

Others who were not known to be in Hampton Roads still boasted in their later years that they served on the *Monitor*. James H. Carey created a tall tale that during the battle with the CSS *Virginia* he "found hot water more effective than shells, boasting that when the ships were close 'we hooked up a fire hose to our boiler and when they opened the shutters we squirted it in. We put several crews out of commission that way.'"[464] While a few scholars believe that Carey was actually James Corwey, no documentation links the two names. Furthermore, no other documentation exists that this event occurred during the battle.

Lawrence Fay also offered stories about the battle, and in his obituary it was stated that he "insisted until his death in the late 1920s that 'the *Monitor* would have won if we'd had just one more ball.'"[465] Others, like Peter Omer and Andrew Fenton, also made claims that they had served on the *Monitor*; however, there is no indication for these men of any naval service.

The descendants of John Price insisted that their ancestor had served on the *Monitor* under an assumed name. The family has a treasured heirloom that has an inscription on a silver plate that reads: "A sliver of the Pilot house of the War Battery Monitor slivered by a cannon blast Fired from the Rebel War Battery Merman in the fight March 9th 1862 in Hampton Roads."

These sailor stories were all tall tales fabricated by men who, in their later years, all dreamed to have been part of the great duel between ironclads.

BRODIE, PETER H.: b. 3/10/45, Ireland; father, John Brodie; mother, Catherine Dorisey; married to Mary J., 8 children; occ. wire millworker; resided Cuyahoga Falls, Ohio; enl. at the age of sixteen, Brooklyn Navy Yard, as a drummer boy 12/20/61; served as "powder boy" until prmtd. to mate, directing the work of the gunners; WIA, twice; served on the USS *Ohio*, *Undine*, *Maria Denning* and, finally, the *Cumberland* and was onboard when it was sunk by the CSS *Virginia*, 3/8/62; served on gunboat assisting Gen. Grant at Vicksburg and on the Mississippi and Missouri Rivers, transporting troops; disch., 8/20/65; postwar: occ. supervisor, paper mill; resided 184

USS *Monitor* Purported Crew Members

Allen St., Cuyahoga Falls; member and post cmdr., GAR; d. of "lobar" (pleural) pneumonia 3/22/1917; bur. 3/24/1917 St. Joseph's Catholic Church, Cuyahoga Falls.[466]

CAREY, JAMES H. (ALSO POSSIBLY CORWEY, JOHN/JAMES): purported to have served on the USS *Monitor* 3/9/62, during the battle w/ the CSS *Virginia*; claimed he used a fire hose hooked up to the boilers to squirt hot water thru open porthole shutters of the Conf. ironclad, putting several crews out of commission; NFR.[467]

CORWEY, JOHN/JAMES (ALSO POSSIBLY CAREY, JAMES): trnsf. from the USS *Stepping Stones* to the USS *Monitor*; served as second-class fireman; NFR.[468]

FAY, LAWRENCE: claimed to have served on the USS *Monitor* during the battle w/ the CSS *Virginia* 3/9/62; d. late 1920s; NFR.[469]

FENTON, ANDREW: b. c. 1844; married, 1 child; purported to have served on the USS *Monitor*; postwar: d. 4/19/1945, Soldier's Home, Vineland, New Jersey; NFR.[470]

OMER, PETER H.: b. 4/14/34; claimed to have served on the USS *Monitor*; NFR; postwar: lived at Soldier's Home, Hampton, Virginia; member, GAR, Western Steam Fire Engine Co.; d. of cancer of the stomach 1/23/97, Philadelphia; bur. Woodland Cem.[471]

PRINCE, JOHN: family legend has it that he joined the USS *Monitor* when under legal age and under an assumed name, apparently enlisting under his real name when becoming of age; claimed to have a sliver of the *Monitor* taken as a souvenir from the damage incurred during the battle w/ the CSS *Virginia* 3/9/62; NFR.

TAYLOR, THOMAS L.: b. 8/47, a slave at Evanston Plantation, Cole's Point, North Carolina; married to Lillian E., 3 children; originally assgn. to the USS *Roanoke*; purported to have served as an ammunition bearer ("powder monkey") onboard the USS *Monitor* during the battle w/ the CSS *Virginia* 3/9/62; is purported to have been WIA on that date (wounds caused by shrapnel in the turret) and recovered in hosp. at Fortress Monroe; purported to have officially joined USN after recovering from injuries sustained during that conflict; served on "other monitors"; honorably

Appendix VI

disch., 7/13/65; postwar: occ. mariner, later a janitor; resided Putnam, Connecticut; net worth, 1930, $1,500; member, GAR; d. of heart attack due to wounds sustained when beaten and robbed, 3/7/1932; bur. Grove St. Cem., Putnam.

Notes

CHAPTER 2

1. Quarstein, CSS *Virginia*, vi.
2. U.S. Department of the Navy, *Official Records of the Union and Confederate Navies*, series 2, vol. 2, 78 (hereafter cited as ORN, series #, vol. #: page #).
3. ORN, 2, 2: 676–68.
4. Quarstein, *CSS Virginia*, 191–93.

The proper name for the Confederate ironclad that fought the USS *Monitor* to a standstill in Hampton Roads on 9 March 1862 is the CSS *Virginia*. The Confederate ironclad's name, however, is consistently expressed in an inaccurate fashion. The most common usage is *Merrimac*. This reference, used alike by Civil War participants and historians ever since, is incorrect. The steam-powered, forty-gun frigate with a screw propeller built at Charlestown Navy Yard, Boston, Massachusetts, was named the USS *Merrimack* by John Lenthall, chief of the U.S. Bureau of Naval Construction, on 25 September 1854. Naval Constructor E.H. Delano, who designed the frigate, noted in his plans the ship's name as *Merrimack*. The frigate was the first of a class of six frigates built during the 1850s. Each of the ships was named for an American river: *Roanoke*, *Wabash*, *Colorado*, *Minnesota* and *Niagara*. President Franklin Pierce was a native of Concord, New Hampshire, the county seat of Merrimack County and located on the Merrimack River. He signed the act approving the appropriation and ship names on 6 April 1854, and the frigate to be built at the Charlestown Navy Yard in Boston was spelled *Merrimack*.

Even though this evidence clearly documents that the frigate's name should always be spelled with a "k," as it was named in honor of the Merrimack

River, confusion concerning the river's name's spelling is commonplace. The name "Merrimack" is a Native American word said to mean "swift water." By the mid-nineteenth century, many writers, Henry David Thoreau excepted, had begun to drop the "k." It appears that the spelling "Merrimack" is more often used at places along the river above Haverhill, New Hampshire, a town located at the head of navigation. "Merrimac" without the "k" is the popular spelling below Haverhill. The river formed the Merrimack Valley, which was often referred to as "Merrimac" Valley. This region was a major textile manufacturing area. One town in the region is named Merrimac, but it was not established until 1876. This circumstance and the fact that it is easier to spell Merrimack with just a "c" rather than a "ck" is why so many Civil War contemporaries use "Merrimac" when writing about the frigate.

Once the Confederates raised the burned hull of the frigate, she was reconfigured into an ironclad and christened on 17 February 1862 as the CSS *Virginia*. Confederate Secretary of the Navy Stephen R. Mallory and Flag Officer Franklin Buchanan, the ironclad's commander, both referred to the ironclad in all of their correspondence after this date as the *Virginia*. Consequently, from February 1862, the ironclad should always be called the *Virginia*. Few, unfortunately, recognize this technicality. Southern journalists usually referred to the vessel by her rechristened name CSS *Virginia*; however, Northern newspapers consistently used the name *Merrimac* without the "k."

Since this volume is all about the *Monitor* Boys, I used the name *Merrimac* when referencing the Confederate ironclad from their point of view. When discussing the Federal opinions about their enemy's warship, I opted to use the correct spelling *Merrimack*. The name *Virginia* is used when the Confederates discuss their ironclad.

5. *Mobile Register*, August 11, 1861.
6. C.S. Bushnell to Gideon Welles, March 9, 1877, Gideon Welles's Papers, Henry E. Huntington Library, San Marino, California.
7. Beale, *Diary of Gideon Welles*, vol. 1, 214; Ericsson, "Negotiations for the Building of the '*Monitor*,'" 750.
8. Ericsson, "The Building of the *Monitor*," 731–32.
9. USS *Monitor* Log, The Mariners' Museum, Newport News, Virginia (hereafter cited as *Monitor* Log).

Chapter 3

10. ORN, 1, 6 :517.
11. Ibid., 518.
12. Ericsson, "The Building of the '*Monitor*,'" 733.
13. Daly, *Aboard the USS* Monitor, 2.

14. ORN, 2, 1: 148.
15. Daly, *Aboard the USS* Monitor, 7.
16. Ibid., 8.
17. Ibid., 9.
18. Ibid., 10.
19. *Brooklyn Daily Eagle*, March 29, 1908.
20. Butts, "The Loss of the *Monitor*," 741.
21. ORN, 1, 6: 671.
22. Stodder, "Aboard the *Monitor*," 31.
23. ORN, 2, 1: 163.
24. Berent, *Crewmen of the USS* Monitor, 30.
25. Daly, *Aboard the USS* Monitor, 10.
26. Ibid., 11.
27. ORN, 1, 25: 757.
28. Lewis, "Life on the *Monitor*," 258.
29. *Monitor* Log.
30. Marvel, *The* Monitor *Chronicles*, 21.
31. Berent, *Crewmen of the USS* Monitor, 29.
32. Daly, *Aboard the USS* Monitor, 17.
33. Ibid.
34. Hawthorne, "Chiefly about War Matters," 58.
35. Daly, *Aboard the USS* Monitor, 20.
36. Ibid.
37. *Norfolk Day Book*, January 7, 1862.
38. ORN, 1, 6: 679.
39. Daly, *Aboard the USS* Monitor, 18.
40. ORN, 1, 1: 338.
41. Ericsson, "The Building of the *Monitor*," 731.
42. Ellis, *The* Monitor *of the Civil War*, 22.
43. ORN, I, 6: 684.

Chapter 4

44. Berent, *Crewmen of the USS* Monitor, 33.
45. Daly, *Aboard the USS* Monitor, 27.
46. Stimers, "An Engineer Aboard the *Monitor*," 29.
47. Daly, *Aboard the USS* Monitor, 28.
48. Greene, "In the *Monitor*'s Turret," 720.
49. Daly, *Aboard the USS* Monitor, 28.
50. Mendell, *Technology and Experience aboard the USS* Monitor, 67.
51. Stimers, "An Engineer Aboard the *Monitor*," 29.
52. Berent, *Crewmen of the USS* Monitor, 34.
53. Greene, "In the *Monitor*'s Turret," 721.
54. Daly, *Aboard the USS* Monitor, 29.
55. Greene, "In the *Monitor*'s Turret," 721.

56. Ibid.
57. Daly, *Aboard the USS Monitor*, 30.
58. U.S. War Department, *War of Rebellion*, series I, vol. 5: 42 (hereafter cited as ORA, series #, vol. #: page #).
59. Thompson and Wainwright, *Confidential Correspondence of Gustavus Vasa Fox*, vol. 1, 285.
60. ORN, 1, 6: 333–34.
61. Goldsborough, "Narrative of Rear Admiral Goldsborough," 1028.
62. ORN, 1, 6: 554.
63. ORN, 1, 6: 672.
64. ORN, 1, 6: 776–77.
65. Ibid.
66. ORN, 1, 6: 778.
67. Phillips, "The Career of the Iron-Clad *Virginia*," 201.
68. Littlepage, "The Career of the *Merrimac-Virginia*," 44.
69. *Norfolk Day Book*, March 10, 1862.
70. Wood, "The First Fight of the Ironclads," 696.
71. Norris, "The Story of the Confederate States' Ship '*Virginia*,'" 205. (Page citations refer to the reprint edition.)
72. Littlepage, "The Career of the Iron-Clad *Virginia*," 44.
73. Wood, "The First Fight of the Ironclads," 696.
74. Reaney, "How the Gun-Boat *Zouave* sided the *Congress*," 714–15.
75. U.S. Department of the Navy, "Subject File of the Confederate States' Navy" (hereafter cited as Ramsay Narrative).
76. ORN, 1, 7: 23.
77. Reaney, "How the Gun-Boat *Zouave* sided the *Congress*," 715.
78. Ramsay, "The Most Famous of Sea Duels," 11.
79. O'Neil, "Engagement Between the *Cumberland* and *Merrimack*," 893.
80. Jones, "Services of the *Virginia*," 69.
81. Quoted in "The Sinking of the *Cumberland* by the Ironclad *Merrimac* off Newport News, Virginia, March 8, 1862."
82. Jones, "Services of the *Virginia*," 68.
83. Norris, "The Story of the Confederate States' Ship '*Virginia*,'" 217.
84. Jones, "Services of the *Virginia*," 68.
85. Colston, "Watching the *Merrimac*," 714.
86. Cannon, *Personal Reminiscences of the Rebellion*, 85–86.
87. Sears, *The Civil War Papers of George B. McClellan*, 198–99.
88. Beale, *Diary of Gideon Welles*, 51–52.

Chapter 5

89. Greene, "In the *Monitor*'s Turret," 720.
90. John Worden to "My darling wife," March 8, 1862, John Worden Papers, Lincoln Memorial University, Harrodsville, Tennessee.
91. Greene, "In the *Monitor*'s Turret," 720.
92. Daly, *Aboard the USS Monitor*, 31.

93. Ibid.
94. Rae, "The Little *Monitor* Saved Our Lives," 34.
95. Greene, "I Fired the First Gun," 13.
96. Berent, *The Crewmen of the USS* Monitor, 33.
97. Ibid.
98. Ibid.
99. Daly, *Aboard the USS* Monitor, 34.
100. Wood, "The First Fight of the Ironclads," 700.
101. Phillips, *The Career of the Iron-Clad* Virginia, 200.
102. Cline, "The Ironclad *Virginia*," 246.
103. Curtis, *History of the Famous Battle*, 10.
104. Ramsay, "The Most Famous of Sea Duels," 12.
105. Eggleston, "Captain Eggleston's Narrative," 174.
106. Curtis, *History of the Famous Battle*, 11.
107. Wood, "The First Fight of the Ironclads," 700.
108. Daly, *Aboard the USS* Monitor, 38.
109. Greene, "In the *Monitor*'s Turret," 722.
110. ORN, 2, 1: 157.
111. White, *The First Iron-Clad Naval Engagement*, 6.
112. Greene, "In the *Monitor*'s Turret," 722.
113. Ibid.
114. Ibid.
115. Daly, *Aboard the USS* Monitor, 36.
116. Greene, "In the *Monitor*'s Turret," 722.
117. Norris, "The Story of the Confederate States' Ship '*Virginia*,'" 219.
118. Wood, "The First Fight of the Ironclads," 701.
119. Daly, *Aboard the USS* Monitor, 36.
120. Stimers, "An Engineer Aboard the *Monitor*," 35.
121. Marvel, *The* Monitor *Chronicles*, 33.
122. Phillips, *The Career of the Iron-Clad* Virginia, 202.
123. Ramsay, "The Most Famous of Sea Duels," 12.
124. Eggleston, "Captain Eggleston's Narrative," 176.
125. Daly, *Aboard the USS* Monitor, 37.
126. Ibid.
127. Berent, *The Crewmen of the USS* Monitor, 33.
128. Wood, "The First Fight of the Ironclads," 703.
129. Greene, "In the *Monitor*'s Turret," 722.
130. ORN, 2, 1: 157.
131. Norris, "The Story of the Confederate States' Ship '*Virginia*,'" 219.
132. Daly, *Aboard the USS* Monitor, 38.
133. Greene, "In the *Monitor*'s Turret," 722.
134. Daly, *Aboard the USS* Monitor, 38.
135. Jones, "Services of the *Virginia*," 72.
136. Greene, "In the *Monitor*'s Turret," 722.
137. Daly, *Aboard the USS* Monitor, 39.

Chapter 6

138. Berent, *The Crewmen of the USS* Monitor, 9.
139. Greene, "Manuscript," 44.
140. Stimers, "An Engineer Aboard the *Monitor*," 35.
141. Daly, *Aboard the USS* Monitor, 39.
142. Ibid.
143. Marvel, *The* Monitor *Chronicles*, 40.
144. Greene, "Manuscript," 45.
145. Berent, *The Crewmen of the USS* Monitor, 9.
146. White, *The First Iron-Clad Naval Engagement*, 7.
147. Greene, "I Fired the First Gun," 104.
148. Daly, *Aboard the USS* Monitor, 39.
149. Ibid.
150. Greene, "Manuscript," 45.
151. ORA, 2, 6: 333–34.
152. ORN, 1, 7: 99–100.
153. ORA, 1, 40, part 2: 381.
154. Daly, *Aboard the USS* Monitor, 39.
155. ORA, 1, 40, part 2: 387.
156. Thompson and Wainwright, *Confidential Correspondence of Gustavus Vasa Fox*, 2, 248–49.
157. Daly, *Aboard the USS* Monitor, 63.
158. ORN, 1, 7: 101.
159. Worden Papers.
160. Daly, *Aboard the USS* Monitor, 46–47.
161. Ibid., 66.
162. Ibid.
163. Ibid.
164. Ransom, "The *Monitor* and the *Merrimac*."
165. *Norfolk Day Book*, March 11, 1862.
166. *London Times*, May 4, 1862.
167. Greene, "Manuscript," 48.
168. Daly, *Aboard the USS* Monitor, 53.
169. Ibid., 56.
170. Marvel, *The* Monitor *Chronicles*, 28.
171. Worden Papers.
172. Daly, *Aboard the USS* Monitor, 77.
173. ORN, 1, 7: 757.
174. Southern Historical Collection. Louis Round, Wilson Library, University of North Carolina, Chapel Hill, Franklin Buchanan Letterbook, 1801–1863.
175. Daly, *Aboard the USS* Monitor, 67.
176. Ibid., 66.
177. Marvel, *The* Monitor *Chronicles*, 47.
178. Daly, *Aboard the USS* Monitor, 67.
179. ORA, 2, 9: 404.

180. ORA, 2, 9: 423.
181. Woodward, *Mary Chesnut's Civil War*, 401.
182. Marvel, *The Monitor Chronicles*, 52.
183. Ibid.
184. Norris, "The Story of the Confederate States' Ship '*Virginia*,'" 212.
185. Daly, *Aboard the USS Monitor*, 74.
186. ORN, 1, 7: 224.
187. Daly, *Aboard the USS Monitor*, 75.
188. Norris, "The Story of the Confederate States' Ship '*Virginia*,'" 213.
189. Foute, "Echoes From Hampton Roads," 247–48.
190. *New York Herald*, April 15, 1862.
191. Marvel, *The Monitor Chronicles*, 51–52.
192. Daly, *Aboard the USS Monitor*, 53.

Chapter 7

193. Marvel, *The Monitor Chronicles*, 55.
194. Daly, *Aboard the USS Monitor*, 85.
195. Wood, "The First Fight of the Ironclads," 706.
196. ORA, 1, 11, part 3: 456.
197. Marvel, *The Monitor Chronicles*, 57.
198. Daly, *Aboard the USS Monitor*, 87.
199. ORA, 1, 11: 473.
200. Ibid., 828.
201. Daly, *Aboard the USS Monitor*, 100.
202. Ibid., 101.
203. Ibid., 107.
204. Ibid.
205. Curtis, *History of the Famous Battle*, 16.
206. Wood, "The First Fight of the Ironclads," 709; Southern Historical Collection, Louis Round Wilson Library, University of North Carolina, Chapel Hill, John Taylor Wood Papers.
207. ORN, 1, 7: 332.
208. Daly, *Aboard the USS Monitor*, 114.
209. ORA, 1, 11, part 3: 153.
210. Daly, *Aboard the USS Monitor*, 115.
211. Ibid.
212. Vick, "A Trip with Lincoln," 819.
213. Franklin, *Memoirs of a Rear Admiral*, 182–83.
214. Marvel, *The Monitor Chronicles*, 66.
215. Daly, *Aboard the USS Monitor*, 122.
216. Ibid., 121.
217. ORA, 1, 11, part 1: 835.
218. Ibid.
219. Ibid.

Chapter 8

220. ORN, 1, 7: 354–55.
221. Daly, *Aboard the USS* Monitor, 123.
222. Ibid., 136.
223. Ibid., 125.
224. Coski, *Capital Navy*, 41.
225. ORN, 2, 1: 636.
226. ORN, 1, 7, 799.
227. Mann, "New Light on the Great Drewry's Bluff Fight," 89.
228. Wood Papers.
229. Soley, "The Navy in the Peninsular Campaign," 269–70.
230. Daly, *Aboard the USS* Monitor, 126.
231. Soley, "The Navy in the Peninsular Campaign," 269.
232. ORN, 1, 7: 369.
233. Daly, *Aboard the USS* Monitor, 126.
234. Marvel, *The* Monitor *Chronicles*, 75.
235. Daly, *Aboard the USS* Monitor, 126.
236. Mann, "New Light on the Great Drewry's Bluff Fight," 92–93.
237. Daly, *Aboard the USS* Monitor, 128.
238. ORN, 1, 7: 370.
239. Mann, "New Light on the Great Drewry's Bluff Fight," 92.
240. Daly, *Aboard the USS* Monitor, 128.
241. ORN, 1, 1: 370.
242. Scharf, *History of the Confederate States Navy*, 764.
243. ORN, 1, 7: 357.
244. Daly, *Aboard the USS* Monitor, 129–30.
245. Mann, "New Light on the Great Drewry's Bluff Fight," 95.
246. Daly, *Aboard the USS* Monitor, 129.
247. Marvel, *The* Monitor *Chronicles*, 72.

Chapter 9

248. Daly, *Aboard the USS* Monitor, 132–33.
249. Ibid., 133–34.
250. Ibid., 135.
251. Ibid., 139–40.
252. Marvel, *The* Monitor *Chronicles*, 72.
253. Ibid.
254. Ibid., 84.
255. Daly, *Aboard the USS* Monitor, 150.
256. Ibid., 151.
257. Marvel, *The* Monitor *Chronicles*, 172.
258. Ibid., 133.

259. Daly, *Aboard the USS* Monitor, 141.
260. Ibid., 160.
261. Marvel, *The* Monitor *Chronicles*, 103.
262. Daly, *Aboard the USS* Monitor, 160.
263. ORN, 2, 18: 736.
264. Ibid., 737.
265. Marvel, *The* Monitor *Chronicles*, 72.
266. New York Historical Society, Ericsson, Box 2.
267. Ibid.
268. Marvel, *The* Monitor *Chronicles*, 72.
269. Daly, *Aboard the USS* Monitor, 163.
270. Ibid., 184.
271. Ibid., 155.
272. Ibid., 205.
273. Ibid., 209.
274. Ibid., 213–14.
275. Marvel, *The* Monitor *Chronicles*, 159.
276. Ibid., 182.

Chapter 10

277. Stodder, "Aboard the *Monitor*," 35.
278. Daly, *Aboard the USS* Monitor, 228.
279. Ibid., 233.
280. Marvel, *The* Monitor *Chronicles*, 202.
281. Daly, *Aboard the USS* Monitor, 232.
282. Marvel, *The* Monitor *Chronicles*, 202.
283. Daly, *Aboard the USS* Monitor, 232.
284. Letters of Jacob Nicklis, The Mariners' Museum, Newport News, Virginia (hereafter cited as Nicklis Letters).
285. Daly, *Aboard the USS* Monitor, 232.
286. ORN, 1, 9: 726–27.
287. Marvel, *The* Monitor *Chronicles*, 200.
288. Berent, *The Crewmen of the USS* Monitor, 34.
289. Ibid., 9–10.
290. Nicklis Letters.
291. Berent, *The Crewmen of the USS* Monitor, 52, 66.
292. Marvel, *The* Monitor *Chronicles*, 206.
293. Daly, *Aboard the USS* Monitor, 241.
294. Ibid., 243.
295. Ibid., 244.
296. Marvel, *The* Monitor *Chronicles*, 207.
297. Daly, *Aboard the USS* Monitor, 239.
298. Marvel, *The* Monitor *Chronicles*, 212–13.
299. Nicklis Letters.

300. Daly, *Aboard the USS* Monitor, 233.
301. Ibid., 251.
302. The Mariners' Museum, Newport News, Virginia. Jacob Nicklis Papers.

Chapter 11

303. New York Historical Society, Greene Papers.
304. Nicklis Letters.
305. Weeks, "The Last Cruise of the *Monitor*," 367.
306. Marvel, *The* Monitor *Chronicles*, 235.
307. Ibid.
308. Daly, *Aboard the USS* Monitor, 251.
309. Marvel, *The* Monitor *Chronicles*, 225.
310. Daly, *Aboard the USS* Monitor, 251.
311. ORN, 82: 349.
312. Marvel, *The* Monitor *Chronicles*, 235.
313. Ibid.
314. Daly, *Aboard the USS* Monitor, 254.
315. ORN, 8: 350.
316. Ibid.
317. Butts, "The Loss of the '*Monitor*,'" 745.
318. Daly, *Aboard the USS* Monitor, 254.
319. Marvel, *The* Monitor *Chronicles*, 235.
320. Daly, *Aboard the USS* Monitor, 255.
321. Ibid., 254.
322. ORN, 8: 352.
323. Stodder, "Aboard the *Monitor*."
324. ORN, 2, 8: 350.
325. Daly, *Aboard the USS* Monitor, 256.
326. Butts, "The Loss of the '*Monitor*,'" 746.
327. Stodder, "Aboard the *Monitor*," 36.
328. Daly, *Aboard the USS* Monitor, 257.
329. Ibid., 257–58.
330. Ibid., 258.
331. Ellis, *The* Monitor *of the Civil War*, 35.
332. Daly, *Aboard the USS* Monitor, 259.
333. Butts, "The Loss of the '*Monitor*,'" 247.
334. Lewis, "Life on the *Monitor*," 261.
335. Weeks, "The Last Cruise of the *Monitor*," 368.
336. Woolen, *Woollen/Woolen Family*.
337. Butts, "The Loss of the '*Monitor*,'" 247.

Chapter 12

338. Marvel, *The* Monitor *Chronicles*, 231.
339. Daly, *Aboard the USS* Monitor, 252.
340. Nicklis Letters.
341. ORN, 7: 341.
342. Marvel, *The* Monitor *Chronicles*, 238–41.
343. ORN, 8: 356.
344. Berent, *The Crewmen of the USS* Monitor, 73.
345. ORN, 9: 247.
346. Worden Papers.
347. Quarstein, *A History of Ironclads*, 186.
348. Berent, *The Crewmen of the USS* Monitor, 11.
349. Ibid., 11–12.
350. Ibid., 65.
351. Ibid., 21.
352. Ibid., 16.
353. Ibid., 79.
354. Ibid.
355. Ibid., 44.

Appendix V

356. Ibid, 8.
357. Ibid., 8–12; "Hans Anderson, 'Shot Man' on the Famous *Monitor*, Dead," *Syracuse (NY) Herald*, April 23, 1909, 17; ORN, 1, 11: 571; pension records, Hans Anderson, Library at The Mariners' Museum, Newport News, Virginia; State of New York Certificate and Record of Death #7764, Hans Anderson, April 20, 1909.
358. Berent, *The Crewmen of the USS* Monitor, 12–13.
359. Ibid., 13.
360. Berent Files, The Mariners' Museum, Ms 164: Box 3; Callahan, *List of Officers*; New Haven, CT, Trinity PE Church records, 2:5, CSL microfilm used; New Haven, CT City Directory, 1860–61, database online, Ancestry.com.
361. Marvel, *The* Monitor *Chronicles*, 165; Daly, *Aboard the USS* Monitor, 244; ORN, I, 3: 472; "The Navy," Niles' National Register (1837–1849), June 1, 1844–16, 14—APS Online; "Our Navy—Loss of the U.S. Brig Traxton," *Cincinnati (OH) Weekly Herald and Philanthropist (1843–1846)*; Sept. 23, 1846, 11, 2, APS Online; It has been reported that Bankhead commanded the *Florida* in 1862 during the capture of Fernandina, Florida. Actually, he was in command of the USS *Pembina* then and had, indeed, been ordered to participate in the attack on Fernandina, but his vessel never made it there, running aground along with several other vessels in the flotilla approximately three miles from the town. See Still, *Ironclad Captains*, 65; also see Post, *The Post Family*, 182, in which it is stated that the USS *Florida*, commanded by John P. Bankhead, was stationed off Wilmington, North

Carolina, not Florida; Obituary, "DIED," *New York Times* (1857–current file); June 2, 1867; ProQuest Historical Newspapers, *New York Times* (1851–2003).

362. Berent, *The Crewmen of the USS* Monitor, 13–14; desertion date of 11/6/62 comes from Pension #54355, dated 9/1/1909, in which his pension is rejected because of the desertion; *Our Brothers Gone Before*, Vol. I, New Jersey Civil War Burials (New Jersey State Archives).

363. Berent, *The Crewmen of the USS* Monitor, 14.

364. Ibid., 15–16; National Archives, M1469-6509, Navy Cert. No. 7224, Can no. 157, Bundle no. 40, Department of Interior Bureau of Pensions #8309, Edmund Brown.

365. Berent, *The Crewmen of the USS* Monitor, 16; information based on New York State Archives Naval Enlistment Record #3916. This John Brown has been confused with John Brown, b. 1838, Haverstraw, New York; occ. none; enl. 10/1/62 in New York for a 1-year term as landsman on the USRS *North Carolina*; desc. blue eyes, brown hair, florid comp., 5'8" tall; later served on the USS *Memphis* and USS *Ceres*, but not on the USS *Monitor*; disch. 10/3/63, per New York State Archives Naval Enlistment Record #3929, Brown, John; private letter dated 9/29/1983 to USS *Monitor* Crew Project, 23 A, from W.M. Forsyth, Clan Forsyth of America Inc., Atlanta, Georgia.

366. Berent, *The Crewmen of the USS* Monitor, 17; New York State Archives Naval Enlistment Record #4386, Bryan, William.

367. Berent, *The Crewmen of the USS* Monitor, 17.

368. Ibid., 17–18; (1) State of Rhode Island, Copy of the Record of a Marriage, during the month of Dec. 1867, full name of groom: Francis B. Butts, age in years: 23. 1867-23=1844. 1843, his birth year according to Berent, *The Crewmen of the USS* Monitor, 17–18, is the year stated by his widow, Helen F. Butts, in an affidavit made when applying for his pension. (See following note.) (2) Wentworth, *The Wentworth Genealogy*, 422. (3) Family Search, Church of Jesus Christ of Latter Day Saints Archive, Provo, Utah. Source information: census place: East Providence, Providence, Rhode Island; family history library film: 1255214; NA film number: T9-1214; page number: 252C; Service record listed is based on official correspondence from the Treasury Department, auditor for the Navy Department #103661-EAP, dated 4/15/1910. Frank Butts's service is often found listed as: *N. Carolina*, Washington Station, *Monitor*, *Stepping Stones* and *Flag* apparently based on statements made by Helen F. Butts, widow of Frank Butts, in a Declaration for Widow's Pension, dated 2/11/1910. However, in a general affidavit dated 5/19/1910, Mrs. Butts admits she cannot remember her husband's birthday; it is possible she misremembered her husband's service record as well.

369. Callahan, *List of Officers*; New York State Archives Naval Enlistment Record #89, Campbell, Albert B.

370. Berent, *The Crewmen of the USS* Monitor, 18.

371. Berent, *The Crewmen of the USS* Monitor, 18–19; *Niphon* is sometimes spelled *Nyphon*, but in official correspondence from the Treasury Department, auditor for the Navy Department #30955-HAO, dated 12/7/1905, it was spelled *Niphon*;

Eleventh Census of the United States, Special Schedule, Surviving Soldiers, Sailors, and Marines, and Widows, Etc., Iola, Grimes Co., Texas, June 1890.

372. Berent, *The Crewmen of the USS* Monitor, 19–21; *1880 United States Federal Census*, New York (Manhattan), New York City-Greater, New York: Roll: T9_886, Family History Film: 1254886, p. 493,000, enumeration district 400, image 0336; Ancestry.com, online database, Church of Jesus Christ of Latter-Day Saints; "His name is on the rolls of the *Monitor* as Thos. Carroll no. 2," official correspondence, Treasury Department Auditor Report to Commissioner of Pensions, dated 7/24/89.

373. Berent, *The Crewmen of the USS* Monitor, 21–27: enlistment date could be May 11, 15 or 19 of 1862; enlistment data reported height of 5'6½", while Carter himself said he was 5'8½"; Declaration for Original Invalid Pension, dated 1/29/85; Declaration for Widow's Pension, dated 4/13/92; Philadelphia, Pa., Health Office Registration Department Death Certificate #3105, Siah Hulett, dated 4/29/92.

374. Berent, *The Crewmen of the USS* Monitor, 27: James was most likely the man listed as "William Coleman," coal heaver, on the *Monitor*'s payroll; New York State Archives Naval Enlistment Record #7184, Coleman, James.

375. Berent, *The Crewmen of the USS* Monitor, 27, listed 3-year term and desertion on 11/7/62; New York State Archives Naval Enlistment Record #7542, Conklin, John P.

376. Berent, *The Crewmen of the USS* Monitor, 27–28.

377. Ibid., 28.

378. Ibid., 28; New York State Archives Naval Enlistment Record #8821, Crown, Joseph, states that he served on the USS *Princeton* between the *North Carolina* and the *Monitor*.

379. Berent, *The Crewmen of the USS* Monitor, 28–29; Cuddeback, *Caudebec in America*, 132; New York State Archives Naval Enlistment Record #8850, Cuddeback, David.

380. Berent, *The Crewmen of the USS* Monitor, 29–34: at enlistment, he claimed his birthplace was Baltimore, Maryland. There is also a question concerning an alias ("John White") he later claimed was his actual name, that he had adopted the use of Driscoll, his mother's maiden name. It is believed Driscoll was his real name.

381. Berent, *The Crewmen of the USS* Monitor, 34–42; official correspondence from the secretary of the navy, Navy Dept. #2809-D, dated 12/4/1907; National Archives and Records, *Ninth Census of the United States, 1870*, M593, 1761 rolls, Ancestry.com, database online, Church of Jesus Christ of Latter-Day Saints; Obituary, "Last Monitor Survivor Dies," *Washington Post*, 1/5/1916.

382. Berent, *The Crewmen of the USS* Monitor, 42–43.

383. Ibid., 43–44; inscription, monument, Mount Annville Cemetery, North Annville Twp., Lebanon Co., Pennsylvania; Obituary, "Rev. David R. Ellis," *Somerset (PA) Herald*, pub. 9/9/1908; Obituary, "Rev. D.R. Ellis to the Grave," *Lebanon (PA) Daily News*, front page, 9/1/1908.

384. Berent, *The Crewmen of the USS* Monitor, 44; New York State Archives Naval Enlistment Record #12804, Feeney, Thomas.

385. Berent, *The Crewmen of the USS* Monitor, 44–45; Marvel, *The* Monitor *Chronicles*, 228.
386. Berent, *The Crewmen of the USS* Monitor, 45–46; New York State Archives Naval Enlistment Record #13173, Fisher, Hugh.
387. Berent Files, The Mariners' Museum, Ms 164: Box 3, Officers Box: Monitor Flye, William; Daly, *Aboard the USS* Monitor, 231; individual record AFN:MSW9-GQ, Family Search, Church of Jesus Christ of Latter-Day Saints, online database, Provo, Utah; official correspondence, Navy Dept., Bureau of Navigation & Office of Detail, dated 2/7/83; Declaration for Original Pension of an Invalid, dated 4/28/75; official correspondence, Treasury Dept., Auditor's Office, dated 3/7/84.
388. Berent Files, The Mariners' Museum, Ms 164: Box 3, Officers Box: Monitor Frederickson, George; affidavit, Magdalena E. Frederickson.
389. Berent Files, The Mariners' Museum, Ms 164: Box 3, Officers Box: Monitor Gager, Edwin V.; No documentation exists that noted the exact date of Gager's assignment to the *Monitor*. Keeler notes (Daly, *Aboard the USS* Monitor, 91) in a letter to his wife on April 29, 1862, that Gager was not aboard during the fight but was assigned to the vessel shortly thereafter.
390. Berent Files, The Mariners' Museum, Ms 164: Box 3, Officers Box: Monitor Gallagher, Darius F.; Obituary, *New York Times*, 6/18/1905.
391. Berent, *The Crewmen of the USS* Monitor, 46; official correspondence, Treasury Department, Fourth Auditor's Office, to Commissioner of Pensions, dated 7/6/91; Deposition A, Case of Bridget Garety, No. 9793, dated 5/13/92; Pension file #9793, Navy Widow, Bridget Garity, widow of John Garity, 1st Cl Fire, USS *Monitor*, dated 3/7/94; State of New York Corrected Certificate and Record of Death, John Garety, signed by D.J. Quirk, MD, dated 6/2/82.
392. Berent, 46; Pedigree Chart and Family Group Record, Family Search Ancestral File, George Spencer Geer (AFN:1F6T-C29), Martha Clark Hamilton (AFN:1F6T-C3H), Gilbert Geer (AFN:1F6T-B6X), Adna Spencer (AFN:1F6T-B75), database online, Church of Jesus Christ of Latter-Day Saints, Provo, Utah; Obituary, "Death of George S. Geer at Charleston—One of the *Monitor*'s Crew," *Troy (NY) Daily Times*, 10/10/92.
393. Berent Files, The Mariners' Museum, Ms 164: Box 3, Officers Box: Monitor Greene, Samuel Dana; Gorton, *The Life and Times of Samuel Gorton*, 612; "Commander S. Dana Greene," *The Records of Living Officers of the U.S. Navy and Marine Corps* (Nimitz Library, U.S. Naval Academy, Special Collections and Archives); official correspondence, from Navy Dept. to Commissioner of Pensions, dated 1/7/55; "Samuel Dana Greene," *Dictionary of American Biography* Base Set, American Council of Learned Societies, 1928–1936; Certificate of Death #7362, S. Dana Greene, dated 12/11/84; Senate Bill #2484, passed 12/20/84.
394. Berent, *The Crewmen of the USS* Monitor, 47.
395. Berent Files, The Mariners' Museum, Ms 164: Box 3, Officers Box: Monitor Hands, Robinson W.; individual record, Robinson Woollen Hands, Family Search, Church of Jesus Christ of Latter-Day Saints, Provo, Utah.

396. Berent, *The Crewmen of the USS* Monitor, 47–49; State of New York Certificate and Record of Death #8029, Patrick Hannan, dated 3/17/92.
397. Berent, *The Crewmen of the USS* Monitor, 49.
398. Ibid., 49; New York State Archives Naval Enlistment Record #17382, Harrison, Henry.
399. Berent, *The Crewmen of the USS* Monitor, 49–50: the *Monitor*'s final transfer lists Robert H. Howard without a ship's no. and indicates he had been on the payroll since 11/7/62. However, Paymaster Keeler listed a "W.H. Howard, ship's no. 74, a negro who joined…on Nov. 8…received from the Washington Navy Yard" (Berent, *The Crewmen of the USS* Monitor, 50); *Black Sailors*.
400. No. 7472, W.A.G., 2nd endorsement, Navy Dept., dated 5/31/98; official correspondence, Dept. of the Interior, Bureau of Pensions, from Acting Commissioner to Samuel Howard, dated 3/21/98; 1860 United States Federal Census Record, 92; Quarstein, *The Battle of the Ironclads*, 78; #7472, 1st Endorsement, Navy Dept., dated 5/11/98; Navy Invalid Pension #43410; U.S. Pension Agency, Notification of Pension dropped, dated 2/9/1900; Obituary, "The *Monitor*'s Pilot," *Newport (RI) Mercury*, January 27, 1900.
401. Berent, *The Crewmen of the USS* Monitor, 50, has the promotion to acting ensign on 11/25/62; Callahan; ORN, 1, 26:80; there is no explanation for a dismissal date of 4/22/64 and documentation of participation in an engagement on 4/26/64.
402. Berent Files, The Mariners' Museum, Ms 164: Box 3, Officers: Monitor Jeffers, William N.; Callahan; *Gloucester County in the Civil War*, reprinted from *The Constitution*, 9/20/1939 to 6/4/1943 (Woodbury, New Jersey: Gloucester County Historical Society), 23–24; McIntire, *Annapolis Maryland Families*, 363–64, entry on William N. Jeffers; Lewis R. Hamersly, *The Records of Living Officers*, 450–51.
403. Berent, *The Crewmen of the USS* Monitor, 50–51, lists 5'5" tall; National Park Service Civil War Soldiers and Sailors System, William H. Jeffrey; *1870 United States Federal Census*, 77.
404. Berent, *The Crewmen of the USS* Monitor, 51.
405. Ibid.
406. Berent Files, The Mariners' Museum, Ms 164: Box 3, Officers: Monitor Keeler, William F.; Daly, 5; Cope, *Genealogy of the Dutton family* (Ancestry.com); Declaration for Increase of an Invalid Pension B, dated 4/4/85, handwritten document appended; Affidavit by William F. Keeler to Edwin Higgins, Clerk, Circuit Court, Duval Co., Florida, dated 11/4/75; Invalid Pension #2177, dated 6/2/75; examining Surgeon's Certificate, No. of Application 3731, dated 7/21/75, signed by A.J. Wanefield, MD; Certificate of Death, Mayport, Florida, William F. Keeler, dated 2/27/86, signed by L.R. Quackenbush, MD.
407. Berent, *The Crewmen of the USS* Monitor, 51–52, apparently used New York State Archives Naval Enlistment Record #23571, Leonard, Matthew, w/ different date of enlistment, term, vessels served on, height and hair color; based on New York State Archives Naval Enlistment Record #23570, Leonard, Mathew; Widow's Claim for Navy Pension, dated 5/5/66.
408. Naval Records, Library at The Mariners' Museum, Newport News, Virginia.

409. Berent, *The Crewmen of the USS* Monitor, 52.
410. Berent Files, The Mariners' Museum, Ms 164: Box 3, Officers Box: Monitor Logue, Daniel C.; obituary, "Dr. Logue, Last Survivor, Died Tuesday," *Middletown (NY) Daily Times-Press*, February 5, 1914.
411. Berent, *The Crewmen of the USS* Monitor, 52; *Massachusetts Soldiers*, Ancestry.com.
412. Berent, *The Crewmen of the USS* Monitor, 52; based on New York State Archives Naval Enlistment Record #24932, Malone, James; there is also New York State Archives Naval Enlistment Record #24933, Malone, James, enlistment 9/10/62, disch. 9/9/63, apparently also served on the *North Carolina, Monitor* and *Ceres* but had light comp. rather than dark. It is unknown if these are the same person.
413. Berent, *The Crewmen of the USS* Monitor, 52–53; *Massachusetts Soldiers*, Ancestry.com.
414. Berent, *The Crewmen of the USS* Monitor, 53, names New London, Connecticut, as birthplace, 2/21/62 as enlistment date, rating coal heaver; private letter from Kenneth S. Carlson, Reference Archivist, dated 5/25/2006, re: Research Project USS Monitor, sources including Records of the RI Adjutant General; ORN, 1, 16: 180; Enclosure Part (C.), dated 1/24/65, to Report of Rear-Adm. Dahlgren; John Mason's rank is listed as first-class fireman.
415. Berent, *The Crewmen of the USS* Monitor, 53.
416. Callahan; *History of Lycoming County, Illustrated*, 1876, 132; Original data: *1880 United States Federal Census Record, Special Schedules of the Eleventh Census (1890) Enumerating Union Veterans and Widows of Union Veterans of the Civil War*, M123, 118 rolls. (National Archives and Records Administration, Washington, D.C.); Ancestry.com, database online, Church of Jesus Christ of Latter-Day Saints, Provo, Utah; *Medical and Surgical Reporter (1858–1898)*, Philadelphia: June 3, 1865. Vol. 12, Issue 34, p. 552; Declaration for the Increase of an Invalid Pension, dated 3/4/89, signed by Thomas W. Meckly; official correspondence from Chief of Bureau, Navy Dept., Bureau of Navigation and Office of Detail, dated 8/24/80; Declaration for Original Invalid Pension A, dated 6/29/80, signed by T.W. Meckly, MD; Death announcement, *Daily Gazette and Bulletin*, Williamsport, Pennsylvania, February 4, 1990.
417. Berent, *The Crewmen of the USS* Monitor, 53.
418. Ibid., 53–54; Declaration for Pension, dated 9/30/07, signed by Michael Mooney; ORN, 1, 14: 324; New York State Archives Naval Enlistment Record #29061, Mooney, Michael; Obituary, "Monitor Veteran Passes," *Los Angeles (CA) Times*, July 15, 1911; AncestryLibrary.com online database—U.S. Veterans Cemeteries, ca. 1800–2006.
419. Berent, *The Crewmen of the USS* Monitor, 54; Affidavit of Sarah Moore, dated 11/12/66, p. 9 ("No Issue").
420. Berent, *The Crewmen of the USS* Monitor, 54; New York State Archives Naval Enlistment Record #29095, Moore, Edward.
421. Berent, *The Crewmen of the USS* Monitor, 54.
422. Ibid., 55; New York State Archives Naval Enlistment Record #30428, Murray, Lawrence.
423. Isaac Newton Papers, Library at The Mariners' Museum, Newport News, Virginia; Callahan; Original Data: *1850 United States Federal Census*, M432,

1009 rolls (National Archives and Records Administration, Washington, D.C.), Ancestry.com, *1850 United States Federal Census* [database online], Provo, Utah: Church of Jesus Christ of Latter-Day Saints; Bennett, *The Steam Navy*, 296.

424. Berent, *The Crewmen of the USS* Monitor, 55; National Archives, Navy 44468, can no. 128, bundle no. 16, Invalid pension declaration, dated 1/25/99.

425. Berent, *The Crewmen of the USS* Monitor, 56–58; Original Data: United States of America, Bureau of the Census, *Eighth Census of the United States, 1860* (National Archives and Records Administration: Washington, D.C.), 1860, M653, 1,438 rolls; Ancestry.com, database online, *1860 United States Federal Census*, Church of Jesus Christ of Latter-Day Saints, Provo, Utah.

426. Callahan, *List of Officers*; 3AE, Library at The Mariners' Museum: M330 or M1328; ORN, 1, 20: 803.

427. Berent, *The Crewmen of the USS* Monitor, 59–60; *United States Federal Census 1870—Town of Russia, County of Herkimer, State of New York*, 6, dated 8/3/70; *United States Federal Census 1880—Town of Russia, County of Herkimer, State of New York*, 38, dated 6/22/80.

428. Berent, *The Crewmen of the USS* Monitor, 60, lists his age as twenty-two, when the Record (see following citation) lists it as thirty-two; based on New York State Archives Naval Enlistment Record #33683, Price, Christy; Ancestry.com: Death Notice, "State News," *Daily Northwestern*, Oshkosh, Wisconsin, September 29, 1916.

429. Berent, *The Crewmen of the USS* Monitor, 61-62, states his height as 5'7½"; based on New York State Archives Naval Enlistment Record #34006, Quinn, Robert.

430. Berent, *The Crewmen of the USS* Monitor, 62–63, states b. 6/7/44, Syracuse, New York, son of William T. and Caroline, enl. in 40th New York Infantry Volunteers, Co. I, 101st Regt., 11/1/61, held seven months as POW, Andersonville, Georgia; due to the timeline, it is likely that these are two different people; information compiled by D.E. Bareilles, Paradise, California, from original document: Homestead certificate #4239, Osceola Co., Michigan, application #6748; *1850 and 1870 United States Federal Census*, ancestry.com database online; McDowell Funeral Home Death Records 1910–1950, Reed City, Michigan (Kinseeker Publications: Grawn, MI, 1998).

431. Berent, *The Crewmen of the USS* Monitor, 63.

432. Ibid., 63–64; Certificate of Death #8720, F.A. Riddey, dated 1/11/99, Health Office Registration Dept., Philadelphia, Pennsylvania.

433. Berent, *The Crewmen of the USS* Monitor, 64, had 1835 as birth date, Ireland as birthplace, ranking as capt.'s clerk, place of enlistment as Brooklyn, 5'7½" tall; based on New York State Archives Naval Enlistment Record #35660, Rooney, John; Declaration for Widow's Pension, signed by Catherine Rooney, dated 7/9/90.

434. Berent, *The Crewmen of the USS* Monitor, 64–66; Isaac Scott did not know his exact birthday. 1837 was given on the New York State Dept. of Health Standard Certificate of Death #3079, dated 5/18/27, and in affidavits for pension, he estimated it as Oct. 16, 1836, and Oct. 1839.

435. Berent, *The Crewmen of the USS* Monitor, 66.
436. Ibid., 66; based on New York State Archives Naval Enlistment Record #36932, Seery, James.
437. Daly, *Aboard the USS* Monitor, 42; the USS *Alligator*, proving a failure for want of speed, was assigned to duty in the Mississippi Squadron, *Records of Living Officers of the U.S. Navy and Marine Corps*, Nimitz Library, Special Collections and Archives Division, United States Naval Academy, Annapolis, Maryland; *1880 U.S. Census Index* provided by Church of Jesus Christ of Latter-Day Saints, Image source: Year: *1880, Census Place: Newport, Newport, Rhode Island*, Roll: T9_1210, Family History Film: *1255210*, Page: *161.3000*, Enumeration District: *92*; Selfridge, *Memoirs of Thomas O. Selfridge Jr.*; Ancestry.com: *The Fitchburg (MA) Sentinel*, Mar. 22, 1897.
438. Berent, *The Crewmen of the USS* Monitor, 66.
439. Berent Files, The Mariners' Museum, Ms 164: Box 3, Officers Box: Monitor Slover, James T.; *Tombstones of Talbot County, Maryland, Including Some Interment Records*, Vol. 1, 1989 (Upper Shore Genealogical Society of Maryland, Easton, Maryland); Questionnaire, Dept. of the Interior, to James T. Slofer, dated 3/24/92; official correspondence from Navy Auditor to Commissioner of Pensions, dated 6/15/96; Declaration for an Original Invalid Pension, signed James T. Slover, dated 3/18/93; Deposition of James T. Slofer, alias Slover, to William B. Wilson, Justice of the Peace, dated 4/14/91.
440. Based on New York State Archives Naval Enlistment Record #38061, Smith, Charles. There is also New York State Archives Naval Enlistment Record #38023, Smith, Charles: b. c. 1840, New York, New York; enl. in New York 9/3/62 for a one-year term as landsman; desc. hazel eyes, black hair, light comp., 5'5½" tall; served on the *North Carolina*, Washington Navy Yard, *Monitor*, *Brandywine* and *Stepping Stones*; disch. 9/30/63. Berent, *The Crewmen of the USS* Monitor, 66, apparently used the vessels from this service record.
441. Based on New York State Archives Naval Enlistment Record #38309, Smith, James; Berent, *The Crewmen of the USS* Monitor, 66–67, lists birth date as 1841, birthplace as Troy, New York, light comp., 5'5" tall, service at Washington Station, *Monitor* (as ship's no. 67), *Brandywine*, *Stepping Stones* and *Alleghany*.
442. Berent, *The Crewmen of the USS* Monitor, 67–68; official correspondence from U.S. Pension Agency, Boston, Mass., to Commissioner of Pensions, dated 11/30/97.
443. Daly, *Aboard the USS* Monitor, 208; dates birth at 5/27/19; Andrews, *The Hamlin Family*; Arlington National Cemetery online database; Rear-Adm. T.H. Stevens, USN, *United Service: A Quarterly Review of Military and Naval Affairs (1879–1905)*, Vol. 5, Issue 5 (New York: May 1891), 545–49; "Army and Navy News," *New York Times*, Feb 24, 1881; "Rear-Adm. Thomas H. Stevens, U.S.N.," *Washington (D.C.) Star*, Friday, May 15, 1896.
444. Callahan, *List of Officers*; official correspondence from Chief of Bureau, Navy Dept., Bureau of Navigation & Office of Detail to Commissioner of Pensions, dated 3/30 /83; "Important from Port Royal—The Bombardment of Ft. McAllister," *New York Times*, Mar 13, 1863, 1; "Local Intelligence—Chief Engineer Stimers in Trouble," *New York Times*, May 29, 1863, 8; Wegner,

"Ericsson's High Priest," 26; Obituary, "Death of Commander A.C. Stimers, United States Navy," *New York Herald,* June 6, 1876, 2.
445. Berent, *The Crewmen of the USS* Monitor, 68; Wentz, *Record of the Descendents of Johann Jost Wentz.*
446. Berent Files, The Mariners' Museum, Ms 164: Box 3, Officers Box: Monitor Stodder, Louis N.; "Fought on the Monitor," *Daily Herald,* Delphos, Ohio, April 30 1902; official correspondence, 1st Endorsement, Navy Dept., Library and Naval War Records, Washington, D.C., Widow Or. 974923, dated 11/27/1911; "Adela," navyhistory.com; Standard Certificate of Death #121948, State of New York, Dept. of Health of the City of New York, Bureau of Records, dated 10/9/1911; Obituary, "Capt. Louis N. Stodder," *New York Times,* 10/10/1911.
447. Callahan, *List of Officers*; Death certificate #2161, Dept. of Public Safety, Sub-Dept. of Health, City of Baltimore, dated 10/14/1908.
448. Berent, *The Crewmen of the USS* Monitor, 68–69, states his height as 5'7½" and did not list the *Princeton*; the New York State Archives Naval Enlistment Record did not list service on the *Philadelphia*; based on New York State Archives Naval Enlistment Record #40319, Syllwester, Charles F.
449. Berent, *The Crewmen of the USS* Monitor, 69; 1870 United States Federal Census, Brooklyn, New York, 196; Office of the Department of Health, Record of Death, City of Brooklyn, dated 9/11/90.
450. Berent, *The Crewmen of the USS* Monitor, 69–70; De Riemer, *The De Riemer Family*, 30; *1860 United States Federal Census, New Jersey*, 7; Obituary, "Daniel Toffey," *Evening Journal*, Feb 10, 1893; Obituary, *New York Times*, Feb 11, 1893, 3.
451. Berent, *The Crewmen of the USS* Monitor, 70–71, states his height as 5'4½", did not have his service on the *Princeton* but did say he served on the *Philadelphia*, which was not on the New York State Archives Naval Enlistment Record (see final citation); King, "Life On the Monitor"; Based on New York State Archives Naval Enlistment Record #41532, Truscott, Peter.
452. Berent, *The Crewmen of the USS* Monitor, 71–73; obituary, *New York Sun*, Nov. 14, 1905.
453. Berent, *The Crewmen of the USS* Monitor, 73; *Massachusetts Soldiers*, 806, ancestry.com; *1880 United States Federal Census*, Lawrence, Massachusetts, 33, ancestry.com; *1890 Veterans Schedules, Special Schedule—Surviving Soldiers, Sailors, and Marines, and Widows, etc.*, 2, ancestry.com; New England Historic Genealogical Society website, Vital Records 1841–1910.
454. *1860 United States Federal Census*, Bordentown, New Jersey, 86; official correspondence from Judge Advocate General to Commissioner of Pensions, dated 3/23/86; Certificate of Death, Joseph Watters, dated 9/13/66; Widow's Pension #953.
455. Daly, *Aboard the USS* Monitor, 22; obituary, "Capt. John J.N. Webber Dead," *New York Times*, October 10, 1909; *1870 United States Federal Census*, Brooklyn, New York, p. 24.
456. Berent Files, The Mariners' Museum, Ms 164: Box 3, Officers Box: Monitor Weeks, Greenville M.; *Who Was Who in America*; "Dr. Grenville M. Weeks Dies at Soldier's Home," *Newark (NJ) Evening News*, April 26, 1919, 6.

457. From the file entitled "research aid" in the Berent Collection, The Mariners' Museum, Newport News, Virginia; U.S. Department of the Navy, *Register of the Commissioned, Warrant, and Volunteer Officers*; ORN, 1, 14: 17–18; ORN, 1, 16; 309.
458. Berent, *The Crewmen of the USS* Monitor, 73–74, has discharge on 4/3/63; although his enlistment papers, as well as the *1864 Naval Register*, showed his place of birth as Germany, his Medal of Honor records consider him to have been Norwegian; based on Naval Historical Center, 805 Kidder Breese St. SE, Washington Navy Yard, D.C.
459. Berent, *The Crewmen of the USS* Monitor, 74–75; based on New York State Archives Naval Enlistment Record #44284, Williams, Robert.
460. Carpenter, *The Reverend John Graham*, 352, online database, ancestry.com; "John Lorimer Worden," *Dictionary of American Biography* Base Set, galenet.galegroug.com/servlet/BioRC; Worden, "The Naval Career of Rear Adm. John Lorimer Worden," 124; Unknown, *The Melrose Memorial: The Annals of Melrose, County of Middlesex, Massachusetts, in the Great Rebellion of 1861–'65* (priv. print by subscription, 1868), 186; "Local Intelligence—The Sword to Lieut. Worden," *New York Times*, Dec 21, 1862; Obituary, "Rear Adm. Worden's Funeral," *New York Times*, Oct 20, 1897; House of Representatives Report #1776, to accompany S. 2919, dated January 19, 1899; Dept. of the Navy, Naval Historical Center, 805 Kidder Breese SE, Washington Navy Yard, Washington D.C.

APPENDIX VI

461. Daly, *Aboard the USS* Monitor, 89.
462. *New York Times*, April 19, 1945.
463. Charles Carillo, "The Cheesebox on a Raft," *New York Post*, July 28, 1979.
464. Ibid.
465. Berent Files, The Mariners' Museum, Ms 164: Box 4, Crewmembers: Monitor Brodie, Peter H.; State of Ohio, Bureau of Vital Statistics, Certificate of Death, registered #47, Peter Brodie; Obituary, "P.H. Brodie Was Naval Veteran of the Civil War," *Cuyahoga Falls (OH) Reporter*, March 30, 1917.
466. Berent, *The Crewmen of the USS* Monitor, 18; Carillo, "The Cheesebox on a Raft," *New York Post*, July 28, 1979.
467. Berent, *The Crewmen of the USS* Monitor, 28.
468. Ibid., 44: Carillo, "The Cheesebox on a Raft," *New York Post*, July 28, 1979.
469. Berent, *The Crewmen of the USS* Monitor, 44.
470. Ibid., 58–59.
471. Ibid., 60–61.

Bibliography

REFERENCE WORKS

Black Sailors: The Howard University Research Project. National Park Service Civil War Soldiers and Sailors System, www.itd.nps.gov/cwss.

Callahan, Charles W., ed. *List of Officers of the United States and of the Marine Corps from 1775–1900: Compromising a Complete Register of All Present and Former Commissioned of the United States Navy, and of the Marine Corps, Regular and Volunteer, Compiled from the Official Records of the Navy Dept.* New York: L.R. Hamersly & Co., 1901.

Dictionary of American Biography. Base Set. American Council of Learned Societies, 1928–1936. Reproduced in Biography Resource Center. Farmington Hills, MI: Thomson Gale, 2006.

Gloucester County in the Civil War. Reprinted from *The Constitution, 9/20/1939 to 6/4/1943.* Woodbury, NJ: Gloucester County Historical Society, n.d.

Hamersly, Lewis R. *The Records of Living Officers of the U.S. Navy and Marine Corps.* N.p., n.d.

Hamersly, Thomas H.S., ed. *Complete General Navy Register of the United States of America from 1776 to 1887.* New York: T.H.S. Nemmersley, 1888.

Massachusetts Soldiers, Sailors, and Marines in the Civil War. Vol. 8. Boston: Adjutant General, 1931–1935.

Our Brothers Gone Before. Vol. 1. New Jersey Civil War Burials, New Jersey State Archives.

BIBLIOGRAPHY

Roberts, Robert B. *Encyclopedia of Historic Ports: The Military, Pioneer, and Trading Posts of the United States.* New York: MacMillan Publishing Co., 1988.

Scharf, J. Thomas. *History of the Confederate States Navy from Its Organization to the Surrender of Its Last Vessel.* New York: Rogers & Sherwood, 1887; reprint, New York: Crameray Books, Random House Publishing, Inc., 1996.

Silverstone, Paul H. *Civil War Navies, 1855–1883.* Annapolis, MD: Naval Institute Press, 2001.

U.S. Department of the Navy. *Dictionary of American Fighting Ships.* 8 vols. Washington, D.C.: U.S. Government Printing Office, 1959.

———. *Official Records of the Union and Confederate Navies in the War of the Rebellion.* 30 vols. Washington, D.C.: Government Printing Office, 1894–1922.

———. *Records of Living Officers of the U.S. Navy and Marine Corps.* Nimitz Library, U.S. Naval Academy, Special Collections and Archives, Annapolis, Maryland.

———. *Register of the Commissioned, Warrant, and Volunteer Officers of the Navy of the United States, Including Officers of the Marine Corps and Others, to January 1, 1865.* Washington, D.C.: Government Printing Office, 1865.

———. "Subject File of the Confederate States' Navy, 1861–1865, File HA, Miscellaneous, Box 160, Narrative of H. Ashton Ramsay, Chief Engineer, Confederate States Steamer *Merrimack*, during her engagements in Hampton Roads, 1862," Microfilm Series, M1091, Roll 13, RG 45, NA.

U.S. War Department. *The War of Rebellion: A Compilation of the Official Records of the Union and Confederate Armies.* 128 vols. Washington, D.C.: Government Printing Office, 1880–1902.

Who Was Who in America. A component volume of *Who's Who in American History.* Vol. I, 1897–1942. Chicago: A.N. Marquis Co., 1943.

PRIMARY

Barnard, J.G. *Notes on Sea Coast Defense.* New York: D. Van Nostrand, 1861.

Beale, Howard K., ed. *Diary of Gideon Welles, Secretary of the Navy Under Lincoln and Johnson.* 3 vols. New York: W.W. Norton Company, 1960.

Butts, Francis B. "The Loss of the '*Monitor.*'" In *Battles and Leaders of the Civil War*, vol. 2, edited by Robert Underwood Johnson and Clarence Clough Buel. New York: Century Co., 1887.

BIBLIOGRAPHY

Cannon, LeGrand Bouton. *Personal Reminiscences of the Rebellion, 1861–1865.* New York: 1895.

Cline, William R. "The Ironclad *Virginia.*" *Southern Historical Society Papers* 32 (December 1904): 243–49.

Colston, Raleigh E. "Watching the *Merrimac.*" In *Battles and Leaders of the Civil War*, vol. 1, edited by Robert Underwood Johnson and Clarence Clough Buel. New York: Century Co., 1887.

Crockett, Albert Stevens. "Aboard the USS '*Monitor.*'" *Civil War Times Illustrated*, January 1963.

Curtis, Richard. *History of the Famous Battle Between the Iron-clad* Merrimac, *C.S.N, and the Iron-clad* Monitor *and the* Cumberland *and the* Congress *of the U.S. Navy, March the 8th and 9th, As Seen by a Man at the Gun.* Repr., Hampton, VA: Houston Print and Publishing Co., 1957.

Dahlgren, Madeline V. *Memoirs of John A. Dahlgren.* Boston: J.R. Osgood & Company, 1882.

Daly, Robert W., ed. *Aboard the USS* Monitor*: 1862: The Letters of Acting Paymaster William Frederick Keeler, US Navy, to his wife Anna.* Annapolis, MD: United States Naval Institute Press, 1964.

Delafield, Richard, Alfred Mordecai and George McClellan. *Report Published by Secretary of War of Military Commission to Europe, 1854–1856.* 3 vols. Washington, D.C.: Government Printing Office, 1857–1860.

Eggleston, John R. "Captain Eggleston's Narrative of the Battle of the *Merrimac.*" *Southern Historical Society Papers* 40 (1966): 166–78.

Ellis, David Robert. *The* Monitor *of the Civil War.* Annville, PA: 1900.

Ericsson, John. "The Building of the *Monitor.*" *Battles and Leaders of the Civil War*, vol. 1. Edited by Robert Underwood Johnson and Clarence Clough Buel. New York: Century Co., 1887.

———. "The Monitors." *Century Illustrated Monthly Magazine*, December 1885, 280–99.

———. "Negotiations for the Building of the '*Monitor.*'" In *Battles and Leaders of the Civil War*, vol. 1, edited by Robert Underwood Johnson and Clarence Clough Buel. New York: Century Co., 1887.

BIBLIOGRAPHY

Foute, Robert Chester. "Echoes From Hampton Roads." *Southern Historical Society Papers* 19 (January 1891): 246–51.

Franklin, S.R. *Memoirs of a Rear Admiral.* New York: Harper & Brothers, 1892.

Goldsborough, Louis M. "Narrative of Rear Admiral Goldsborough, US Navy." U.S. Naval Institute Proceedings 54 (July 1933): 1022–34.

Greene, Samuel Dana. "I Fired the First Gun and Thus Commenced the Great Battle." *American Heritage*, June 1957, 103.

———. "In the *Monitor*'s Turret." In *Battles and Leaders of the Civil War*, vol. 1, edited by Robert Underwood and Clarence Clough Buel. New York: Century Co., 1887.

———. "Manuscript." *U.S. Naval Academy Trident*, Spring 1942, 42–44.

Griffiths, Oliver W. "The New War Steamers." *The United States Nautical Magazine*, April 1855, 298–310.

Hawthorne, Nathaniel. "Chiefly about War Matters by a Peaceable Man." *Atlantic Monthly*, July 1862, 58.

Isherwood, Benjamin F. *Experimental Researches in Steam Engineering.* 2 vols. Philadelphia: J.B. Lippincott, 1860.

Jones, Catesby ap Roger. "Services of the *Virginia*." *Southern Historical Society Papers* 11 (January 1883): 65–75.

Lewis, Samuel. "Life on the *Monitor*: A Seaman's Story of the Fight with the *Merrimac*; Lively Experiences Inside the Famous 'Cheesebox on a Raft.'" In *Campfire Sketches and Battlefield Echoes of '61–'65*, edited by William C. King and William P. Derby. Springfield, MA: 1883.

Littlepage, Hardin Beverly. "The Career of the *Merrimack-Virginia*: With Some Personal History." In *Voices of the Civil War: The Peninsula*, edited by Paul Mathless. Alexandria, VA: Time-Life Books, 1997.

———. "*Merrimac* vs. *Monitor*: A Midshipman's Account of the Battle with the 'Cheese Box.'" In *Campfire Sketches and Battlefield Echoes of '61–'65*, edited by William C. King and William P. Derby. Springfield, MA: 1883.

———. "A Midshipman Aboard the *Virginia*." With an introduction by Jon Nielson. *Civil War Times Illustrated*, April 1974.

BIBLIOGRAPHY

———. "With the Crew of the *Virginia*." *Civil War Times Illustrated*, May 1974, 36–43.

Lull, Edward P. *History of the United States Navy Yard at Gosport*. Washington, D.C.: Government Printing Office, 1874.

Mann, Samuel A. "New Light on the Great Drewry's Bluff Fight." *Southern Historical Society Papers* 34 (1906).

The Mariners' Museum, Newport News, Virginia. Jacob Nicklis Papers.

Marvel, William, ed. *The* Monitor *Chronicles*. New York: Simon & Schuster, 2000.

McClellan, George Brinton. *McClellan's Own Story*. New York: Charles L. Webster Company, 1887.

New York Historical Society. Ericsson Papers, Box 2.

———. Greene Papers.

Norris, William. "The Story of the Confederate States' Ship '*Virginia*' (once *Merrimac*): Her Victory Over the *Monitor*; Born March 7th, Died May 10th, 1862." *Southern Historical Society Papers* 41 (September 1916): 204–33.

Phillips, Dinwiddie Brazier. *The Career of the Iron-clad* Virginia, *(formerly the* Merrimac*), Confederate States Navy, March–May 1862*. Collection of the Virginia Historical Society. Richmond: Virginia Historical Society, 1887, 193–231.

Porter, David Dixon. *Incidents and Anecdotes of the Civil War*. New York: D. Appleton & Company, 1885.

Rae, Thomas W. "The Little *Monitor* Saved Our Lives." *American History Illustrated*, July 1960, 34.

Ramsay, Henry Ashton. "The Most Famous of Sea Duels, The Story of the *Merrimac*'s Engagement with the *Monitor*, and the Events That Preceded and Followed the Fight, Told by a Survivor." *Harper's Weekly*, February 10, 1912, 11–12.

———. "Wonderful Career of the *Merrimac*." *Confederate Veteran*, July 1907, 310–13.

Ransom, Thomas. "The *Monitor* and the *Merrimac* in Hampton Roads." *Hobbies*, September 1959.

Reaney, Henry. "How the Gun-Boat *Zouave* sided the *Congress*." In *Battles and Leaders of the Civil War*, vol. 1, edited by Robert Underwood Johnson and Clarence Clough Buel. New York: Century Co., 1887.

Bibliography

Rodman, J.T. *Reports of Experiments on the Properties of Metals for Cannons, and the Qualities of Cannon Powder.* Washington, D.C.: Government Printing Office, 1801.

Sears, Stephen W., ed. *The Civil War Papers of George B. McClellan.* New York: Da Capo Press, 1992.

Selfridge, Thomas O., Jr. *Memoirs of Thomas O. Selfridge Jr.: Rear Admiral USN.* New York: G.P. Putnam's Sons, 1924.

———. "The *Merrimac* and the *Cumberland.*" *Cosmopolitan,* June 1893, 180.

"The Sinking of the *Cumberland* by the Ironclad *Merrimac* off Newport News, Virginia, March 8, 1862." New York: Currier and Ives, 1862.

Soley, James Russell. "The Navy in the Peninsular Campaign." In *Battles and Leaders of the Civil War,* vol. 2, edited by Robert Underwood Johnson and Clarence Clough Buel, 269–70. New York: Century Co., 1887.

Southern Historical Collection. Louis Round Wilson Library. University of North Carolina, Chapel Hill. Franklin Buchanan Letterbook, 1801–1863.

———. John Taylor Wood Papers.

Stimers, Alban C. "An Engineer Aboard the *Monitor.*" *Civil War Times Illustrated,* April 1970.

Stodder, Louis N. "Aboard the *Monitor.*" *Civil War Times Illustrated* 1, January 1963, 31.

Stuyvesant, Moses S. "How the *Cumberland* Went Down." *War Papers and Reminiscences, 1861–1865, Read Before the Missouri Commandery, Military Order of the Loyal Legion of the United States (MOLLUS).* St. Louis: Missouri Commandery MOLLUS, 1892.

Thompson, C.M., and R. Wainwright, eds. *Confidential Correspondence of Gustavus Vasa Fox, Assistant Secretary of the Navy 1861–1865.* 2 vols. New York: Naval History Society, 1918–1919.

Vick, Egbert L. "A Trip with Lincoln, Chase and Stanton." *Scribner's Monthly,* October 1878, 819.

Weeks, Grenville M. "The Last Cruise of the *Monitor.*" *Atlantic Monthly* 11 (March 1863): 367.

Welles, Gideon. "The First Iron-Clad *Monitor.*" *Annals of War.* Philadelphia: J.B. Lippincott, 1879.

Bibliography

White, Elsberry Valentine. *The First Iron-Clad Naval Engagement in the World.* New York: J.S. Ogilvie Publishing Co., 1906.

Wood, John Taylor. "The First Fight of the Ironclads; March 9, 1862." In *Battles and Leaders of the Civil War*, vol. 1, edited by Robert Underwood Johnson and Clarence Clough Buel. New York: Century Co., 1887.

Secondary

Anderson, Bern. *By Sea and By River: The Naval History of the Civil War.* New York: Alfred A. Knopf Co., 1962.

Andrews, H. Franklin. Th*e Hamlin Family: A Genealogy of Capt. Giles Hamlin of Middletown, Connecticut, 1654–1900.* Exira, IA: H.F. Andrews, 1900.

Barthell, Edward E., Jr. *The Mystery of the* Merrimack. Muskegon, MI: Dana Printers Company, 1959.

Baxter, James Phenney, III. *The Introduction of the Ironclad Warship.* Cambridge, MA: Harvard University Press, 1933.

Bearss, Edwin C. *River of Lost Opportunities: The Civil War on the James River 1861–1862.* Lynchburg, VA: H.E. Howard Inc., 1995.

Beese, Sumner B. *C.S. Ironclad* Virginia *and U.S. Ironclad* Monitor. Newport News, VA: The Mariners' Museum, 1996.

Bennett, Frank M. *The Steam Navy of the United States: A History of the Steam Vessel of War in the U.S. Navy and of the Naval Engineer Corps.* Pittsburgh, PA: Warren & Company, 1897.

Berent, Irwin Mark. *The Crewmen of the USS* Monitor*: A Biographical Directory.* Raleigh: North Carolina Department of Cultural Resources, 1985.

Brooke, George M., Jr. *John M. Brooke, Naval Scientist and Educator.* Charlottesville: University Press of Virginia, 1980.

Carpenter, Helen Graham. *The Reverend John Graham of Woodbury, Connecticut and His Descendents.* Chicago: Monastery Hill Press, 1942.

Catton, Bruce. "When the *Monitor* Met the *Merrimac.*" *New York Times Magazine*, March 4, 1962.

Cope, Gilbert. *Genealogy of the Dutton family of Pennsylvania—Preceded by a history of the family in England from the time of William the Conqueror to the year 1669: with an appendix containing a short account of the Duttons of Conn., West Chester, Pa.* N.p, n.d.

BIBLIOGRAPHY

Coski, John M. *Capital Navy: The Men, Ships and Operations of the James River Squadron.* Campbell, CA: Savas Woodbury Publishers, 1996.

Cuddeback, William Louis. *Caudebec in America—A Record of Descendents of Jacques Caudebec 1700 to 1920.* New York: Amereon House, n.d.

Daly, Robert W. *How the* Merrimac *Won: The Strategic Story of the CSS* Virginia. New York: Crowell Inc., 1957.

Davis, William C. *Duel Between the First Ironclads.* Garden City, NJ: Doubleday & Company, 1975.

De Riemer, W.E. *The De Riemer Family: A.D. 1640(?)–1903.* New York: T.A. Wright, 1905.

Gorton, Adelos. *The Life and Times of Samuel Gorton: The Founders and the Founding of the Republic: A Section of Early United States History and a History of the Colony of Providence and Rhode Island Plantations in the Narragansett Indian Country.* Philadelphia: 1907.

Halsy, Ashley, Jr. "The Plan to Capture the *Monitor*: Seal the Turtle in Its Shell." *Civil War Times Illustrated,* June 1966, 28–31.

Hochling, A.A. *Thunder at Hampton Roads.* New York: Da Capo Press, 1993.

Hogg, Ian V. *A History of Artillery.* London: Hamlyn Publishing Group, 1974.

Johnson, Robert Erwin. *Rear Admiral John Rodgers, 1812–1882.* Annapolis, MD: United States Naval Institute Press, 1967.

Jones, Virgil Carrington. *The Civil War at Sea.* 3 vols. New York: Holt, Rinehart and Winston, 1960–1962.

McIntire, Robert Harry. *Annapolis Maryland Families.* Baltimore, MD: Gateway Press, Inc., 1979.

Mendell, W.N. *Technology and Experience aboard the USS* Monitor. Baltimore: Johns Hopkins University Press, 2000.

Niven, John. *Gideon Welles, Lincoln's Secretary of the Navy.* New York: Oxford University Press, 1973.

O'Neil, Charles. "Engagement Between the *Cumberland* and *Merrimack*." *U.S. Naval Institute Proceedings* 48, June 1922, 893.

Post, Marie Caroline de Trobriand. *The Post Family.* New York: Sterling Potter, 1905.

Bibliography

Preston, Robert L. "Did the *Monitor* or *Merrimac* Revolutionize Naval Warfare?" *William & Mary Quarterly*, July 1915, 58–66.

Quarstein, John V. *The Battle of the Ironclads*. Charleston, SC: Arcadia Publishing, 2000.

———. *CSS Virginia: Mistress of Hampton Roads*. Appomattox, VA: H.E. Howard, Inc., 2001.

———. *Hampton and Newport News in the Civil War: War Comes to the Peninsula*. Lynchburg, VA: H.E. Howard, Inc., 1998.

———. *A History of Ironclads: The Power of Iron Over Wood*. Charleston, SC: The History Press, 2006.

Quarstein, John V., and Dennis Mroczowski. *Fort Monroe: The Key to the South*. Charleston, SC: Arcadia Publishing, 2000.

Rae, Thomas W. "The Little *Monitor* Saved Our Lives." *American History Illustrated*, July 1966.

Ripley, William. *Artillery and Ammunition of the Civil War*. New York: Promontory Press, 1970.

Still, William H., Jr. *Iron Afloat: The Story of the Confederate Armourclad*. Nashville, TN: Vanderbilt University Press, 1971.

———. *Ironclad Captains: The Commanding Officers of the USS* Monitor. Monitor National Marine Sanctuary, Historical Report Series. N.p.: n.d.

Veinart, Richard P., Jr., and Robert Arthur. *Defender of the Chesapeake: The Story of Fort Monroe*. Shippensburg, PA: White Mane Publishing Company, 1989.

Wegner, Dana. "Ericsson's High Priest." *Civil War Times Illustrated*, February 1975.

Wentworth, John. *The Wentworth Genealogy: Comprising the origin of the name, the family in England, and a particular account of Elder William Wentworth, the emigrant, and of his descendents*. Boston, MA; Press of A. Mudge & Son, 1870.

Wentz, Richard W. *Record of the Descendents of Johann Jost Wentz*. Brooklyn, NY, Binghamton: *Daily Republican*, 1884.

Woodward, C. Vann, ed. *Mary Chesnut's Civil War*. New Haven, CT: Yale University Press, 1981.

Bibliography

Woolen, Edward A., ed. *Woollen/Woolen Family Biographical and Historical Records and Genealogy of Edmond/Edward Woollen of Dorchester County Maryland and Richard Woolen of Maryland*. Decorah, IA: Anundsen Publishing Company, 1984.

Worden, Robert L. "The Naval Career of Rear Adm. John Lorimer Worden (1818–1897)." *Worden's Past* 4, no. 2, October 1983.

Periodicals

Baltimore American, 11 May 1862.
Boston Journal, 13 March 1862.
Cincinnati (OH) Weekly Herald and Philanthropist, 23 September 1846.
Civil War Times Illustrated, February 1975.
Cuyahoga Falls (OH) Reporter, 30 March 1917.
Daily Gazette and Bulletin (Williamsport, PA), 4 February 1990.
Daily Herald (Delphos, OH), 30 April 1902.
Daily Northwestern (Oshkosh, WI), 29 September 1916.
Daily Star (Long Island City), March 3, 1906.
Evening Journal, 10 February 1893.
Fitchburg (MA) Sentinel, 22 March 1897.
Lebanon (PA) Daily News, 1 September 1908.
London Times, 4 May 1862.
Los Angeles (CA) Times, 15 July 1911.
Lynchburg Virginian, 12 September 1861.
Medical and Surgical Reporter (Philadelphia), 3 June 1865.
Middletown (New York) Daily Times-Press, 5 February 1914.
Mobile Register, 11 August 1861.
Newark (NJ) Evening News, 26 April 1919.
Newport (RI) Mercury, 27 January 1900.
New York Herald, 15 April 1862, 6 June 1876.
New York Post, 28 July 1979.
New York Sun, 14 November 1905.
New York Times, 21 December 1862, 13 March 1863, 29 May 1863, 2 June 1867, 24 February 1881, 11 February 1893, 20 October 1897, 18 June 1905, 10 October 1909, 10 October 1911.
Niles' National Register, 1 June 1844.
Norfolk Day Book, 7 January 1862, 10 March 1862, 11 March 1862.
Somerset (PA) Herald, 9 September 1908.
Syracuse (NY) Herald, 23 April 1909.
Troy (NY) Daily Times, 10 October 1892.
Washington (D.C.) Post, 5 January 1916.
Washington (D.C.) Star, 15 May 1896.

Index

A

Abbott & Sons 31, 32, 217
A. Colby 179
Adams centrifuge pump 171
Adams Centrifuge Pump 49, 58, 247
Aden, Arabia 267
Albany Ironworks 31, 217
Alene 195
Allen, William 162, 210, 242, 265
Amazon River 186, 288
American Merchant Marine 43, 265
Anderson, Hans 89, 91, 181, 195, 198, 265
Anderson, Priscilla Gladden (or Gladding) 265
Andrew's centrifugal pump 144, 171, 172, 247
Anjier, Richard 181, 206, 222, 266
Annapolis, Maryland 61, 63, 96, 100, 188
Anna Reynolds 35
Annville, Pennsylvania 254, 272
Antioch, Texas 269
Appomattox Manor 138
Appomattox River 124, 146, 152, 237, 275, 289
Aquia Landing, Virginia 150

Arlington National Cemetery 189, 192
Army of the Potomac 61, 94, 106, 150, 158, 232, 282
Ashland, Kentucky 253, 273
Asiatic Squadron 186, 190, 288, 289
Atkins, John 43, 207, 221, 266
Atwater, Charlotte 266
Atwater, John Knox 266
Atwater, Norman Knox 159, 262, 266

B

Baker, Benjamin Franklin 187
Baltimore, Maryland 32, 40, 159, 166, 207, 208, 212, 216, 217, 250, 266, 276, 280, 289, 293
Bankhead, Anne Pyne 266
Bankhead, James 153, 266
Bankhead, John Pyne 152, 153, 154, 156, 161, 162, 165, 166, 167, 168, 169, 170, 171, 172, 173, 174, 175, 177, 180, 181, 189, 190, 191, 206, 240, 241, 245, 246, 247, 248, 250, 261, 266
Barnard, John Gross 94, 106
Barrios, Justo Rufino 189, 292
Barry, Joseph N. 109

INDEX

Basting, Anton 43, 161, 202, 209, 222, 242, 255, 267
Basting, Ellen Spriggs 267
Bath, New York 252, 268
Battle of Antietam 158, 282
Battle of Bull Run 158, 296
Battle of Drewry's Bluff 128
Battle of Eltham's Landing 150, 291
Battle of Hampton Roads 75, 147, 160
Battle of Lake Erie 150, 290
Battle of Malvern Hill 146, 291
Battle of Mobile Bay 188, 291
Battle of Oak Grove 145
Beaufort, North Carolina 167, 244, 245, 267
Beaumont, John C. 124, 130
Bedford Artillery 126, 131
Bellmore, Long Island, New York 281
Bermuda Hundred, Virginia 200
Binghamton, New York 39, 207, 292
Blair, Franklin 148, 238
Blair's Landing, Louisiana 185
Bordentown, New Jersey 169, 209, 295
Boston, Massachusetts 38, 43, 70, 96, 168, 208, 209, 266, 268, 269, 270, 272, 276, 277, 281, 288, 290, 293, 295
Boston Navy Yard 289
Braithwaite, John 16
Brazil Squadron 101, 266, 278, 290, 297
Breese, K. Randolph 195
Bremen, Germany 208, 267
Bridgeport, Connecticut 35, 280
Bringman, Derick 43, 161, 208, 222, 242, 267
Bristol, Rhode Island 210, 295
Brodie, Peter H. 200, 299, 300
Brooke bolts 26, 91, 105, 230
Brooke, George Mercer 230
Brooke, John Mercer 24, 25, 91, 105, 230
Brooke rifle 26, 67, 75, 86, 228
Brooklyn Navy Yard 33, 36, 42, 46, 52, 55, 56, 59, 68, 158, 161, 189, 191, 198, 220, 269, 277, 283, 297, 300
Brooklyn, New York 31, 32, 38, 44, 52, 92, 207, 208, 211, 212, 216, 250, 251, 254, 256, 266, 268, 272, 274, 284, 286, 287, 293, 294, 296, 298
Brown, Edmund 196, 224, 252, 268
Browne, D. Rodney 177
Brown, John 197, 211, 242, 254, 268
Bryan, William 42, 207, 222, 268
Buchanan, Franklin 62, 65, 68, 74, 104, 105, 224
Buchanan, Thomas McKean 68
Buffalo, New York 32, 161, 196, 211, 271, 284, 286
Bureau of Ordnance 188, 279
Burnside, Ambrose Everett 61, 112
Burrows, George W. 150, 239, 268
Burwell's Bay, Virginia 117
Bushnell, Cornelius 28, 29, 30, 35, 112, 214, 215, 216
Butler, Benjamin Franklin 45
Butts, Francis Banister 39, 161, 171, 174, 177, 178, 183, 201, 212, 242, 247, 253, 268
Butts, Helen Francis Battey 269
Butts, John Wood 268
Butts, Mehitable Wentworth 268

C

Calcasieu Pass, Louisiana 182
California Gold Rush 35
Caloric engine 22
Calypso 189, 267
Cambridge 61
Campbell, Albert Bogart 40, 56, 99, 158, 169, 191, 220, 245, 249, 261, 269
Camp Butler 66, 68, 97, 105
Camp Groce 182, 249, 280
Camp Gross. *See* Camp Groce
Camp Hamilton 105
Cann, Edward 160, 200, 243, 269
Cannon, LeGrand B. 69, 99

INDEX

Canonicus class 192, 292
Cape Charles, Virginia 60
Cape Fear River 167, 186, 244, 288
Cape Hatteras Lighthouse 170, 246
Cape Hatteras, North Carolina 179, 265, 266, 267, 268, 269, 270, 271, 272, 273, 274, 276, 277, 280, 281, 282, 283, 284, 285, 286, 287, 293, 294, 297
Cape Henry 223, 245
Cape Henry Lighthouse 223
Carey, James H. 300, 301
Caribbean Sea 33, 291
Carmarthen, England 211, 272
Carroll #1, Thomas 43, 208, 220, 253, 269
Carroll #2, Thomas 140, 196, 212, 222, 251, 269
Carroll, Eliza Stanley 269
Carroll, Margaret Walsh 269
carronades 16, 18
Cartagenera 102, 278
Carter, Hill 137, 270
Carter, Siah Hulett 137, 189, 200, 210, 236, 242, 252, 270
Casco class 193, 292
Case, A. Ludlow 188
Catinet 65
Cavalli, Giovanni 19
Chaffin's Bluff, Virginia 125, 128
Charles City Co., Virginia 210, 270
Charleston Harbor, South Carolina 23, 153, 162, 182, 184, 188, 193, 197, 206, 266, 274, 284, 286, 293, 296
Charleston Navy Yard 295
Charleston, South Carolina 240, 249, 252, 275, 276, 279, 282, 291, 298
Charlestown, Massachusetts 209, 269, 287
Charlestown Navy Yard 96, 168
Charlotte, North Carolina 114
Chase, Salmon P. 115, 122
Chelsea, Massachusetts 253, 290

Chesapeake & Albemarle Canal 103, 112
Chesapeake Bay 28, 60, 70, 94, 107, 120, 162, 168, 223, 233
Chesnut, Mary 106
Chesterfield Co., Virginia 270
Chincoteague Island, Virginia 59
Chiriqui Isthmus 279
Christinia, Norway 208, 285
City Point, Virginia 125, 126, 137, 138, 139, 144, 146, 149, 236, 238, 240, 291
Clemson class 194
Cleveland, Ohio 202, 253, 269
Cline, William 75
Clute Brothers Foundry 32, 217
Coatzacoalcos 278
Coleman, James 162, 183, 211, 242, 270
Coles, Cowper 213
Cole's Point, North Carolina 213, 301
Columbiad 17, 126, 131, 133, 235
Congress 19, 28, 70, 102, 152, 164, 212, 214, 238, 244, 294, 296, 298
Conklin, John P. 161, 208, 222, 242, 270
Connoly, Anthony 180, 208, 222, 270
Constable, David C. 118
Constitution 106
Continental Iron Works 31, 32, 216, 217, 218, 219
contraband 45, 137, 200
Cook, Robert 160, 200, 212, 242, 271
Cornelius Creek, Virginia 130
Corwey, John/James 300, 301. *See* Carey, James H.
Cotton for Cannon 23
County Cork, Ireland 42, 209
Craney Island, Virginia 65, 104, 107, 122, 233
Cranston, Rhode Island 161, 268
Crimean War 20, 29
Crown, Joseph 39, 41, 181, 206, 222, 271

INDEX

CSS *Albemarle* 190
CSS *America* 150, 291
CSS *Beaufort* 64, 65, 112, 128
CSS *Elmira* 185, 288
CSS *General Price* 185, 288
CSS *Hampton* 115
CSS *Jamestown* 75, 95, 109, 112, 113, 115, 119, 128
CSS *Louisville* 185, 288
CSS *Merrimac*. *See* CSS *Virginia (Merrimack)*
CSS *Merrimac II*. *See* CSS *Richmond*
CSS *Merrimack*. *See* CSS *Virginia (Merrimack)*
CSS *Nashville* 183, 298
CSS *Patrick Henry* 75, 112, 115, 128, 225
CSS *Raleigh* 64, 112
CSS *Rattlesnake* 183
CSS *Richmond* 112, 114, 115, 146, 162, 238
CSS *Shenandoah* 190
CSS *Teaser* 75, 95, 112, 128, 146, 147, 148, 238, 291
CSS *Virginia (Merrimack)* 26, 27, 31, 42, 61, 62, 65, 67, 68, 74, 75, 79, 81, 85, 88, 93, 94, 95, 99, 106, 109, 113, 117, 119, 297
  casemate 25
  command change, post-battle 104
  fight or flight 121
  first day of battle 66
  first shot fired in the battle 75
  launch and commission 62
  scuttled and burned 121
CSS *Yorktown* 95, 113, 119
Cuddeback, David 45, 211, 222, 250, 271
Cumberland Landing 150, 291
Cumberland, Maryland 38, 211, 275
Curtis, Richard 75

D

Dabney's Battery 128
Dahlgren, John A.B. 46, 102, 225, 229, 278
Dahlgren rifle 154, 188
Dahlgren smoothbores 26, 31, 37, 47, 48, 78, 79, 85, 91, 112, 124, 129, 130, 132, 138, 146, 154, 157, 164, 174, 181, 188, 189, 214, 219, 227, 235, 257, 279
Davidson, Hunter 75, 147
David, Walter. *See* Durst, William
Davis, Charles Henry 28, 214
Davis, Jefferson 23, 186
Deckerman, Allan 216
Delafield Report 20
Delafield, Richard 20
Delameter, Cornelius H. 21, 29, 214
Delameter Ironworks 21, 29, 31
de Lesseps, Ferdinand 186, 288
Deloach's Bluff, Louisiana 191, 278
de Montaignac, Marquis 66
Dévastation 20
de Villeroi, Brutus 146
Dismal Swamp Canal, North Carolina/Virginia 102
Dixon, Peter 130
Drayton, Percival 163, 291
Drewry, Augustus Herman 126, 131, 133
Drewry's Bluff, Virginia 125, 126, 128, 129, 130, 134, 137, 141, 146, 235, 236, 273, 275, 279, 291
Driscoll, Abigail Sweeney 271
Driscoll, John Ambrose 42, 49, 55, 57, 73, 74, 85, 161, 181, 209, 222, 271
Dublin, Ireland 206, 277
DuPont, Samuel Francis 154, 181, 183, 184, 188, 193, 197, 267, 274, 276, 286, 290, 291
Durst, Anna Goronozy Neuman 272
Durst, Ester Wallensteine 272
Durst, William 43, 73, 87, 89, 160, 181, 198, 199, 209, 220, 242, 255, 271
Dutchess Co., New York 208, 274
Dutch Gap, Virginia 125
Dutton, Henry 35, 280

INDEX

E

Eads, James 181
East Gulf Blockading Squadron 293
East India Squadron 104, 267
East River 31, 42, 46, 52
Egan, William H. 162, 272
Eggleston, John Randolph 75, 83
Elizabeth City, North Carolina 102, 279
Elizabeth River 25, 42, 60, 64, 65, 88, 107, 113, 115, 117, 121, 228, 234
Ellis, David Roberts 43, 53, 189, 199, 211, 222, 254, 272
Ellis, Martha Jane Keck 272
Ellis, Roberts 272
Ericsson, John 8, 16, 18, 20, 21, 29, 30, 31, 34, 36, 37, 42, 46, 49, 50, 52, 56, 78, 90, 143, 163, 192, 205, 209, 211, 212, 213, 214, 215, 216, 217, 218, 219, 221, 257, 284, 292, 298
Ericsson's Battery 15, 33, 34, 215, 216, 217, 218, 219
European Squadron 185, 186, 187, 188, 288, 289, 291, 298

F

Fairfax, D.M. 296
Farragut, David Glasgow 102, 279
Farrand, Ebenezer 126, 130, 132
Fay, Lawrence 300, 301
Feeney, Thomas 43, 211, 220, 272
Fenton, Andrew 256, 299, 300, 301
Fenwick, Charles 272
Fenwick, Elizabeth 272
Fenwick Island, Delaware 58
Fenwick, James R. 43, 174, 210, 222, 241, 272
Fenwick, Mary Ann Duffy 241, 273
Fernandina, Florida 291
Fisher, Hugh 43, 206, 222, 273
Flye, Mary Elizabeth Perkins 273
Flye, William P. 92, 158, 205, 229, 233, 242, 253, 261, 273

Fort Beauregard 150, 290
Fort Boykin 117, 123
Fort Clinch 150
Fort Darling. *See* Drewry's Bluff
Fort Fisher 167, 186, 187, 195
Fort Gregg 282
Fort Huger 123
Fort Itapiru 102
Fort Johnston 153, 206, 266
Fort McAllister 183, 184, 193
Fort Monroe 45, 60, 63, 65, 66, 69, 90, 92, 94, 100, 105, 106, 107, 108, 112, 115, 117, 120, 166, 225, 229, 230, 232, 233, 234, 284
Fort Moultrie 282
Fort Pickens 33, 213
Fort Simpson 193, 293
Fort Sumter 162, 282, 291, 292, 296, 298
Fort Wagner 282
Fort Walker 150, 290
Fort Wool 94, 106, 107, 119
Foster, John Gray 168
Foute, R.C. 108
Fox, Gustavus Vasa 36, 42, 61, 89, 90, 92, 94, 103, 219, 229, 233
Francis B. Ogden 16, 209
Franklin, S.R. 121
Franklin, William Buell 111
Frederickson, George 39, 41, 43, 48, 175, 208, 218, 221, 237, 261, 273
Frederickson, Magdalena E. Heobst 273
Fugitive Slave Act 45

G

Gager, Edwin Velie 32, 92, 145, 191, 208, 231, 236, 238, 255, 261, 274
Gager, Hanna M. Velie 274
Gager, Joseph 274
Gager, Julia P. Werner 274
Gager, Rose A. Morley 274
Gallagher, Darius F. 274

337

INDEX

Gallagher, Darius Farrington 191, 253, 274
Gallagher, Margaret A. Lynch 274
Gallagher, Marshal W. Higgins 274
Gardiner, David 19
Garety, Bridget Davis 274
Garety, John 44, 181, 209, 221, 251, 274
Garety, Mary 274
Garvin, Benjamin F. 53, 221
Gassendi 65
Geer, Adna Spencer 274
Geer, George Spencer 13, 44, 49, 82, 90, 103, 105, 106, 109, 111, 112, 121, 125, 131, 135, 138, 139, 143, 152, 154, 156, 159, 162, 165, 166, 168, 170, 172, 175, 177, 179, 180, 181, 209, 220, 223, 236, 238, 242, 246, 247, 252, 274
Geer, Gilbert 274
Geer, Martha Clark Hamilton 274
Gibson, James F. 148
Gilmer, Thomas 19, 212
Gloucester Co., Virginia 212, 271
Gloucester Point, Virginia 95, 106, 287
Goldsborough, Louis Malesherbes 61, 95, 102, 103, 107, 113, 115, 116, 117, 118, 119, 122, 123, 124, 137, 141, 145, 146, 148, 149, 185, 234, 238, 279, 288
Gosport Navy Yard 24, 27, 51, 64, 96, 102, 109, 114, 117, 120, 180, 234, 287
Göta Canal 16
Gothenberg, Sweden 43, 206
Gottenberg, Sweden. *See* Gothenberg, Sweden
Graham, Andrew 297
Greek War of Independence 15
Greene, George Sears 37, 275
Greene, Martha Barrett Dana 275
Greene, Mary Abby Babbitt 275
Greene, Mary Willis Dearth 275
Greene, Samuel Dana 37, 40, 56, 58, 59, 71, 72, 76, 77, 78, 79, 85, 87, 88, 89, 90, 92, 95, 103, 129, 168, 177, 189, 194, 211, 219, 223, 224, 226, 228, 230, 244, 251, 261, 275
Greenpoint Ship Yard 31, 34, 35, 41, 46, 216, 219, 257, 298
Green, Thomas 185
Gregory, Francis H. 53, 221, 291, 298
Griswold, John A. 30, 215, 216
grog 139, 152, 238
Guttenberg, Sweden. *See* Gothenberg, Sweden

H

Halman, William S. 242, 276, 280
Hamlin, Hannibal 232
Hands, Jane Woollen 276
Hands, Robinson Woollen 40, 57, 208, 220, 261, 276
Hands, Washington 276
Hannan, Bridget Kenery 276
Hannan, Mary 276
Hannan, Nora 276
Hannan, Patrick 44, 73, 89, 181, 196, 207, 220, 252, 276
Hardy, John 150, 210, 239, 240, 276
Harrison, Henry 180, 206, 239, 240, 277
Harrison's Bar 121
Harrison's Landing, Virginia 146, 148, 149, 202, 238, 239, 295
Hartford, Connecticut 27, 29
Hartt, Edward 53, 221
Harvest Queen 38
Hasker, Charles H. 129, 235
Hatteras Inlet, North Carolina 92
Hattie 189, 267
Haverstraw, New York 161, 207, 290
Hawthorne, Nathaniel 51
HMS *Ariadne* 166
HMS *Ironside* 100
HMS *Warrior* 20, 100
Hoboken, New Jersey 106
Hog Island, Virginia 124
Holdane and Company 31

INDEX

Home Squadron 33, 150, 154, 266, 267, 287, 290, 297
Howard Co., Virginia 206, 277
Howard, Mary Dugan 277
Howard, Robert H. 159, 200, 206, 243, 277
Howard, Samuel 71, 76, 87, 191, 206, 224, 253, 261, 277
howitzers 102, 124, 129, 130
Hubbell, Charles Benjamin, Jr. 277
Hubbell, Mary Adeline Knox 277
Hubbell, Robert Knox 45, 181, 191, 222, 261, 277
Hudson River 287
Huger, Benjamin 112, 120
Hulett, Eliza Tarrow 200, 270
Hulett, John 270
Hulett, Molly 270
hydrostatic guns 37
hydrostatic javelins 31

I

Indian Ocean 190
Industrial Revolution 15
International Canal Congress 186, 289
Inter-Oceanic Canal 186, 288
Inter-Oceanic Railway 101, 278
Intrepid 106
Iola, Texas 253, 269
Ironclad Board 28, 29, 34, 68, 112, 214, 215
Iron Witch 21
island of Møn, Denmark 39, 208, 273
Isthmus of Darien 186, 288
Isthmus of Honduras 278

J

Jacksonville, Florida 150, 291
James Gordon Bennett 195
James River 60, 62, 67, 93, 94, 95, 111, 112, 113, 115, 116, 117, 119, 121, 123, 124, 125, 126, 145, 146, 148, 149, 150, 151, 162, 165, 181, 185, 186, 204, 236, 238, 240, 270, 275, 279, 287, 288, 291, 295
James River Flotilla 149, 234, 275
James River Squadron 128, 147
James River Task Force. *See* James River Flotilla
Jamestown Island, Virginia 117, 120, 124
Jeffers, John Ellis 278
Jeffers, Lucie LeGrand Smith 278
Jeffers, Ruth Westcott 278
Jeffers, William Nicholson, III 100, 101, 102, 103, 104, 113, 117, 118, 120, 130, 131, 134, 137, 138, 141, 143, 145, 150, 152, 187, 188, 206, 231, 234, 235, 236, 251, 261, 278
Jeffrey, William H. 209, 241, 279
Jersey City, New Jersey 203, 252, 255, 267, 294
Jersey Shore, Pennsylvania 191, 252, 282
Johnston, Joseph Eggleston 113, 114, 124
Joice, Thomas 210, 222, 280
Jones, Catesby ap Roger 64, 67, 68, 74, 75, 82, 83, 85, 88, 102, 104, 126, 127, 132, 224, 227, 228, 278
Jones, Jessie M. 160, 229, 280
Jones, Thomas ap Catesby 100, 278

K

Katrina 15
Kearny, New Jersey 191, 255, 296
Keeler, Anna Elizabeth Dutton 35, 280
Keeler, Mary Eliza Plaut 280
Keeler, Roswell 280
Keeler, William Frederick 35, 36, 40, 44, 46, 50, 52, 56, 57, 58, 60, 72, 74, 75, 78, 80, 83, 85, 87, 88, 90, 92, 94, 95, 97, 98, 99, 103, 104, 105, 106, 107, 108, 109, 111, 112, 113, 115, 119, 120, 121, 125, 128, 130, 131,

INDEX

132, 133, 134, 135, 137, 138, 139, 147, 148, 149, 151, 152, 155, 156, 159, 162, 165, 166, 168, 169, 171, 172, 173, 174, 175, 179, 189, 190, 191, 206, 218, 224, 226, 233, 239, 241, 243, 244, 247, 252, 261, 280, 299
Kennon, Beverly 19
Keyes, Erasmus Darwin 106
Key West, Florida 186
Kinburn Peninsula, Ukraine 20
Kingsland Creek, Virginia 125, 235
Kings, New York 296

L

La Gloire 20, 213
Laguna de Terminos, Mexico 100, 278
Lamb, William 120
Langbanshyttan, Sweden 16, 205
La Plata River 101, 278
La Salle, Illinois 35, 280
LaSalle, Illinois 193
La Salle Iron Works, Founders & Machinists 36
Lave 20
Lave class 20
Lawrence, Massachusetts 254, 295
Lee, George Washington Custis 126
Lee, Robert Edward 112, 126, 145
Lee, Samuel Phillips 149, 189
Lee, Sidney Smith 114
Leonard, Catherine Dailey 280
Leonard, Mathew 182, 207, 222, 249, 280
Leo XIII 186, 289
Le Pacificateur 17
Lewis, Samuel. *See* Truscott, Peter
Lewis, Samuel Augee 159, 174, 241, 261, 280
Lincoln, Abraham 23, 27, 28, 29, 61, 69, 97, 99, 108, 115, 116, 117, 119, 122, 123, 148, 187, 191, 214, 215, 225, 231, 233, 234, 238, 244, 279, 293, 296, 298

Lincoln Gun 94
Littlefield, George 161, 209, 242, 281
Littlepage, Hardin Beverly 64, 65
Liverpool, England 16, 40, 211, 284
Logue, Daniel C. 40, 49, 56, 75, 87, 90, 91, 158, 161, 191, 208, 220, 241, 255, 261, 281
Logue, Elizabeth A. Cassidy 281
Logue, John 281
Logue, Ruth Otis 281
Long Island, New York 191, 216, 255, 257, 281, 298
Lopez, Carlos Antonia 101
Los Angeles, California 254, 283
Loughran, Thomas 76, 160, 207, 222, 242, 281
Lowe, Thaddeus S.C. 106
Lynch, William F. 102

M

Madeira River 186, 288
Magruder, John Bankhead 94, 106, 111, 112, 113, 145, 153, 188, 191, 291
Mahone, William 126
Mallory, Stephen Russell 23, 24, 26, 27, 63, 104, 112, 113, 114, 115
Malone, James 161, 208, 242, 281
Malvern Hill, Virginia 238
Manhatten, New York 274
Mann, Thomas 132, 133
Mansfield, Joseph King Fenno 66, 94, 99, 120
Mare Island Navy Yard, California 296
Maria Helena 150, 290
Marion, William 180, 207, 222, 281
Marston, John 71, 223
Martin boilers 31, 49, 218, 257
Mason, Charles 125
Mason, James 149
Mason, John 182, 211, 222, 249, 282
Mason, Thomas 127
Matagorda Peninsula, Texas 182
Maury, Matthew Fontaine 147, 230

INDEX

Maxson Ship Yard 112, 220
Mayport, Florida 193, 252, 280
McClellan, George Brinton 20, 60, 69, 94, 95, 100, 105, 106, 107, 111, 113, 115, 117, 122, 123, 137, 144, 146, 149, 150, 229, 232, 238, 239, 291
McKinley, William 298
McKinley, William, Jr. 185
McPherson, Norman 43, 221, 282
Meckly, Elizabeth E. Frederick 282
Meckly, John 282
Meckly, Rebecca Martin 282
Meckly, Thomas W. 158, 191, 211, 241, 252, 262, 282
Medal of Honor 181, 297
Mediterranean Sea 33, 279
Mediterranean Squadron 154, 267, 297
Meigs, Montgomery 116
Melville, Herman 100
Merrick and Sons 28
Mersey Iron Works 18, 211
Mexican-American War 19, 33, 61, 63, 100, 104, 153, 165, 240, 266, 297
Miantonomoh class 203
Middletown, Connecticut 150, 205, 290
Militia Act 200
Milton, Pennsylvania 158, 211, 282
Mississippi Squadron 277
Mobile, Alabama 192, 277
Monaghan, John 162, 211, 242, 282
Monterey, California 100, 278
Montgomery, Alabama 34, 213, 217, 298
Mooney, James 282
Mooney, Mary Delany 282
Mooney, Michael 44, 181, 210, 242, 254, 282
Moore, Daniel 177, 200, 243, 283
Moore, Edward 161, 209, 222, 283
Moore, Henry 283
Moore, Sarah 283
Mordecai, Alfred 20
Morris, George Upham 67, 124, 130

Morris Island, Charleston, South Carolina 181, 276
Morrison, William 161, 183, 211, 242, 283
Mosquito Fleet 278
Mount Pleasant Township, New York 33, 297
Murray, Lawrence 45, 152, 207, 222, 240, 283
Mystic, Connecticut 112, 220

N

Nakhimov, Pavel Stepanovich 20
Nansemond River 60, 183
Napoleon, Louis 20, 37, 213
Natchez, Mississippi 288
Naval Observatory 33, 92, 188, 213, 273, 279, 297
Naval Retiring & Examining Board 185, 298
Navarino Bay 15
Newark, New Jersey 191, 255, 274
New Bedford, Massachusetts 295
Newcastle, Maine 205
New Haven, Connecticut 159, 207, 266
New Inlet, North Carolina 167
New Orleans, Louisiana 192, 250, 277, 278, 291, 295
Newport News Point 65, 66, 97, 109, 113, 151, 162, 240
Newport News, Virginia 61, 92, 94, 230, 240, 243, 283, 287
Newport, Rhode Island 277
Newport, Rhode Island, Torpedo Station 186, 289
Newton, Anna H. 284
Newton, Isaac, Jr. 40, 57, 76, 97, 143, 150, 158, 194, 210, 220, 239, 251, 261, 284
Newton, Isaac, Sr. 284
New York City, New York 31, 39, 44, 70, 158, 160, 162, 190, 206, 207, 210, 211, 251, 252, 253, 254, 267, 268, 269, 270, 271,

INDEX

274, 276, 277, 281, 282, 283, 284, 285, 296
Niagara Steam Forge 32
Nicholas II 186, 289
Nichols, William H. 44, 180, 284
Nicklis, Catherine 284
Nicklis, Jacob 158, 160, 161, 162, 165, 166, 180, 211, 242, 243, 245, 249, 284
Nicklis, William, Sr. 284
Norfolk Navy Yard 188, 279, 285, 291
Norfolk, Virginia 42, 52, 60, 88, 93, 95, 105, 107, 109, 112, 113, 114, 115, 118, 119, 120, 121, 122, 154, 185, 188, 214, 231, 234, 277, 288, 292
Norris, William 65, 79, 86
North Atlantic Blockading Squadron 27, 61, 150, 167, 180, 181, 186, 266, 267, 275, 288, 291, 293
Novelty 16
Novelty Iron Works 31, 32, 48, 217

O

Ocracoke Inlet, North Carolina 102
office of engineer in chief 34
Office of Engineer in Chief 292
Ogden, Francis 16
Ogeechee River 183
Old Inlet, North Carolina 167
Old Point Comfort Lighthouse 94
Omer, Peter H. 202, 300, 301
O'Neil, Charles 66
Orator 18, 211
Oregon 18, 211
Osceola, Michigan 255, 285
Otisville, New York 208, 281

P

Pacific Squadron 33, 100, 213, 239, 267, 276, 278, 287, 292, 297
Paget, Clarence 100
Page, Thomas Jefferson 101
Paixhans, Henri-Joseph 17, 18, 20

Pamunkey River 291
Panama 186
Parana River 101, 278
Park, William Dunlap 192, 220, 249, 261, 284
Parliament 100
Parrott rifle 106, 111, 112, 124, 125, 129, 130, 154, 214, 230, 233, 235
Passaic class 167
Paulding, Hiram 28, 52, 53, 214, 215
Pawling, New York 44, 210, 294, 298
Peacemaker 18, 19, 212
Pendergrast, Austin 68
Peninsula Campaign 100, 147, 150, 158, 185, 239, 240, 282, 291
Peninsula, Virginia 61, 94, 105, 107, 113, 115, 124, 229
Pensacola, Florida 213
Pensacola Squadron 33, 297
Petersburg, Virginia 125, 150, 237, 287
Peterson, Charles/Philip 43, 91, 208, 251, 285
Peterson, Emma Jean Newman 285
Philadelphia Naval School 33, 150, 154, 290, 297
Philadelphia Navy Yard 16, 271, 289, 296
Philadelphia, Pennsylvania 28, 39, 43, 146, 160, 200, 207, 209, 252, 253, 255, 265, 270, 272, 273, 275, 279, 282, 286
Phillips, Dinwiddie 64, 74, 82
pivot gun 18, 26, 147, 160, 189, 228
Pivot gun 85
pivot rifle 130
Pook, Samuel 29, 112, 214
Pope, John 150
Porter, David Dixon 46, 186, 195, 288
Porter, Fitz John 145
Porter, John Luke 24, 112
Port Hudson, Louisiana 192, 249, 284
Port Jervis, New York 45, 211, 271
Portland, Maine 210, 276

INDEX

Port Royal Sound, South Carolina 150, 154, 184, 267
Port Royal, South Carolina 189, 275, 284, 290
Portsmouth (NH) Navy Yard 194, 251, 276
Portsmouth, Virginia 24, 51, 60, 74, 114, 188, 292
Post, Charles 189, 190
Potomac Flotilla 183, 293
Potomac River 18, 63, 70, 162, 212, 240, 279
Price, Christy 180, 207, 222, 249, 255, 285
Prince, John 301
Prince William, Virginia 283
Profit Island, Mississippi River 192, 284
Providence, Rhode Island 211, 212, 268, 282
Putnam, Connecticut 200, 256, 302

Q

Quackenbush, Stephen P. 182
Quebec, Canada 161, 210, 286
Quinn, Robert 180, 211, 222, 285

R

Rae, Thomas 72
Rainhill Trials 16
Ramsay, Ashton 66, 75, 82, 227
Rappahannock River, Virginia 287
Ready, Thomas 131
Reaney, Henry 66
Red River 278
Red River Campaign 185, 191
Red River Expedition. *See* Red River Campaign
Reed City, Michigan 285
Remington, Emoline Cook 285
Remington, Mary Bates McCartney 285
Remington, William Henry 161, 180, 212, 242, 255, 285

Remmington, Caroline 285
Remmington, William T. 285
Reno, Jesse Lee 112
Rensselaer Iron Works 31, 217
Richardson, William 181, 209, 222, 286
Richmond Howitzers 96
Richmond, New York 296
Richmond & Petersburg Railroad Bridge 146
Richmond, Virginia 24, 60, 61, 72, 93, 94, 95, 105, 106, 111, 113, 114, 116, 121, 123, 124, 125, 126, 129, 134, 135, 137, 144, 145, 150, 204, 229, 234, 236
Riddey, Francis A. 43, 207, 253, 286
Riddey, Margery Buchanan 286
Ritchfield Spring, New York 269
Rives, Alfred 125
Roanoke Island, South Carolina 61, 102, 279
Roanoke River 190
Roberts, Ellis. *See* Ellis, David Roberts
Rockville, Maryland 252, 292
Rodgers, John 116, 117, 119, 123, 124, 125, 126, 128, 129, 132, 134, 137, 146, 188, 234, 235, 236, 291
Rodman Gun 94
Rolfe, Iowa 198, 268
Rome, New York 161, 208, 289
Rooney, Catherine Mitchell 286
Rooney, John 41, 44, 180, 208, 222, 251, 286
Rowan, S.C. 102, 103, 279
Rowland, Thomas Fitch 217
Rumford Prize 22
Russia, New York 251, 285
Ryeday, Frank. *See* Riddey, Francis A.

S

Sabine Pass, Texas 182
Sacketts Harbor 290
Sacketts Harbor Naval Station 150
Saco, Maine 161, 209, 281

INDEX

Sandy Hook, New Jersey 43, 53, 55, 222
San Francisco, California 283
San Juan d'Ulloa, Mexico 100, 278
Savannah, Georgia 183, 192, 277, 291
Savannah Squadron 104
Schenectady, New York 32
Scientific American 51, 108
Scott, Catherine Sausse 287
Scott, Isaac H. 161, 196, 203, 210, 242, 256, 286
Scott, John 286
Scott, Mary Rye 286
Scott, William 180, 200, 287
Scott, Winfield 153, 154, 266
Seery, James 210, 222, 244, 287
Selfridge, Ellen F. Shepley 288
Selfridge, Gertrude Wildes 289
Selfridge, Louisa Cary Soley 287
Selfridge, Thomas Oliver, Jr. 96, 97, 100, 103, 121, 185, 186, 191, 195, 203, 209, 231, 255, 261, 278, 287
Selfridge, Thomas Oliver, Sr. 287, 289
Sevastopol, Crimea 20
Seward, William 29, 215
Sewell's Point 69, 73, 88, 107, 113, 115, 117, 118, 119, 120, 121, 225, 228, 234, 273
Shirley Plantation 137, 200, 210, 270
Simla 267
Simms, John D. 128
Sinclair, Henry 222, 289
Sinope, Turkey 20
Slidell, John 149
Slover, James T. 205, 234, 253, 289
Slover, Sarah Ann Hopkins 289
Smith, Charles 161, 182, 208, 242, 289
Smithfield, New York 34, 207, 292
Smith, James 161, 183, 207, 242, 290
Smith, Joseph 28, 34, 68, 214, 215, 218
Smith, Joseph B. 62, 68
Smith, Kirby 188, 291
Smith, William 137, 236

soap box navy 28
South Atlantic Blockading Squadron 150, 154, 189, 267, 269, 282, 290, 293, 298
South Mills, North Carolina 112
Southside Artillery 126, 127, 132
Squires, E.G. 101
Stanton, Edwin 69, 115, 119, 122, 225
Staten Island, New York 193, 250, 292
Stearns, Moses M. 82, 91, 161, 206, 221, 253, 290
Stearns, Ruth Gardiner 290
Stevens, Anna Maria Christie 290
Stevens Battery 34. *See* USRMS *Naugatuck*
Stevens, Elizabeth Read Sage 290
Stevens, Robert L. 106
Stevens, Thomas Holdup, Jr. 150, 152, 188, 205, 238, 239, 240, 252, 261, 267, 290
Stevens, Thomas Holdup, Sr. 290
Stimers, Alban Crocker 34, 40, 46, 47, 52, 55, 56, 57, 76, 81, 87, 90, 91, 97, 192, 207, 216, 217, 218, 220, 221, 226, 229, 232, 233, 250, 292
Stimers, Imogene 292
Stimers, James 292
Stimers, Julia Ann Appleby 292
St. Louis, Missouri 45, 277
St. Mary, Maryland 270
St. Mary's, Florida 291
St. Michaels, Maryland 205, 253, 289
Stockbridge, Wisconsin 255, 285
Stocking, John 39, 41, 76, 174, 207, 222, 292
Stockton, Robert F. 16, 18, 212
Stodder, Almira Fuller 293
Stodder, John Low 293
Stodder, Louis Napoleon 38, 40, 41, 76, 81, 91, 155, 172, 174, 193, 209, 220, 221, 223, 226, 233, 234, 238, 241, 254, 261, 293
Stodder, Rose B. Champlin 293
Stodder, Watie Howland Alderich 293

INDEX

Stringham, Silas Horton 27, 61
Suffolk, Virginia 120, 183
Sunboat Squadron 290
Sunstrom, Mark Trueman 40, 212, 220, 250, 261, 293
Sunstrom, Rachel 293
Sunstrom, Rhoda M. Bullock 293
Sunstrom, Robert C. 293
Surratt, John Harrison 187
Swedesboro, New Jersey 100, 206, 278
Swift Creek 237
Sylvester, Charles F. 91, 211, 222, 294
Syracuse, New York 161, 212, 285

T

Tabasco, Mexico 100, 278
Talcott, Andrew 123
Tarnow, Austria 209, 271
Tattnall, Josiah 104, 107, 108, 109, 112, 113, 117, 120, 127, 233, 234, 287
Taylor, Thomas L. 200, 213, 256, 299, 301
Terry, Alfred 195
Tester, Abraham 183, 189, 210, 222, 250, 294
Tester, Bridget Graham 294
Toffey, Adeline S. Wilson 294
Toffey, Daniel 41, 44, 87, 201, 210, 226, 233, 252, 294, 299
Toffey, George A. 294
Toffey, Mary De Riemer 294
Tonnante 20
Treadwell, David 18
Tredegar Iron Works 24, 25
Trenchard, Stephen Decauter 168, 169, 249
Trent Affair 149, 238
Troy, New York 31, 44, 209, 274
Truscott, Peter 39, 41, 46, 81, 91, 177, 181, 210, 222, 226, 294
Tucker, John Randolph 112, 115, 126, 127, 128, 131
Turkey Point, Virginia 146
Turner, Thomas 154

Tuspan, Mexico 100, 278
Tyler, John 18, 150, 290
Tyler, John, Jr. 212

U

Union Gun 94
Upshur, Abel 19, 212
USF *United States* 24, 100, 278
U.S. Naval Academy 38, 61, 63, 96, 100, 185, 186, 194, 219, 230, 275, 276, 278, 288, 298
U.S. Naval Academy Board of Visitors 189, 292
U.S. Revenue Cutter Service 193, 293
U.S. Revenue Marine 192, 277
USRMS *Miami* 115
USRMS *Naugatuck* 106, 107, 109, 113, 117, 118, 123, 124, 125, 128, 130, 135, 233, 234, 235
USRS *North Carolina* 39, 42, 43, 44, 49, 55, 160, 180, 198, 265, 266, 267, 268, 269, 270, 271, 272, 273, 274, 275, 276, 277, 278, 280, 281, 282, 283, 284, 285, 286, 287, 289, 294, 295, 297
USS *Adela* 293
USS *Allegheny* 101, 290
USS *Alligator* 146, 185, 204, 288
USS *Amanda* 71, 191, 277
USS *Arctic* 34, 292
USS *Arizona* 270
USS *Aroostook* 117, 124, 128, 130, 234, 291
USS *Ashuelot* 293
USS *August Dinsmore* 283
USS *Baltimore* 122, 234
USS *Belmont* 200, 270
USS *Ben Morgan* 287
USS *Benton* 273
USS *Big Bells* 295
USS *Boston* 286
USS *Brandywine* 61, 158, 180, 200, 269, 270, 276, 277, 281, 283, 285, 286, 287, 290, 296
USS *Brooklyn* 102, 278

Index

USS *Cairo* 185, 204, 288
USS *Calypso* 293
USS *Catherine A. Dix* 109
USS *Catskill* 181, 195, 198, 265, 271, 283, 295
USS *Ceres* 281
USS *Chattanooga* 169, 295
USS *Chippewa* 295
USS *Clematis* 181, 297
USS *Coeur de Lion* 137
USS *Colorado* 150, 166, 290, 296
USS *Columbia* 154, 267
USS *Commodore Barney* 200, 270, 289
USS *Commodore Morris* 289
USS *Concord* 154, 266
USS *Conestoga* 185, 288
USS *Congress* 43, 61, 62, 65, 66, 67, 68, 69, 72, 224, 240, 265, 273, 277, 278
USS *Connecticut* 161, 271
USS *Constellation* 154, 188, 267, 279
USS *C.P. Williams* 183
USS *Cranford* 154, 267
USS *Crusader* 295
USS *Cumberland* 33, 61, 66, 67, 68, 69, 74, 83, 96, 124, 151, 200, 203, 231, 240, 273, 277, 287, 297, 299
USS *Currituck* 55, 222, 285
USS *Cyane* 33, 297
USS *Dacotah* 37, 47, 117, 219, 234
USS *Dai Ching* 269
USS *Dale* 33, 297
USS *Dawn* 183
USS *Despatch* 276
USS *Dragon* 61, 85, 228, 289
USS *Enterprise* 186, 288
USS *Erie* 33, 297
USS *Ewing* 150, 290
USS *Falmouth* 43, 265
USS *Flag* 183, 269
USS *Florida* 181, 183, 189, 193, 200, 267, 270, 272, 275, 280, 294, 297
USS *Fulton* 15

USS *Galena* 28, 29, 111, 112, 113, 115, 116, 117, 123, 124, 125, 128, 129, 130, 131, 132, 133, 134, 135, 146, 164, 181, 214, 215, 220, 233, 234, 235, 236, 238, 244, 275
USS *Germantown* 101, 278, 290
USS *Gettysburg* 265
USS *Granite City* 182, 280
USS *Guerriere* 188, 291
USS *G.W.P. Custis* 106
USS *Hartford* 38, 275, 288
USS *Home* 265
USS *Huron* 186, 288
USS *Idaho* 185, 298
USS *Illinois* 288
USS *Independence* 96, 150, 154, 266, 287, 290, 297
USS *Iosco* 267
USS *Iroquois* 275
USS *Island Belle* 137, 146
USS *Jacob Bell* 146, 268
USS *Jamestown* 92, 273
USS *John Adams* 273
USS *Juniata* 276
USS *Kensington* 273
USS *Keokuk* 181, 197, 274, 276, 286
USS *Kickapoo* 295
USS *King Phillip* 156
USS *Levant* 33, 297
USS *Lexington* 273
USS *Lodona* 158, 191, 282
USS *Macedonian* 153, 186, 266, 276, 278, 288
USS *Madawaska* 284
USS *Mahaska* 146, 238, 294
USS *Manitou* 185, 288
USS *Maratanza* 137, 146, 147, 150, 238, 291
USS *Marblehead* 276
USS *Marcus* 109
USS *Mary Haley* 237
USS *Merrimack* 24, 25, 27, 34, 203, 216, 292
USS *Michigan* 34, 150, 286, 290, 292

INDEX

USS *Mingoe* 295
USS *Minnesota* 61, 62, 66, 68, 71, 72, 73, 75, 80, 82, 85, 88, 89, 93, 160, 224, 225, 227, 228, 229, 233
USS *Mississippi* 192
USS *Monitor*
  arrival at Hampton Roads 71
  battle at Drewry's Bluff 129
  beginning construction 30
  captain's cabin and stateroom 50
  commissioning 46
  engine 49
  enlisted berth deck 49
  first sea voyage 55
  first time in battle 76
  galley 49, 140, 144, 156
  greatest weakness in battle 97
  hull specifications 48
  launching 41
  magazine and shell room 50
  officers' quarters and wardroom 50
  original design 31
  pilothouse 45
  reason for name 36
  sinking 178
  specifications 45, 257
  toilets 49
  turret 31, 32, 46
USS *Monongahela* 276
USS *Montauk* 164, 167, 183, 184, 298
USS *Montgomery* 183, 294
USS *Monticello* 92, 274
USS *Morris* 278
USS *Mount Vernon* 277, 286
USS *Nantucket* 159, 189, 296
USS *Nautilus* 96, 287
USS *Neosho* 181, 192, 277
USS *New Ironsides* 28, 153, 154, 164, 215, 244
USS *Niphon* 269, 277, 293
USS *Nipsic* 186, 288, 294
USS *Nyack* 276
USS *Ohio* 42, 268, 269, 270, 271, 272, 274, 276, 281, 290, 295, 297, 300

USS *Oliver Wolcott* 193, 293
USS *Omaha* 186, 289
USS *Oneida* 291
USS *Onondaga* 203
USS *Onward* 202, 295
USS *Osage* 181, 185, 191, 273, 278, 288
USS *Ossipee* 276, 295
USS *Otsego* 190, 267
USS *Ottawa* 150, 267, 286, 290, 296
USS *Passaic* 163, 167, 168, 171, 193, 244, 245, 292
USS *Patapsco* 167, 182, 188, 282, 291
USS *Pembina* 154, 267, 279
USS *Pennsylvania* 289
USS *Pensacola* 185, 189, 192, 276, 277, 284, 292, 298
USS *Philadelphia* 102, 181, 275, 279, 294
USS *Plymouth* 102, 278
USS *Pocahontas* 286
USS *Pontoosuc* 293
USS *Port Royal* 117, 124, 125, 128, 130, 146, 150, 234, 239, 268
USS *Powhatan* 292
USS *Princeton* 16, 18, 19, 43, 212, 265, 271, 278, 286, 294, 295
USS *Raritan* 289
USS *R.B. Forbes* 92, 273
USS *Release* 293
USS *Relief* 33, 297
USS *Rhode Island* 168, 169, 170, 171, 172, 173, 174, 176, 177, 179, 245, 246, 247, 248, 249, 266, 267, 293
USS *Richmond* 284
USS *Roanoke* 34, 40, 61, 66, 68, 71, 92, 102, 200, 203, 223, 268, 273, 279, 284, 290, 292, 299, 301
USS *Sabine* 39, 42, 44, 213, 268, 269, 270, 272, 281, 290, 293, 295
USS *Sachem* 55, 222
USS *Saginaw* 276
USS *Salbea* 109
USS *San Francisco* 289

INDEX

USS *San Jacinto* 34, 117, 234, 292
USS *Saranac* 269, 276
USS *Saratoga* 278
USS *Sassacus* 272
USS *Satelitte* 146
USS *Savannah* 33, 276, 286, 297
USS *Seminole* 117, 234
USS *Seneca* 183, 295
USS *Seth Low* 55, 58, 60, 222, 223
USS *Sonoma* 291
USS *Southampton* 33, 297
USS *Southfield* 146
USS *State of Georgia* 168, 171, 245
USS *Stepping Stones* 137, 146, 182, 269, 270, 283, 289, 290, 301
USS *St. Lawrence* 61, 68
USS *Susquehanna* 117, 154, 234, 267
USS *Swatara* 187, 279, 296
USS *Truxtun* 154, 267
USS *Tunxis* 193, 272, 292
USS *Tuscarora* 158, 282
USS *Unadilla* 293
USS *Underwriter* 102, 158, 242, 273, 279
USS *Union* 289
USS *Valley City* 158, 296
USS *Vandalia* 154, 267
USS *Vanderbilt* 108, 233
USS *Vermont* 192, 274, 275, 276, 277, 295
USS *Vincennes* 96, 287, 295
USS *Vindicator* 288
USS *Vixen* 100, 278
USS *Wabash* 193, 200, 265, 270, 283, 292
USS *Wachusetts* 137, 150, 277
USS *Walker* 34, 292
USS *Warren* 297
USS *Water Witch* 34, 101, 278, 292
USS *Wave* 182
USS *Weehawken* 167
USS *Whitehall* 277
USS *Whitehead* 268
USS *Winnebago* 188, 291
USS *Winona* 295
USS *Wissahickon* 183
USS *Wyoming* 190, 267
USS *Yankee* 96, 287
USS *Young America* 232
USS *Zouave* 61, 66, 68
Utica, New York 35, 206, 280

V

Van Brunt, Gershom Jacques Henry 62, 72, 76, 85, 88, 160, 224, 225
Vanderbilt, Cornelius 108
Vera Cruz, Mexico 100, 154, 266, 278
Viall, Ezra B. 295
Viall, Julia (or Juliette) 295
Viall, Margaret Pearce Phillips 295
Viall, Thomas Brown 202, 210, 220, 239, 254, 295
Vicksburg, Mississippi 185, 288, 301
Viele, Egbert L. 115, 120
Vineland, New Jersey 256, 301

W

Wales 272, 296, 297
Walsh, Catherine Moran 295
Walsh, Daniel 43, 180, 208, 222, 254, 295
Walsh, Ellen McAuliff 295
Walsh, John 295
Walsh, Margaret O'Donnell 295
Walton, James 130
War of 1812 38
Warwick, Rhode Island 254, 295
Warwick River 106, 112, 147
Washington Artillery 128
Washington, D.C. 18, 29, 33, 61, 63, 69, 70, 71, 90, 119, 145, 155, 160, 185, 187, 188, 189, 192, 200, 213, 214, 215, 217, 222, 223, 225, 231, 251, 253, 255, 267, 268, 276, 277, 279, 285, 289, 292, 293, 297, 298, 299
Washington Navy Yard 145, 154, 156, 158, 160, 161, 162, 164, 165, 173, 188, 196, 200, 240, 241,

INDEX

242, 243, 265, 267, 268, 270, 272, 279, 280, 281, 282, 283, 284, 285, 286, 290
Watertown, New York 208, 270
Watters, Caroline Kelly 295
Watters, John 295
Watters, Joseph 169, 172, 177, 209, 245, 248, 250, 262, 295
Watters, Mary 295
Webber, Edward 296
Webber, John Joshua Nathaniel 38, 39, 40, 76, 207, 221, 254, 261, 296
Webber, Nancy 296
Webb, William 147
Weber, Max 120
Weeks, Cyrus 296
Weeks, Grenville Mellen 13, 158, 168, 177, 180, 191, 210, 242, 244, 248, 249, 255, 262, 296
Weeks, Helen Campbell Stuart 296
Weeks, Maria L. Child 296
Weeks, Maria Oberg 296
Weeks, Pauline M. Sauer 296
Welles, Gideon 27, 28, 29, 33, 39, 42, 52, 61, 62, 69, 71, 78, 95, 97, 108, 123, 146, 149, 167, 180, 214, 215, 220, 222, 223, 225, 231
Wentz, Wells. *See* Stocking, John
West Gulf Blockading Squadron 181, 182, 188, 291, 297
White, E.V. 76, 91
White, George H. 159, 239, 261, 296
White House Landing 150, 291
White, John. *See* Driscoll, John Ambrose
Wilkes, Charles 149, 238, 240, 275, 291
Williamson, William Price 24
Williams, Peter 39, 41, 76, 83, 86, 175, 181, 189, 208, 222, 296
Williamsport, Pennsylvania 282
Williams, Robert 297
Willoughby Spit, Virginia 120
Wilmington, North Carolina 162, 167, 186, 189, 193, 244, 267, 280, 286, 288
Wilson, Thomas Woodrow 200
Winslow, John F. 30, 31, 32, 215, 216
Wise, Henry A. 90, 97, 231
Wood, John Taylor 65, 66, 67, 74, 75, 79, 85, 86, 111, 118, 127, 128, 134, 228, 236
Woodward crank pump 144
Wool, John Ellis 66, 69, 89, 92, 94, 115, 119, 120, 122, 230, 232
Worden, Ananias 297
Worden, Harriet Graham 297
Worden, John Lorimer 33, 37, 39, 40, 42, 44, 46, 52, 53, 55, 56, 58, 71, 72, 75, 76, 79, 81, 83, 85, 86, 88, 90, 91, 97, 108, 164, 169, 183, 185, 204, 213, 217, 218, 220, 223, 224, 226, 227, 228, 229, 231, 244, 261, 297
Worden, Olivia Aiken Toffey 44, 97, 297
Worthington steam pump 144
Worthington Steam Pump 49, 58, 171, 172, 216, 247
Wynne, Thomas 126

Y

Yazoo River 185, 204, 288
York River 94, 106, 112
Yorktown, Virginia 95, 106, 108, 111, 112, 113
Yorktown-Warwick Siege 111, 113

About the Author

John V. Quarstein is an award-winning historian, preservationist and author. He presently serves as historian for the City of Hampton. He previously worked as the director of the Virginia War Museum and as consultant to The Mariners' Museum's Monitor Center.

Quarstein is the author of ten books, including *Fort Monroe: The Key to the South*, *CSS Virginia: Mistress of Hampton Roads* and *A History of Ironclads: The Power of Iron Over Wood*. He also has produced, narrated and written several PBS documentaries, including *Jamestown: Foundations of Freedom* and *Hampton: From the Sea to the Stars*.

John Quarstein is the recipient of the National Trust for Historic Preservation's 1993 President's Award for Historic Preservation; the Civil War Society's Preservation Award in 1996; the United Daughters of the Confederacy's Jefferson Davis Gold Medal in 1999; and a 2007 Silver Telly for his *Civil War in Hampton Roads* film series.

Besides his lifelong interest in Tidewater Virginia's Civil War experience, Quarstein is also an avid duck hunter and decoy collector. He lives on Old Point Comfort in Hampton, Virginia, and on his family's Eastern Shore farm near Chestertown, Maryland.

In 1987, The Mariners' Museum was designated by NOAA, on behalf of the federal government, as the repository for artifacts and archives from the USS *Monitor*. Working jointly with NOAA and the U.S. Navy, the museum has received more than 1,200 artifacts from the *Monitor*, including the steam engine, propeller and revolving gun turret, all now permanently housed in the state-of-the-art USS *Monitor* Center.

See www.marinersmuseum.org

www.ingramcontent.com/pod-product-compliance
Lightning Source LLC
Chambersburg PA
CBHW070403100426
42812CB00005B/1614